THEODORE L. BROWN
University of Illinois

H. EUGENE LEMAY, JR.
University of Nevada

SOLUTIONS TO EXERCISES IN

Chemistry

The central science

2ND EDITION

Prentice-Hall, Inc., Englewood Cliffs, New Jersey 07632

Printed in the United States of America

ISBN 0-13-128538-6

10 9 8 7 6 5 4 3 2 1

PRENTICE-HALL INTERNATIONAL, INC., London
PRENTICE-HALL OF AUSTRALIA PTY. LIMITED, Sydney
PRENTICE-HALL OF CANADA, LTD., Toronto
PRENTICE-HALL OF INDIA PRIVATE LIMITED, New Delhi
PRENTICE-HALL OF JAPAN, INC., Tokyo
PRENTICE-HALL OF SOUTHEAST ASIA PTE. LTD., Singapore
WHITEHALL BOOKS, LTD., Wellington, New Zealand

Contents

Introduction

Chemistry: The Central Science, 2nd edition, contains about 1600 end-of-chapter exercises. We have given much attention to these exercises because we believe that one of the best ways for students to master chemistry is by solving problems. Because of their limited time, students will normally be able to work only a portion of the available exercises. By grouping them according to subject matter, we have aided the student in selecting and recognizing particular types of problems. This grouping also aids the instructor in assigning an appropriate range of problems to be worked. For those who prefer general exercises, we have provided in each chapter a substantial number of these to supplement those grouped by type. Providing brief answers in the text to about 600 of these exercises has helped to make the text a useful self-contained vehicle for learning.

This manual, Solutions to Exercises in Chemistry: The Central Science, 2nd edition, was written to further facilitate the use of the end-of-chapter exercises. In providing complete, worked-out solutions to all of these exercises, this manual should give further assistance to both instructor and student. For the instructor, it will save time spent in working out solutions for assigned problem sets. For students, it gives a means of checking their understanding of the material. Most of the solutions have been worked in the same detail as the in-chapter sample exercises in order to guide the students in their studies.

We have tried to keep this book as error-free as possible. All exercises have been worked by at least two persons to ensure not only that the answer arrived at is correct but also that the steps in arriving at the answer are shown clearly and completely. However, it is probably too much to expect that we have caught all typographical errors in our proofreading. Should you find any errors, we would appreciate hearing from you. We hope, whether you are an instructor or a student, that you will find this manual helpful and instructive.

Theodore L. Brown
School of Chemical Sciences
University of Illinois
Urbana 61801

H. Eugene LeMay, Jr.
Department of Chemistry
University of Nevada
Reno 89557

1

Introduction: some basic concepts

1.1 Chemistry is an experimental science. The systems on which experiments are performed are relatively simple collections of matter. The experiments the chemist performs should be reproducible, and should reveal some new aspect of nature. Normally, there is no sense of an adversary relation toward nature. The laws formulated by the chemist are the logical outgrowth of experimental observations. These laws should be independent of the social or political origins of the chemist who formulates them.

By contrast, the lawyer and judge are intensely concerned with human values. The laws of society are based on judgments of what is good or bad for society. The laws rest mainly on intellectual judgments and are the logical outgrowth of a particular view of human society. They have less basis in "experiments" than do scientific laws. Evidence in the law is used to judge human conduct in relationship to laws, but not as a test of them as it is in science.

1.2 Science advances largely by showing that one or more hypotheses or theories is wrong because of a conflict with reliable experimental results. When conflicting hypotheses are advanced to account for a body of experimental data, it is usually possible to design an experiment that will prove one or the other hypothesis incorrect.

1.3 It should be called Coulomb's law, because it simply summarizes experimental observations regarding the forces between charged objects. A theory would advance an explanation for why the charged objects behave as they do.

1.4 Wöhler, by preparing a substance found in living systems from inorganic starting materials, showed that the compound, urea, did not need the agency of a living system for its formation. Thus, the basic assumption behind the vitalist theory was demolished.

1.5 (a) A hypothesis is an informed guess or tentative explanation about some aspect of nature, based on a relatively few experimental facts. A theory is a more comprehensive intellectual construct, or model, based on a large number of observations, which "explains" those observations and permits predictions regarding the results of new experiments. (b) A paradigm is a generally accepted set of beliefs about how things are based on wide experience on the part of an entire group of persons working in the same area. For example, among geologists it is generally accepted that the earth is about 4.5 billion years old. A law is a concise verbal statement or equation that summarizes a particular well-established set of observations. (c) Experiments are organized observations of how nature behaves under controlled conditions. A theory is an intellectual construct, an "explanation," that accounts for a number of related experiments in a single, unifying manner. (d) Quantitative experiments involve making accurate measurements; of quantities, temperature, times, and so forth. Qualitative observations are simply reports of what seemed to happen in an experiment (boiling, a yellow flame, a solid formed, etc).

1.6 (a) meters, m; (b) cubic meters, m^3; (c) kilograms, kg; (d) square meters, m^2; (e) second, sec or simply s

1.7 (a) kilo; (b) deci; (c) mega; (d) micro; (e) pico

1.8 (a) $(1.00 \text{ kg gold})\left(\frac{1000 \text{ g}}{1 \text{ kg}}\right)\left(\frac{0.0518 \text{ cm}^3 \text{ gold}}{1.00 \text{ g gold}}\right) = 51.8 \text{ cm}^3$

(b) $(1.00 \text{ dg gold})\left(\frac{0.1 \text{ g}}{1 \text{ dg}}\right)\left(\frac{0.0518 \text{ cm}^3 \text{ gold}}{1.00 \text{ g gold}}\right) = 5.18 \times 10^{-3} \text{ cm}^3$

(c) $5.18 \times 10^{-5} \text{ cm}^3$ (d) 0.518 cm^3 (e) $5.18 \times 10^{-9} \text{ cm}^3$

(f) $5.18 \times 10^{-6} \text{ cm}^3$

1.9 (a) 650 cm or 6.50×10^2 cm; (b) $33 \text{ kg}\left(\frac{1000 \text{ g}}{1 \text{ kg}}\right)\left(\frac{1000 \text{ mg}}{1 \text{ g}}\right) = 3.3 \times 10^7$ mg;
(c) 1.2×10^4 msec; (d) 0.043 g; (e) 2.35×10^3 g; (f) 2.25×10^4 μm

1.10 (a) 5×10^5m; (b) 0.326 kg; (c) $476 \text{ nm}\left(\frac{1 \text{ m}}{10^9 \text{ nm}}\right)\left(\frac{100 \text{ cm}}{1 \text{ m}}\right) = 4.76 \times 10^{-5}$ cm;
(d) 8.2×10^{-2} msec

1.11 (a) volume; (b) area; (c) mass; (d) length

1.12 3.08×10^{16}m. 3.08×10^{10}m per microparsec

$$\left(\frac{3.08 \times 10^{13} \text{km}}{1 \text{ parsec}}\right)\left(\frac{1000 \text{ m}}{1 \text{ km}}\right)\left(\frac{100 \text{ cm}}{1 \text{ m}}\right)\left(\frac{1 \text{ parsec}}{10^9 \text{ nparsec}}\right) = \frac{3.08 \times 10^9 \text{cm}}{1 \text{ nparsec}}$$

1.13 2.205 lb; 39.39 in.; 28.35 g; 0.4717 L;

$$55 \text{ cm}\left(\frac{1 \text{ m}}{100 \text{ cm}}\right)\left(\frac{1.094 \text{ yd}}{1 \text{ m}}\right)\left(\frac{3 \text{ ft}}{1 \text{ yd}}\right) = 1.80 \text{ ft}$$

1.14 $\left(\frac{70{,}664 \text{ mi}^2}{649{,}000 \text{ people}}\right)\left(\frac{1760 \text{ yd}}{1 \text{ mi}}\right)^2\left(\frac{1 \text{ m}}{1.094 \text{ yd}}\right)^2 = \frac{281{,}802 \text{ m}^2}{\text{person}}$

Round off to 3 significant figures, because the population is estimated to 3 significant figures. Thus, 282,000 m^2/person, or $2.82 \times 10^5 m^2$/person.

1.15 (a) four; (b) three; (c) three; (d) one; (e) two; (f) two, three or four - can't be sure without knowing the context; (g) six

1.16 (a) three; (b) three; (c) three; (d) from one to five - can't be sure; (e) three; (f) two

1.17 (a) 4.568×10^6; (b) 2.358×10^3; (c) 4.256×10^4; (d) 2.389×10^{-3}; (e) 9.876

1.18 (a) 1.245×10^3; (b) 6.50×10^4; (c) 5.975×10^4; (d) 4.56×10^{-3}

1.19 (a) 0.55; (b) 3.5×10^3; (c) 5.74×10^{-9}; (d) 2×10^{-6}; (e) 1.5; (f) 1.82×10^3

1.20 (a) 6.71×10^{-3}; (b) 9.0×10^{-3}; (c) 18.6; (d) 5.74

1.21 (a) 1.78 m; (b) 2.20×10^3kg; (c) 3282 ft; (d) 43.4 L

1.22 (a) $1302 \text{ ft} \left(\frac{1 \text{ yd}}{3 \text{ ft}}\right)\left(\frac{1 \text{ m}}{1.094 \text{ yd}}\right)\left(\frac{1 \text{ fathom}}{1.8288 \text{ m}}\right) = 216.9 \text{ fathoms}$

(b) $10.000 \text{ furlong} \left(\frac{201.17 \text{ m}}{\text{furlong}}\right)\left(\frac{1.094 \text{ yd}}{1 \text{ m}}\right)\left(\frac{1 \text{ mi}}{1760 \text{ yd}}\right) = 1.250 \text{ mi}$

1.23 (a) 1.2×10^2L; (b) 15.432 grains; (c) 1.62×10^3 pennywt.; (d) 1016 kg

1.24 (a) 86.62 dm^3; (b) 5.24 L; (c) 10^6 m^2; 0.386 mi^2

1.25 mass of toluene is 87.127 g - 57.832 g = 29.295 g.

$$(29.295\text{g toluene})\left(\frac{1 \text{ cm}^3 \text{ toluene}}{0.866\text{g toluene}}\right) = 33.8 \text{ cm}^3$$

The number of significant figures is limited by the accuracy with which we know the density of toluene.

1.26 (a) 10 miles is 1.760×10^4 yds; add 120 yds, obtain 1.772×10^4 yds.

$$1.772 \times 10^4 \text{ yds}\left(\frac{1 \text{ m}}{1.094 \text{ yd}}\right)\left(\frac{1 \text{ km}}{1000 \text{ m}}\right) = 16.20 \text{ km}$$

(b) $2.74026 \times 10^8 \text{ yd}^3 \left(\frac{1 \text{ m}}{1.094 \text{ yd}}\right)^3 = 2.09286 \times 10^8 \text{ m}^3 = 0.2093 \text{ km}^3$

(c) $11{,}647 \text{ ft} \left(\frac{1 \text{ yd}}{3 \text{ ft}}\right)\left(\frac{1 \text{ m}}{1.094 \text{ yd}}\right) = 3549 \text{ m}$

(The four-place precision is limited by the precision of the conversion, 1.094 yd = 1m, Table 1.3. Using the value on the back inside cover, 1.0936, we obtain for (b), 0.20952 km^3; for (c), 3.5500 km.)

(d) 147 m high; 230 m on each side; 2.3×10^3 kg each

(e) 804 km race;

$$\frac{99.482 \text{ mi}}{1 \text{ hr}}\left(\frac{1760 \text{ yd}}{1 \text{ mi}}\right)\left(\frac{1 \text{ m}}{1.094 \text{ yd}}\right)\left(\frac{1 \text{ km}}{1000 \text{ m}}\right)\left(\frac{1 \text{ hr}}{3600 \text{ sec}}\right)$$

$$= 0.04446 \text{ km/sec} = 44.46 \text{ m/sec}$$

1.27 (a) $1 \text{ mi} \left(\frac{5280 \text{ ft}}{1 \text{ mi}}\right)\left(\frac{12 \text{ in.}}{1 \text{ ft}}\right)\left(\frac{1 \text{ link}}{7.92 \text{ in.}}\right)\left(\frac{1 \text{ chain}}{100 \text{ links}}\right) = 80.0 \text{ chains}$

(b) We can use the result from (a) to help answer part (b):

$$1 \text{ rod} \left(\frac{1 \text{ chain}}{4 \text{ rods}}\right)\left(\frac{1 \text{ mi}}{80 \text{ chains}}\right)\left(\frac{1760 \text{ yd}}{1 \text{ mi}}\right)\left(\frac{1 \text{ m}}{1.094 \text{ yd}}\right) = 5.027 \text{ m}$$

(c) We have that 80 chains = 1 mi. Thus $(80 \text{ chain})^2 = 1 \text{ mi}^2$ = 6400 chain^2. Using the answer from (b), we have

(d) $1 \text{ yd}^3 \left(\frac{1 \text{ m}}{1.094 \text{ yd}}\right)^3\left(\frac{1 \text{ rod}}{5.027 \text{ m}}\right)^3 = 5.13 \times 10^{-3} \text{ rod}^3$

1.28 (a) $\left(\frac{62 \text{ mi}}{1 \text{ hr}}\right)\left(\frac{1760 \text{ yd}}{1 \text{ mi}}\right)\left(\frac{1 \text{ m}}{1.094 \text{ yd}}\right)\left(\frac{1 \text{ km}}{1000 \text{ m}}\right) = 100 \text{ km/hr}$

(b) 28 m/sec (c) $\left(\frac{\$2.45}{1 \text{ lb}}\right)\left(\frac{2.205 \text{ lb}}{1 \text{ kg}}\right) = \$5.40/\text{kg}$

$$\left(\frac{\$5.40}{\text{kg}}\right)\left(\frac{1 \text{ Mark}}{\$0.46}\right) = 11.7 \text{ Mark/kg}$$

(d) $12 \text{ fl oz} \left(\frac{1 \text{ qt}}{32 \text{ fl oz}}\right)\left(\frac{1 \text{ L}}{1.06 \text{ qt}}\right) = 0.35 \text{ L}$

1.29 (a) $238{,}850 \text{ mi} \left(\frac{1760 \text{ yd}}{1 \text{ mi}}\right)\left(\frac{1 \text{ mi}}{1.094 \text{ yd}}\right)\left(\frac{1 \text{ km}}{1000 \text{ m}}\right) = 3.842 \times 10^5 \text{ km}$

The number of significant figures is limited to the number of significant figures in the conversion from m to yd. If we were to use a more precise number here, we could report the result to five significant figures, since there are that many in the originally given distance to the moon.

$$27.32 \text{ days} \left(\frac{24 \text{ hr}}{1 \text{ day}}\right)\left(\frac{3600 \text{ sec}}{1 \text{ hr}}\right) = 2.360 \times 10^6 \text{ sec}$$

$$1081 \text{ mi} \left(\frac{1760 \text{ yd}}{1 \text{ mi}}\right)\left(\frac{1 \text{ m}}{1.094 \text{ yd}}\right)\left(\frac{1 \text{ km}}{1000 \text{ m}}\right) = 1739 \text{ km}$$

$$1.62 \times 10^{23} \text{ lb} \left(\frac{1 \text{ kg}}{2.205 \text{ lb}}\right) = 7.35 \times 10^{22} \text{ kg}$$

1.30 $4.84 \times 10^8 \text{ mi} \left(\frac{1760 \text{ yd}}{1 \text{ mi}}\right)\left(\frac{1 \text{ m}}{1.094 \text{ yd}}\right)\left(\frac{1 \text{ km}}{1000 \text{ m}}\right) = 7.79 \times 10^8 \text{ km}$

$$11.86 \text{ yr} \left(\frac{365 \text{ day}}{1 \text{ yr}}\right)\left(\frac{24 \text{ hr}}{1 \text{ day}}\right)\left(\frac{3600 \text{ sec}}{1 \text{ hr}}\right) = 3.740 \times 10^8 \text{ sec}$$

$$\frac{1.330 \text{ g}}{1 \text{ cm}^3}\left(\frac{1 \text{ kg}}{1000 \text{ g}}\right)\left(\frac{100 \text{ cm}}{1 \text{ m}}\right)^3 = 1.330 \times 10^3 \frac{\text{kg}}{\text{m}^3}$$

1.31 $(4.5 \text{ cm})(12.54 \text{ cm})(1.25 \text{ cm}) = 71 \text{ cm}^3$ The result is limited to two significant figures because 4.5 cm contains just two.

$$(71 \text{ cm}^3)\left(\frac{1 \text{ m}}{10^2 \text{ cm}}\right)^3 = 7.1 \times 10^{-5} \text{ m}^3$$

1.32 $\left(\frac{980.66 \text{ cm}}{\text{sec}^2}\right)\left(\frac{1 \text{ m}}{10^2 \text{ cm}}\right) = 9.8066 \text{ m/sec}^2$

1.33 (a) volume $= \pi(6.5 \text{ cm})^2(28.6 \text{ cm}) = 3.8 \times 10^3 \text{ cm}^3$

(b) $\pi 8.0 \text{ ft} \left(\frac{1 \text{ yd}}{3 \text{ ft}}\right)\left(\frac{1 \text{ m}}{1.094 \text{ yd}}\right)(10.0 \text{ in.})^2\left(\frac{2.54 \text{ cm}}{1 \text{ in.}}\right)^2\left(\frac{1 \text{ m}}{100 \text{ cm}}\right)^2 = 0.49 \text{ m}^3$

(c) $0.49 \text{ m}^3 \left(\frac{100 \text{ cm}}{1 \text{ m}}\right)^3\left(\frac{1.00 \text{ g } H_2O}{1 \text{ cm}^3}\right)\left(\frac{1 \text{ kg}}{10^3 \text{ g}}\right) = 4.9 \times 10^2 \text{ kg}$

1.34 $°C = (100.6 - 32)(5/9) = 38.1°C$

1.35 (a) 19.4°C; (b) −22°C; (c) 628 K; (d) −123°C; (e) 322 K; (f) 227 K

1.36 $1 \text{ dm}^3 \left(\frac{10 \text{ cm}}{1 \text{ dm}}\right)^3\left(\frac{13.6 \text{ g}}{1 \text{ cm}^3}\right) = 13{,}600 \text{ g} = 13.6 \text{ kg}$

1.37 density $= \frac{\text{mass}}{\text{volume}} = \frac{4.268 \text{ kg}}{3.47 \text{ L}} = \frac{1.23 \text{ kg}}{\text{L}}\left(\frac{1 \text{ L}}{10^3 \text{ cm}^3}\right)\left(\frac{10^3 \text{ g}}{1 \text{ kg}}\right) = 1.23 \text{ g/cm}^3$

1.38 $\frac{\$0.39}{600 \text{ mL}}\left(\frac{1 \text{ mL}}{1 \text{ cm}^3}\right)\left(\frac{1 \text{ cm}^3}{0.79 \text{ g}}\right)\left(\frac{10^3 \text{ g}}{\text{kg}}\right) = \frac{\$0.82}{\text{kg}}$

1.39 density $= \frac{(251.65 - 204.58) \text{ g}}{58.3 \text{ cm}^3} = 0.807 \text{ g/cm}^3$

1.40 volume of room = (8.2)(13.5)(2.75) = $3.0 \times 10^2 m^3$

$$3.0 \times 10^2 m^3 \left(\frac{100\ cm}{m}\right)^3 \left(\frac{1.19\ g}{10^3\ cm^3}\right) = 3.6 \times 10^5\ g = 360\ kg$$

(Here we use the fact that 1 L = 1000 cm^3.)

1.41 Add enough water to the cylinder so that the marble will be completely immersed when it is added later. Weigh the cylinder as accurately as possible, and note the volume level as accurately as possible. Add the marble, and repeat both measurements. The difference in volume readings gives the volume of the marble; the difference in mass gives the mass of the marble. The density is simply the mass divided by the volume.

1.42 Weight-extensive; color-intensive; density-intensive; volume-extensive; temperature-intensive; melting point-intensive; ease of corrosion-intensive

1.43 (a) inappropriate - The circulation of a widely read publication would vary over a year's time, and could simply not be counted to the nearest single subscriber. Probably about 4 significant figures would be appropriate. (b) appropriate - It might be possible to do better than two significant figures, but an estimate to two significant figures should be easily possible. (c) Rainfall can be measured to within 0.01 in., but it is probably not possible to record an entire year's rainfall to the nearest 0.01 in. Further, the variation from year to year is sufficiently large that it does not make much sense to report the average to this number of significant figures. Probably two significant figures would be appropriate. (d) inappropriate - The population of a city is not constant during a year, and cannot probably even be counted at any one time to an accuracy of one person. Probably 51,000 would be an appropriate estimate.

1.44 -88°C; 185 K

1.45 (a) 5 ft 3 in. = 63 in. $\left(\frac{2.54\ cm}{1\ in.}\right)\left(\frac{1\ m}{100\ cm}\right) = 1.60\ m$

$$100\ lb \left(\frac{1\ kg}{2.205\ lb}\right) = 49.9\ kg \qquad 122\ lb \left(\frac{1\ kg}{2.205\ lb}\right) = 55.3\ kg$$

(b) 5 ft 9 in. = 69 in. $\left(\frac{2.54\ cm}{1\ in.}\right)\left(\frac{1\ m}{100\ cm}\right) = 1.75\ m$

$$154\ lb \left(\frac{1\ kg}{2.205\ lb}\right) = 69.8\ kg \qquad 2.5\ oz \left(\frac{28.35\ g}{1\ oz}\right) = 71\ g\ \text{protein}$$

1.46 volume of room = (2.5)(15)(40) = $1.5 \times 10^3 m^3$

$$1.5 \times 10^3 m^3 \left(\frac{10\ mg\ CO}{1\ m^3}\right)\left(\frac{1\ g}{1000\ mg}\right) = 15\ g\ CO$$

1.47 100°C = 212 - 32 = 180°F Thus, 1°C = $\frac{180}{100}$ °F = 1.80°F

1.48 (a) $\left(\frac{4.5\ g}{cm^3}\right)\left(\frac{1\ kg}{1000\ g}\right)\left(\frac{100\ cm}{1\ m}\right)^3 = 4.5 \times 10^3 \frac{kg}{m^3}$

(b) $248\ cm^2 \left(\frac{1\ m}{100\ cm}\right)^2 = 2.48 \times 10^{-2} m^2$

(c) $\left(289 \frac{g\ cm}{sec^2}\right)\left(\frac{1\ kg}{1000\ g}\right)\left(\frac{1\ m}{100\ cm}\right) = 2.89 \times 10^{-3}\ kg\text{-}m/sec^2$

(d) $2230 \frac{dm^3}{hr} \left(\frac{1\ m}{10\ dm}\right)^3 \left(\frac{1\ hr}{3600\ sec}\right) = 6.19 \times 10^{-4} m^3/sec$

1.49 A most important characteristic of scientific activity is doing experiments under controlled and reproducible conditions. Experiments are designed so as to answer some question about an aspect of nature that interests the scientist. The evaluation of experimental results should be totally objective. That is, all who perform the experiment should be able to agree upon the results.

The interpretations placed on experimental observations are based on the model, or theory, one has about that aspect of nature. Such theories are evaluated on the extent to which they correctly account for experimental facts, and not on the basis of any other factors such as political or religious beliefs. Thus, a theory about a set of chemical observations is based on experimental data. By contrast, any theory one might have about the origins of the characters in Ulysses would necessarily be based on historical information, such as Joyce's letters and notes, and contemporary accounts. One's conclusions would very likely be influenced in part by one's own point of view regarding literary aesthetics, political theory, and religious beliefs. It is much less straightforward to subject such a theory to "experimental" test. All this, of course, does not mean that literary criticism has less validity than scientific research. It merely indicates that the two kinds of activity differ in important ways, in their methods and in their goals.

1.50 $8\ qt \left(\frac{1\ L}{1.06\ qt}\right)\left(\frac{1000\ cm^3}{1\ L}\right)\left(\frac{19.3\ g}{1\ cm^3}\right)\left(\frac{1\ kg}{1000\ g}\right) = 1.5 \times 10^2\ kg$

The thief would need to be exceptionally strong, since the bucket has a mass about twice that of an average man (see problem 1.45(b)).

1.51 $\frac{22.4\ mi}{gal}\left(\frac{1760\ yd}{1\ mi}\right)\left(\frac{1\ m}{1.094\ yd}\right)\left(\frac{1\ km}{1000\ m}\right)\left(\frac{1\ gal}{4\ qt}\right)\left(\frac{1.06\ qt}{1\ L}\right) = 9.55\ km/L$

1.52 $\frac{100\ m}{9.9\ sec}\left(\frac{1.094\ yd}{1\ m}\right)\left(\frac{1\ mi}{1760\ yd}\right)\left(\frac{3600\ sec}{hr}\right) = 22.6\ mi/hr$

1.53 (a) $2.00 \times 10^2\ g \left(\frac{1\ cm^3}{7.86\ g}\right) = 25.4\ cm^3$

volume of a sphere $= (4/3)\pi r^3 = 25.4\ cm^3$

$r = \left(\frac{(25.4)(3)}{4\pi}\right)^{1/3} = 1.82\ cm$

(b) 2.60 cm (c) 1.61 cm

2

Our chemical world: atoms, molecules, and ions

2.1 (a) mercury; (b) sulfur melts at 112°C; phenol (carbolic acid) melts at 41°C; naphthalene (used as moth crystals) melts at 80°C; (c) naphthalene or para-dichlorobenzene, used as moth crystals sublime slowly into the vapor state when allowed to stand. Solid CO_2, "dry ice," sublimes without melting.

2.2 Some distinguishing physical properties: solubility in water, melting point, density, hardness, crystal form. Some distinguishing chemical properties: sugar can be combusted in air, salt cannot; solutions of table salt produce precipitates when mixed with solutions of other ionic substances, whereas sugar does not. This and many other chemical distinctions will become more obvious as we proceed in the text.

2.3 (a) A battery fluid is a solution of sulfuric acid in water. (b) The gold bars should be pure substance. (c) Sand is a heterogeneous mixture of different minerals; this is so even within most individual grains of sand. (d) Seven-up is a solution, that is, a homogeneous mixture, of sugar and other substances, including carbon dioxide. (e) A Bufferin or Anacin tablet is a heterogeneous mixture of two or more substances.

2.4 (a) Burning of the magnesium ribbon is a chemical process; it involves oxidation of magnesium to magnesium oxide. (b) Extrusion of copper into copper wire is a physical process; it does not change the chemical nature of the substance. (c) Heating of a filament is mainly a physical process, though there may be a slow chemical change in the filament while it is hot that eventually causes the bulb to burn out. (d) Distillation is a physical process. (e) Crystallization is a physical process; chemical composition of the substance being crystallized does not change. (f) Tarnishing of silver is a chemical process; silver metal is converted to some other compound of silver, such as silver sulfide, Ag_2S.

2.5 (a) physical change; (b) Hard-boiling an egg is a chemical change; the protein in the egg has been chemically altered. (c) physical change; (d) Burning of coal produces new chemical substances; it is thus a chemical process. (e) Forming a solution of salt in water is a physical change; the salt is readily recovered by evaporating off the water. (f) The "dissolving" of an Alka-Seltzer tablet is in part a chemical reaction; we see gas given off, indicating that a new substance has been formed. (g) physical change.

2.6 (a) Fractional distillation; the methyl alcohol would distill over at lower temperature than butyl alcohol. (b) Place the two salts in water; sodium chloride dissolves, silver chloride does not; filter to remove silver chloride, then evaporate water off to recover sodium chloride. (c) Use column chromatography, as illustrated in Figure 2.6. (d) Use a large distillation apparatus with fractionating column, as shown in Figure 2.4.

2.7 Physical Properties: silver-white color; soft; good conductor of electricity; boils at 883°C; vapor is violet-colored. Chemical Properties: prepared by electrolysis of molten sodium chloride; tarnishes rapidly in air; burns on heating in air or in bromine vapor.

2.8 The molten material present on the surface of the hot planet can be considered a solution of many substances all together. Some of the components of this solution mixture could react with one another or with gases present in the atmosphere to form insoluble products that would separate as either solid precipitates or as liquids. Copper, for example, would not likely be soluble in a molten mineral mixture. Some of the mixture components that were soluble at higher temperatures would begin to crystallize out as the temperature dropped. These would then form blocks of essentially pure solids which would eventually be surrounded by other solid substances as the liquid cooled still further.

2.9 In the early days of the atomic theory, the evidence for an element was that the substance was incapable of being chemically broken down into still simpler substances. Thus the evidence was always negative in character. Only when means were devised for analyzing materials carefully could the elemental nature of a substance be fully established. Of course, there was also much reasoning by analogy. A substance with a bright, shiny metallic appearance was likely to be a metallic element.

2.10 The ink components are adsorbed onto the paper. The most strongly adsorbed components are those which move most slowly up the paper.

2.11 The product of reaction of hydrogen and bromine is a substance with a fixed ratio of one substance to the other, because the molecules of the product each contain the same relative numbers of hydrogen and bromine atoms. The composition of the product is fixed; the identity of the product depends on the fact that the composition of each molecule is the same. If there is an excess of one gas over the other when the two are reacted, molecules of the component present in excess remain unreacted. These observations illustrate the law of constant composition.

2.12 There is no difference in nutritional value. If the two preparations of ascorbic acid were different, it would be because traces of other substances were present. These other substances could be carried along in the purification processes, probably recrystallization. However, careful work should result in negligible levels of any other substances in the ascorbic acid isolated from either source.

2.13 There are no simple ways to be sure that atoms retain their identity in molecules. Various sophisticated techniques are now available which make it possible to learn about the shapes and sizes of molecules, and about how charge and mass are distributed within them. It appears that "atoms" are present in molecules. For example, the water molecules, formula H_2O, appear to consist of a central mass connected to two smaller masses, with an overall boomerang-like shape. We presume that a central oxygen atom is connected to two hydrogen atoms.

When we electrolyze water, as in Figure 2.8, we obtain the two substances oxygen and hydrogen from which the water was formed in the first place. This suggests, but does not prove, that the oxygen and hydrogen atoms retain their identity in the water molecule.

2.14 Calculate the ratio mass oxygen/mass nitrogen. A, 1.14; B, 2.28; C, 1.705. Dividing through by the smallest of these, we obtain A, 1.00; B, 2.00; C, 1.50. To convert these to whole numbers, multiply by 2: A, 2.00; B, 4.00; C, 3.00. These data tell us that the ratios of oxygen to nitrogen in the three compounds are related to one another as small whole numbers. In other words, while we don't know yet, without doing more calculating, just what the ratio of oxygen to nitrogen atoms in compound A is, we know that in compound B there is twice as much oxygen relative to nitrogen; in Compound C there is 1.5 times as much.

2.15 Substance A must be a compound, because it is broken down into other substances on heating. The gas formed is also a compound, because it consists of carbon and oxygen. We do not know enough about B to know whether it is a compound or an element.

2.16

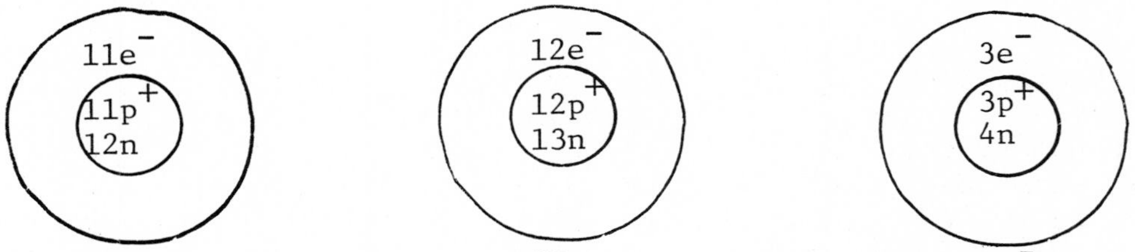

(a) ^{23}Na (b) ^{25}Mg (c) ^{7}Li

2.17 The tracks are long because the α particles encounter only empty space or the outer electrons of atoms in their path. By gradually losing energy via many contacts with the electrons of atoms, the α particles eventually slow down and are stopped. Occasionally, an α particle makes a direct hit on a small but highly charged nucleus of a nitrogen or oxygen atom. Such an encounter causes the α particle to be deflected from its course. However, these encounters are rare because the nuclei have such a small cross-section.

2.18 The droplets contain different charges because there may be 1,2,3 or more excess electrons on the droplet. The electronic charge is likely to be the lowest common factor in all the observed charges. Assuming this is so, we calculate the apparent electronic charge from each drop as follows:

A	$1.60 \times 10^{-19}/1$	=	1.60×10^{-19} coul
B	$3.15 \times 10^{-19}/2$	=	1.58×10^{-19} coul
C	$4.81 \times 10^{-19}/3$	=	1.60×10^{-19} coul
D	$6.31 \times 10^{-19}/4$	=	1.58×10^{-19} coul

The average value from these four observations is 1.59×10^{-19} coul. There appears to be an experimental uncertainty of about ±0.01 coul, so 3 significant figures are justified.

<u>2.19</u> mass of 3 protons = $3 \times 1.67 \times 10^{-24}$ g = 5.01×10^{-24} g
mass of 4 neutrons = $4 \times 1.67 \times 10^{-24}$ g = 6.68×10^{-24} g
mass of 3 electrons = $3 \times 9.1 \times 10^{-28}$ g = 2.7×10^{-27} g

Total mass 11.69×10^{-24} g

proton fraction of mass = 5.01/11.69 = 0.429
neutron fraction of mass = 6.68/11.69 = 0.571
electron fraction of mass = $2.7 \times 10^{-27}/11.69 \times 10^{-24} = 2.3 \times 10^{-4}$

The electrons clearly constitute a very small fraction of the atomic mass, negligible in comparison with the masses of protons and neutrons.

<u>2.20</u> $2.7\ \text{Å}\left(\frac{10^{-10}\ \text{m}}{1\ \text{Å}}\right)\left(\frac{10^{12}\ \text{pm}}{1\ \text{m}}\right) = 270\ \text{pm}$ $\quad 2.7\ \text{Å}\left(\frac{10^{-10}\ \text{m}}{1\ \text{Å}}\right)\left(\frac{10^{9}\ \text{nm}}{1\ \text{m}}\right) = 0.27\ \text{nm}$

$$\frac{1\ \text{Ir}}{0.27\ \text{nm}}\left(\frac{10^{9}\ \text{nm}}{1\ \text{m}}\right)\left(\frac{1\ \text{m}}{10^{3}\ \text{nm}}\right) = 3.7 \times 10^{6}\ \text{Ir atoms/mm}$$

<u>2.21</u> Volume = $(4/3)\pi r^3 = 4/3\pi(0.75\ \text{km})^3 = 1.8\ \text{km}^3$

density = 2×10^{30} kg/1.8 km^3 = 1×10^{30} kg/km^3 = 1×10^{18} g/cm^3
to be compared with 0.07 g/cm^3 for liquid H_2!

<u>2.22</u> (a) Must contain the same number of protons, neutrons and electrons. (b) Must contain the same number of protons and electrons. (c) Must contain the same total number of protons plus neutrons.

<u>2.23</u> (a) ^{13}C 6 protons, 7 neutrons, 6 electrons
(b) ^{55}Mn 25 protons, 30 neutrons, 25 electrons
(c) ^{97}Mo 42 protons, 55 neutrons, 42 electrons

<u>2.24</u> (a) ^{6}Li 3 protons, 3 neutrons, 3 electrons
(b) ^{57}Fe 26 protons, 31 neutrons, 26 electrons
(c) ^{27}Al 13 protons, 14 neutrons, 13 electrons
(d) ^{19}F 9 protons, 10 neutrons, 9 electrons

<u>2.25</u> (a) $^{39}_{19}K$; (b) $^{35}_{17}Cl$; (c) $^{29}_{14}Si$; (d) $^{32}_{16}S$, $^{34}_{16}S$

<u>2.26</u> (a) Mn, a metal; (b) Se, a nonmetal; (c) Br, a nonmetal; (d) Zn, a metal; (e) Cr, a metal; (f) Ge, a metalloid; (g) S, a nonmetal

<u>2.27</u> (a) Hydrogen, a nonmetal; (b) potassium, a metal; (c) nitrogen, a nonmetal; (d) antimony, a metalloid; (e) fluorine, a nonmetal; (f) barium, a metal; (g) cadmium, a metal; (h) cesium, a metal; (i) selenium, a nonmetal.

<u>2.28</u> (a) Ca, calcium, and Sr, strontium, are both elements of group 2A, located in adjacent rows. One expects that they will be closely similar in chemical and physical properties. (b) Arsenic, As, antimony, Sb, and nitrogen, N, are all members of group 5A. However, nitrogen is quite removed from the heavier elements As and Sb. These two are thus the most closely similar pair.

<u>2.29</u> The total number of molecules in the two flasks is the same. In the one, however, half the molecules are H_2, half are Cl_2. In the other, all

the molecules are HCl. If a spark were struck in the first flask, an explosion would result. In the second flask, a spark would have no effect.

2.30 (a) P_2O_5; (b) CH_2; (c) C_3H_7; (d) SiO_2; (e) $Na_2B_4O_7$; (f) C_2O_2H

2.31 (a) K^+; (b) F^-; (c) Sc^{3+}; (d) Ca^{2+}; (e) O^{2-}; (f) At^-

2.32 (a) Li_2O; (b) SrS; (c) $MgCl_2$; (d) ScF_3; (e) CsBr

2.33 (a) Ca; (b) Kr; (c) Br; (d) Zn

2.34 (a) S; forms S^{2-} ion; (b) Kr, a noble gas element; (c) Cs; forms Cs^+ ion; (d) Sb

2.35 (a) H_2CO; (b) CH; (c) CHCl; (d) C_5H_7N; (e) P_2S_5; (f) HgCl Acetic acid, benzene, dichloroethylene and nicotine are organic substances.

2.36 (a) chromium(II), or chromous ion; (b) strontium ion (or strontium(II); the Roman numeral is not really necessary, because strontium is never seen in any other positive oxidation state); (c) bromide ion; (d) sulfide ion; (e) cesium ion; (f) titanium(III) ion

2.37 (a) sodium chlorate; (b) potassium phosphate; (c) titanium(III) bromide; (d) barium hydrogen sulfite or barium bisulfite; (e) zinc cyanide

2.38 (a) $CaCl_2$; (b) KCN; (c) $NaHCO_3$; (d) $Al(HSO_4)_3$; (e) CuBr; (f) $FeCl_3$; (g) $CuCO_3$; (h) $(NH_4)_2SO_4$; (i) NiO_2; (j) $KMnO_4$; (k) BaO_2; (l) $Zn_3(PO_4)_2$.

2.39 (a) $VBr_3(s)$; (b) $Zn(HCO_3)_2(s)$, $ZnO(s)$, $CO_2(g)$, $H_2O(g)$; (c) HF(aq), $SiO_2(s)$, $SiF_4(g)$, $H_2O(g)$; (d) $SO_2(g)$, $H_2O(\ell)$, $H_2SO_3(aq)$, $H_2SO_4(aq)$; (e) PH_3; (f) $HClO_4(aq)$, Cd(s), $Cd(ClO_4)_2(aq)$

2.40 (a) $NaNO_2$; (b) $Ca(ClO)_2$; (c) H_3PO_3 (PO_4^{3-} is phosph<u>ate</u>, thus PO_3^{3-} is phosph<u>ite</u>, and the corresponding acid is phosphor<u>ous</u> acid.); (d) $KHSeO_4$ (Selen<u>ate</u> and biselen<u>ate</u> ions are named by analogy with sulfate, SO_4^{2-} and bisulf<u>ate</u>, HSO_4^-, since both S and Se are members of group 6A.); (e) $LiClO_3$; (f) $NaBrO_4$ (by analogy with the names of the oxyanions of chlorine.); (g) HIO_4 (by analogy with $HClO_4$, perchloric acid.); (h) $Pb_3(AsO_3)_2$ (by analogy with the naming of phosph<u>ate</u>, PO_4^{3-}, and phosph<u>ite</u>, PO_3^{3-}, ions)

2.41 (a) $KClO_3$; O_2; (b) NaClO; (c) NH_3; NH_4NO_3; (d) HF(aq); (e) $H_2S(g)$; (f) HCl(aq); $NaHCO_3$; CO_2.

2.42 (a) strontium nitrate; (b) tin(IV) bromide (or stannic bromide); (c) silver nitrate; (d) potassium cyanide; (e) aluminum hydroxide; (f) sodium hydrogen sulfate, or sodium bisulfate; (g) ammonium sulfate; (h) iodic acid; (i) copper(I) bromide, or cuprous bromide

2.43 (a) calcium hydrogen carbonate, or calcium bicarbonate; (b) sodium carbonate; (c) cobalt(II) chloride; (d) barium hydroxide; (e) sodium peroxide; (f) sodium oxide; (g) potassium sulfide; (h) hydrogen iodide (in aqueous solution it is called hydriodic acid); (i) hydrogen selenate (selenic acid)

2.44 (a) <u>homogeneous</u> indicates the same throughout, as opposed to <u>heterogeneous</u>, which indicates variation throughout a material; (b) a <u>gas</u> expands so as to uniformly occupy all of the volume containing it. A liquid is of fixed volume, but is fluid, i.e., it adapts to the shape of the volume it occupies. Gases are highly compressible, liquids are not; (c) atomic number is a measure of the number of protons in the atomic nucleus; the mass number gives the number of protons plus neutrons; (d) a chemical property is a characteristic of a substance as it undergoes a chemical change; that is, as it is converted into one or more other substances. A physical property, for example the melting point of a solid, is exhibited by a substance while it retains its chemical properties and composition; (e) the symbol Ca refers to the neutral calcium atom, with 20 electrons. The symbol Ca^{2+} refers to the calcium ion, which has two electrons fewer than the neutral atom; (f) hydrochloric acid is HCl; chloric acid is the oxyacid, $HClO_3$; (g) iron(II) and iron (III) are two ionic forms of the element iron. In the iron(II) (or ferrous) form, the element has lost two electrons; in the iron(III) or ferric form, the element has lost three electrons; (h) sodium carbonate is Na_2CO_3; sodium bicarbonate is $NaHCO_3$; (i) H_2O is water; H_2O_2 is hydrogen peroxide; (j) a metal is an element with characteristic metallic properties, found mainly on the left side of the periodic table. Nonmetals are elements with characteristic properties different from those of metals; nonmetals are found on the upper right portion of the periodic table; (k) H represents the element hydrogen with 1 proton in the nucleus and one external electron per atom. He represents helium, which has 2 protons per nucleus, and two electrons in each atom; (l) chloride ion is Cl^-; chlorate ion is ClO_3^-.

2.45 (a) An alpha particle is a helium nucleus, with mass number four and charge of 2+. A beta particle is an electron, which has a very small mass relative to the proton, and a charge of -1; (b) a proton has mass number 1, charge 1+; a neutron has mass number 1, charge zero; (c) an ion may consist of one or more atoms, and it bears a charge. An atom is neutral; (d) the empirical formula gives the ratio of the numbers of atoms present in a substance, reduced to the lowest possible whole numbers. The molecular formula gives the actual numbers of atoms of each element present in each molecule of a substance that exists in the form of discrete molecules; (e) chemical change results in conversion of one or more substances into other substances with different properties. A physical change does not result in a change in the essential chemical character of the substance underoing change; (f) A solid has a rigid structure. It maintains a fixed shape independent of its container, and is relatively incompressible. A liquid is also not easily compressed, but it does flow to follow the shape of its container; (g) O represents the neutral oxygen atom, with 8 electrons; O^{2-} represents the oxide ion, with 10 electrons surrounding the nucleus; (h) C represents the element carbon, atomic number 6; Ca represents calcium, atomic number 20; (i) sodium sulfate is Na_2SO_4; sodium sulfite is Na_2SO_3; (j) sodium hydrogen phosphate is Na_2HPO_4, sodium dihydrogen phosphate is NaH_2PO_4; (k) cuprous chloride is CuCl; cupric chloride is $CuCl_2$; (l) O_2 represents diatomic oxygen, the stable form of the element; O_3 represents ozone, a higher energy form of the element.

2.46 (a) mercury-liquid; (b) iron-solid; (c) aluminum-solid; (d) oxygen-gas; (e) alcohol-liquid; (f) water-liquid; (g) hydrogen-gas; (h) helium-gas; (i) chlorine-gas

2.47 compressibility - much higher for gases than for solids or liquids. ability to retain shape - only solids have it. density - much higher for solids and liquids than for gases.

2.48 (a) lithium, Li; (b) aluminum, Al; (c) neon, Ne

2.49

Symbol	$^{12}_{6}C$	$^{17}_{8}O^{2-}$	$^{25}_{12}Mg$	$^{23}_{11}Na^{+}$	$^{18}_{8}O^{2-}$
Protons	6	8	12	11	8
Neutrons	6	9	13	12	10
Electrons	6	10	12	10	10
Net Charge	0	2-	0	1+	2-

2.50

Symbol	$^{37}_{17}Cl^{-}$	$^{40}_{18}Ar$	$^{44}_{21}Sc^{3+}$
Protons	17	18	21
Neutrons	20	22	23
Electrons	18	18	18
Net Charge	1-	0	3+

2.51

Symbol	$^{9}_{4}Be^{2+}$	$^{80}_{35}Br$
Atomic Number	4	35
Mass Number	9	80
Net Charge	2+	0

2.52 (a) ^{16}O and ^{18}O possess the same number of protons in the nucleus, 8, and the same number of electrons in the atom, also 8 for the neutral atoms. These electrons are the major factor in determining the chemical properties of the elements. ^{16}O and ^{18}O differ only in the number of neutrons in the nuclei, not a chemically important difference. (b) The alpha particle is a helium nucleus, of mass number 4 and charge 2+. The beta particle is identical with an electron, which has a charge of -1, and a very small mass in comparison with 1 amu.

2.53 (a) $CdCl_2$; (b) CdS; (c) ZnO (Here we rely on the expectation that Zn and Cd, as members of the same family of elements, will exhibit similar chemical behavior.)

2.54 (a) Dalton formulated the first modern version of the theory that matter is composed of atoms and that the atoms of each element are essentially identical in their chemical behavior. (b) Rutherford first established the charges and other characteristics of radioactive emissions. Secondly, he and coworkers established the nuclear nature of atoms.

(c) Thomson first established the ratio of charge to mass, e/m, for the electron. (d) Millikan first determined the charge of the electron with fairly high precision. (e) Becquerel was the first to note that radioactivity is a property of substances; that is, that certain substances spontaneously emit radioactivity.

2.55 (a) density - silver is much denser; melting point - 960°C for silver, 660°C for aluminum; electrical conductivity, mechanical properties such as malleability, etc, also differ. Silver is less reactive chemically, though it is not so easy to observe this fact, because aluminum forms a protective oxide coat. (b) density - water is denser; boiling point - water has a higher boiling point; reaction with oxygen - alcohol burns; odor - alcohol has characteristic smell.

2.56 Elements are I_2, S_8, O_3. Compounds are CO_2, NH_3, H_2O_2.

2.57 Gallium and aluminum are in the same family (3A); sulfur and oxygen are both in group 6A. (a) Ga_2O_3; (b) Al_2S_3; (c) Ga_2S_3

2.58 A family is a group of elements located in a vertical column on the periodic table. Elements that are related in this way possess characteristically similar chemical and physical properties, because they have similar arrangements of electrons.

2.59 The term "family" is used to refer to a group of elements that have similar chemical and physical properties. They have such similar properties because the electronic arrangements in the outer periphery of the atoms are the same. We can also say that they bear the same relationship to the nearest rare gas element (Mg as compared with Ne, Ca as compared with Ar, Sr as compared with Kr.) On the other hand, K, Ca and Sc each have differing numbers of electrons as compared with Ar, and thus have different chemical and physical characteristics.

3

Conservation of mass; stoichiometry

3.1 Among the most common examples are combustion processes such as burning of gasoline, oil and coal. Detergents used in cleaning products of all kinds eventually become part of the sewage that may cause undesirable growth of algae. Chemicals applied to control insects or weeds may end up polluting water supplies and harming wildlife. Fertilizers that can improve crop production may contaminate streams and rivers. Lead, added to gasoline to improve burning properties and thus increase octane rating, is deposited in the environment in forms that are a source of lead poisoning to human and animal populations. Iron used in the manufacture of many useful articles is oxidized (rusting).

3.2 Equations (a) and (d) are consistent with the law of conservation of mass, because there are precisely the same numbers of atoms of each element on both sides. In (b), there are two oxygens on the left, only one on the right. In (c), there should be $2H_2O$ on the right to achieve a mass balance.

3.3 (a) $2Al(s) + 3Cl_2(g) \rightarrow 2AlCl_3(s)$
(b) $P_2O_3(s) + 3H_2O(l) \rightarrow 2H_3PO_3(aq)$
(c) $Ca(OH)_2(aq) + 2HBr(aq) \rightarrow CaBr_2(aq) + 2H_2O(l)$
(d) $16Al(s) + 3S_8(s) \rightarrow 8Al_2S_3(s)$
(e) $Mg_2C_3(s) + 4H_2O(l) \rightarrow 2Mg(OH)_2(aq) + C_3H_4(g)$

3.4 (a) $C(s) + 2F_2(g) \rightarrow CF_4(g)$
(b) $N_2O_5(s) + H_2O(l) \rightarrow 2HNO_3(l)$
(c) $C_2H_4(g) + 3O_2(g) \rightarrow 2CO_2(g) + 2H_2O(g)$
(d) $La_2O_3(s) + 3H_2O(l) \rightarrow 2La(OH)_3(s)$
(e) $2HCl(g) + CaO(s) \rightarrow CaCl_2(s) + H_2O(g)$
(f) $Pb(NO_3)_2(aq) + 2NaCl(aq) \rightarrow PbCl_2(s) + 2NaNO_3(aq)$

3.5 (a) $2PH_3(g) + 4O_2(g) \rightarrow P_2O_5(s) + 3H_2O(g)$
(b) $Ba(s) + 2CH_3OH(l) \rightarrow H_2(g) + Ba(OCH_3)_2(alc)$
(c) $B_2S_3(s) + 6H_2O(l) \rightarrow 2H_3BO_3(aq) + 3H_2S(g)$
(d) $Cu(s) + 2H_2SO_4(l) \rightarrow CuSO_4(s) + SO_2(g) + 2H_2O(\ell)$
(e) $2NH_3(g) + 2Na(l) \rightarrow H_2(g) + 2NaNH_2(s)$

3.6 (a) $HCNO(g) + H_2O(l) \rightarrow NH_3(g) + CO_2(g)$
(b) $2Hg(NO_3)_2(s) \rightarrow 2HgO(s) + 4NO_2(g) + O_2(g)$
(c) $PCl_3(l) + 3H_2O(l) \rightarrow H_3PO_3(aq) + 3HCl(aq)$
(d) $2KNO_3(s) \rightarrow 2KNO_2(s) + O_2(g)$
(e) $3H_2S(g) + 2Fe(OH)_3(s) \rightarrow Fe_2S_3(s) + 6H_2O(g)$

3.7 (a) $CaH_2(s) + 2H_2O(l) \rightarrow Ca(OH)_2(s) + 2H_2(g)$
(b) $CO(g) + 3H_2(g) \rightarrow H_2O(g) + CH_4(g)$
(c) $Na_2CuCl_4(aq) + Cu(s) \rightarrow 2NaCuCl_2(aq)$
(d) $10HNO_3(aq) + 4Zn(s) \rightarrow NH_4NO_3(aq) + 3H_2O(l) + 4Zn(NO_3)_2(aq)$
(e) $FeSO_4(s) + H_2S(g) \rightarrow FeS(s) + H_2SO_4(l)$
(f) $2Fe_2S_3(s) + 6H_2O(g) + 3O_2(g) \rightarrow 4Fe(OH)_3 + 6S(s)$

3.8 (a) $C_2H_5OH(l) + 3O_2(g) \rightarrow 2CO_2(g) + 3H_2O(\ell)$
(b) $Pb(NO_3)_2(aq) + Na_2CrO_4(aq) \rightarrow PbCrO_4(s) + 2NaNO_3(aq)$
(c) $2K(s) + 2HNO_3(aq) \rightarrow 2KNO_3(aq) + H_2(g)$
(d) $H_2SO_4(aq) + 2KOH(aq) \rightarrow K_2SO_4(aq) + 2H_2O(l)$ or
$Ca(OH)_2(s) + H_2SO_4(aq) \rightarrow CaSO_4(s) + 2H_2O(l)$
(e) $Ca(OH)_2(s) + 2HNO_3(aq) \rightarrow Ca(NO_3)_2(aq) + 2H_2O(l)$
NOTE: Reaction of $Ca(OH)_2$ with H_2SO_4 leads to formation of insoluble $CaSO_4$.
(f) $H_2SO_4(aq) + BaCl_2(aq) \rightarrow BaSO_4(s) + 2HCl(aq)$

3.9 (a) $4CH_3NO_2(g) + 7O_2(g) \rightarrow 4NO_2(g) + 4CO_2(g) + 6H_2O(g)$
(b) $2K(s) + 2NH_3(l) \rightarrow 2KNH_2(sol) + H_2(g)$
(c) $CH_4(g) + 4F_2(g) \rightarrow CF_4(g) + 4HF(g)$
(d) $Zn(s) + 2HCl(aq) \rightarrow ZnCl_2(aq) + H_2(g)$

3.10 (a) $SOCl_2$, FW = AW(S) + AW(O) + 2AW(Cl) = 118.98; (b) $Tl_2(C_2O_4)$, FW = 496.80; (c) $C_9H_6O_2$, FW = 146.14; (d) $C_{16}H_{22}O_8 \cdot 2H_2O$, FW = 378.4; (e) XeF_4, FW = 207.29

3.11 (a) mol. wt. CO_2 = 12.011 + 2(16.00) = 44.01 g/mol

$$\left(\frac{44.01 \text{ g } CO_2}{1 \text{ mol } CO_2}\right) 2 \text{ mol } CO_2 = 88.02 \text{ g } CO_2$$

(b) $$3.58 \times 10^{22} \text{ atoms Kr} \left(\frac{1 \text{ mol Kr}}{6.022 \times 10^{23} \text{ atoms}}\right)\left(\frac{83.8 \text{ g Kr}}{1 \text{ mol Kr}}\right) = 4.98 \text{ g Kr}$$

(c) $$4.83 \times 10^{24} \text{ molec HCl} \left(\frac{1 \text{ mol Kr}}{6.022 \times 10^{23} \text{ molec}}\right)\left(\frac{36.5 \text{ g HCl}}{1 \text{ mol HCl}}\right) = 293 \text{ g HCl}$$

(d) $$0.0090 \text{ mol } C_2H_4 \left(\frac{28.0 \text{ g } C_2H_4}{1 \text{ mol } C_2H_4}\right) = 0.25 \text{ g } C_2H_4$$

3.12 $$33.4 \text{ g } C_3H_7OH \left(\frac{1 \text{ mol } C_3H_7OH}{60.10 \text{ g } C_3H_7OH}\right)\left(\frac{3 \text{ mol C}}{1 \text{ mol } C_3H_7OH}\right)\left(\frac{6.022 \times 10^{23} \text{ C atoms}}{1 \text{ mol C}}\right)$$

$$= 1.00 \times 10^{24} \text{ atoms C.}$$

wt. percentage C is given by

$$\left(\frac{1 \text{ mol } C_3H_7OH}{60.10 \text{ g } C_3H_7OH}\right)\left(\frac{3 \text{ mol C}}{1 \text{ mol } C_3H_7OH}\right)\left(\frac{12.01 \text{ g C}}{1 \text{ mol C}}\right)(100) = 59.9 \text{ \% C}$$

3.13 No. If we accept Avogadro's hypothesis that equal volumes of gases at the same temperature and pressure contain equal numbers of "particles," then we must conclude that there are 2 HCl molecules for each particle of hydrogen or chlorine that reacts. But if those particles were only single atoms of hydrogen or chlorine, it would be necessary to divide them in half to form two hydrogen chloride particles. An atom, by definition, is not divisible in a chemical reaction. Hence, hydrogen and chlorine must be diatomic; that is, H_2 and Cl_2.

The experiment also tells us that the formula for hydrogen chloride must contain hydrogen and chlorine in the same ratio as the formulas for the two gases. The simplest assumption is that both elements exist as diatomic gases. However, it is conceivable that another ratio could exist. For example, suppose that element X exists as $X_2(g)$ and element Y as $Y_4(g)$. Reaction of 1 volume of $X_2(g)$ with one volume of $Y_4(g)$ to give two volumes of product would then necessarily form $XY_2(g)$:

$$\underset{\text{1 volume}}{X_2(g)} + \underset{\text{1 volume}}{Y_4(g)} \rightarrow \underset{\text{2 volumes}}{2XY_2(g)}$$

Note that the ratio of X to Y in the product is the same as the ratio of X to Y in X_2/Y_4.

3.14 Using the same ideas as outlined in the solution to problem 3.31 above, we have

$$Cl_2(g) + 3F_2(g) \rightarrow 2ClF_3(g)$$

Note that the coefficients before the formulas in the equation correspond to the relative volumes of the gases involved. It follows that the product gas must be ClF_3.

3.15 $$(5 \text{ g } ^{12}C)\left(\frac{6.0220 \times 10^{23} \text{ atoms } ^{12}C}{12.000 \text{ g } ^{12}C}\right) = 2.5092 \times 10^{23} \text{ atoms } ^{12}C$$

$$(5 \text{ g C})\left(\frac{6.0220 \times 10^{23} \text{ atoms C}}{12.011 \text{ g C}}\right) = 2.5069 \times 10^{23} \text{ atoms C}$$

These numbers are different because naturally occurring carbon contains about 1% ^{13}C, a nuclide of higher mass than ^{12}C. Thus, the average carbon atom in nature has higher mass than an atom of ^{12}C.

$$(5 \text{ g Ag})\left(\frac{6.0220 \times 10^{23} \text{ atoms Ag}}{107.87 \text{ Ag}}\right) = 2.7913 \times 10^{22} \text{ atoms Ag}$$

3.16 $$3 \times 10^5 \text{ g Al}\left(\frac{1 \text{ mol Al}}{27.0 \text{ g Al}}\right)\left(\frac{1 \text{ mol Na}}{1 \text{ mol Al}}\right)\left(\frac{23.0 \text{ g Na}}{1 \text{ mol Na}}\right) = 2.56 \times 10^5 \text{ g Na}$$

$$= 256 \text{ kg Na}$$

3.17 We assume that the overall reaction is the same:

$$2Na(s) + 2H_2O(l) \rightarrow 2NaOH(aq) + H_2(g)$$

The volume of H_2 gas formed is proportional to the number of moles of metal reacted. Thus our unknown metal is present to the extent of (0.497/1.14) as many moles as Na.

$$2.1 \text{ g Na}\left(\frac{1 \text{ mol Na}}{23.0 \text{ g Na}}\right)\left(\frac{0.497 \text{ mol X}}{1.14 \text{ mol Na}}\right) = 0.040 \text{ mol X}$$

The atomic weight of X is then $\frac{3.4 \text{ g X}}{0.040 \text{ mol X}} = 85$ g/mol. The element involved is rubidium, Rb.

3.18 $\frac{4.032 \text{ g Zn}}{1 \text{ g O}}\left(\frac{16.00 \text{ g O}}{1 \text{ mol O}}\right)\left(\frac{2 \text{ mol O}}{1 \text{ mol Zn}}\right) = \frac{129.0 \text{ g Zn}}{1 \text{ mol Zn}}$

Berzelius' value for the atomic weight of Zn is a factor of 2 too high. His erroneous assumption was in the assumed formula for the oxide, which is ZnO, not ZnO_2 as he assumed. Correcting this leads to a value of 64.50 g Zn/mol, in good agreement with the currently accepted value.

3.19 We have that the ratio of the masses $KClO_3/KCl$ is experimentally determined to be 1.64382. Let us call the mass of KCl x. Then

$$[x + 3(15.9994)]/x = 1.64382$$

Solving for x we obtain

$$x + 47.9982 = 1.64382x$$

$$x = 47.9982/0.64382 = 74.552$$

The sum of the atomic weights of K and Cl is 74.552.

3.20 Proceeding as in problem 3.19, we write the ratio of masses of Ag to $AgNO_3$, and call the unknown quantity, the atomic mass of nitrogen, y:

$$\frac{Ag}{AgNO_3} = 0.634985 = \frac{107.870}{107.870 + 3(15.9994) + y}$$

Solve for y to obtain y = 14.010. This is to be compared with the currently accepted value of 14.007.

3.21 (a) "a mole of argon" means 6.022×10^{23} argon atoms. (b) "a mole of potassium bromide" means a mass of this salt that corresponds to 6.022×10^{23} KBr units in the solid ionic lattice. (c) "a mole of methyl alcohol" means 6.022×10^{23} molecules of methyl alcohol, CH_3OH.

3.22 (a) The formula weight is simply the sum of the atomic weights. Thus, FW = 3(12.01) + 8(1.008) = 44.09 g; (b) 64.06; (c) 32.12; (d) 65.14; (e) 65.99

3.23 (a) 299.08; (b) 167.00; (c) 151.99; (d) 158.04; (e) 74.10

3.24 We want the average mass of a neon atom which is given by the sum of the mass of each isotope present in the sample times its fractional abundance:

$$19.99(0.9092) + 20.99(0.0025) + 21.99(0.0883) = 20.17 \text{ amu}$$

3.25 We can write the fraction of ^{107}Ag as x; the fraction of ^{109}Ag is then 1-x. Then

$$x(106.905) + (1-x)(108.905) = 107.870$$

Solving for x yields 0.5175. The fraction of silver which is ^{109}Ag is then 1 - 0.5175 = 0.4825.

3.26 It is simplest to obtain the average mass of the PbO^+ ion, then subtract 15.9948 to obtain the average mass of Pb^+; Mass PbO^+ =

$$(220.002)(.0137) + (222.056)(0.2630) + (223.050)(0.2080) + (224.055)(0.5153) = 207.27$$

Mass Pb^+ = 223.264 - 15.9948

In fact, this value for the atomic weight of Pb is good to the full five significant figures, because the effect of an uncertainty of 0.0001 in the fractions is not that serious in the overall average. The mass of the electron missing from Pb^+ can be ignored; it is smaller than the uncertainty in the estimated atomic mass.

3.27 $0.068 \text{ g } C_5H_5N \left(\frac{1 \text{ mol } C_5H_5N}{79.1 \text{ g } C_5H_5N}\right)\left(\frac{6.02 \times 10^{23} \text{ molecules}}{1 \text{ mol}}\right)$

$$= 5.2 \times 10^{20} \; C_5H_5N \text{ molecules}$$

$$5.0 \text{ g ZnO} \left(\frac{1 \text{ mol ZnO}}{81.37 \text{ g ZnO}}\right)\left(\frac{6.02 \times 10^{23} \text{ ZnO}}{1 \text{ mol}}\right) = 3.7 \times 10^{22} \text{ ZnO}$$

There is one C_5H_5N molecule for each $3.7 \times 10^{22}/5.2 \times 10^{20} = 71$ ZnO units.

$$\left(\frac{48 \text{ m}^2}{\text{g ZnO}}\right)(5 \text{ g ZnO})\left(\frac{1}{5.2 \times 10^{20} \text{ } C_5H_5N}\right) = 4.6 \times 10^{-19} \text{ m}^2\text{/molecules}$$

$$\left(\frac{4.6 \times 10^{-19} \text{ m}^2}{C_5H_5N \text{ molecule}}\right)\left(\frac{10^{10} \text{ Å}}{1 \text{ m}}\right)^2 = \frac{46 \text{ Å}^2}{C_5H_5N \text{ molecule}}$$

3.28 $$\frac{2.05 \times 10^{-6} \text{ g } C_2H_3Cl}{1 \text{ L}}\left(\frac{1 \text{ mol } C_2H_3Cl}{62.5 \text{ g } C_2H_3Cl}\right) = \frac{3.28 \times 10^{-8} \text{ mol } C_2H_3Cl}{\text{L}}$$

$$\times \left(\frac{6.02 \times 10^{23} \text{ molecules}}{1 \text{ mol}}\right) = \frac{1.97 \times 10^{16} \text{ molecules } C_2H_3Cl}{\text{L}}$$

3.29 $$2.5 \times 10^{-5} \text{ g } C_{12}H_{30}O_2 \left(\frac{1 \text{ mol } C_{12}H_{30}O_2}{206 \text{ g } C_{12}H_{30}O_2}\right) = 1.2 \times 10^{-7} \text{ mol } C_{12}H_{30}O_2$$

$$1.27 \times 10^{-7} \text{ mol } C_{12}H_{30}O_2 \left(\frac{6.02 \times 10^{23} \text{ molecules}}{1 \text{ mol}}\right) = 7.3 \times 10^{16} \text{ molecules } C_{12}H_{30}O_2$$

3.30 (a) $$4.2 \text{ g } I_2 \left(\frac{1 \text{ mol } I_2}{254 \text{ g } I_2}\right)\left(\frac{6.02 \times 10^{23} \text{ molecules}}{1 \text{ mol}}\right) = 1.0 \times 10^{22} \text{ molecules } I_2$$

(b) 5.2×10^{-5}g; (c) 2.1×10^{22} Ag atoms; (d) 8.1×10^{-3} g $C_6H_{12}O_6$; (e) 2.08 mol H_2O

3.31 $$182{,}000 \text{ mol wt} \left(\frac{1 \text{ } C_2H_4 \text{ unit}}{28.0 \text{ g}}\right) = 6.49 \times 10^3 \text{ } C_2H_4 \text{ units}$$

3.32 (a) Let us assume 1 g sample. Then

$$\text{mol N} = 0.280 \text{ g N}\left(\frac{1 \text{ mol N}}{14.0 \text{ g N}}\right) = 0.0200 \text{ mol N}$$

$$\text{mol Ag} = 0.720 \text{ g Ag}\left(\frac{1 \text{ mol Ag}}{107.9 \text{ g Ag}}\right) = 6.67 \times 10^{-3} \text{ mol Ag}$$

Ratio N/Ag = $0.0200/6.67 \times 10^{-3} = 3.00$. Thus, the formula is AgN_3. Similarly, (b) Mn_3C; (c) $CSCl_2$; (d) ICN

3.33 The ratio involved is

$$Sr(OH)_2\cdot XH_2O(s) \rightarrow Sr(OH)_2(s) + XH_2O(g)$$

Let us first calculate the number of moles of product $Sr(OH)_2$; this is the same as the number of moles of starting hydrate. We can write the formula wt of the hydrate as

$$FW(Sr(OH)_2\cdot XH_2O) = FW(Sr(OH)_2) + XFW(H_2O)$$

$$3.13 \text{ g } Sr(OH)_2 \left(\frac{1 \text{ mol } Sr(OH)_2}{121 \text{ g } Sr(OH)_2}\right)\left(\frac{1 \text{ mol } Sr(OH)_2\cdot XH_2O}{1 \text{ mol } Sr(OH)_2}\right)$$

$$= 0.0259 \text{ mol } Sr(OH)_2 \cdot XH_2O$$

Thus, $$\frac{6.85 \text{ g } Sr(OH)\cdot XH_2O}{0.0259 \text{ mol}} = 264 \text{ g} = \text{FW}$$

$264 = 121 + x(18.0)$. Thus, $x = 7.94$. The formula of the hydrate is $Sr(OH)_2\cdot 8H_2O$. Alternatively, we could calculate the number of moles of water represented by the weight loss:

$$(6.85 - 3.13)\ g\ H_2O \left(\frac{1\ mol\ H_2O}{18.0\ g\ H_2O}\right) = 0.207\ mol\ H_2O$$

$$\frac{mol\ H_2O}{mol\ Sr(OH)_2} = \frac{0.207}{0.0259} = 7.98.$$ Thus the correct formula is $Sr(OH)_2{\cdot}8H_2O$.

3.34 The weight percentage is determined by the relative number of atoms of the element times the atomic weight, divided by the total formula mass. Thus, the weight percentage of bromine in $KBrO_x$ is given by

$$0.5292 = \frac{79.91}{39.10 + 79.91 + x(16.00)}$$. Solving for x, we obtain x = 2.00. Thus, the formula is $KBrO_2$.

3.35 The wt. percentage of P in each case is obtained by dividing the mass of phosphorus in each formula by the total formula mass. We obtain: (a) 18.4%; (b) 24.5%; (c) 23.4%. The weight percentage of P is highest in $Ca(H_2PO_4)_2{\cdot}H_2O$.

3.36 $$1.45\ g\ CO_2 \left(\frac{1\ mol\ CO_2}{44\ g}\right)\left(\frac{12\ g\ C}{1\ mol\ CO_2}\right) = 0.395\ g\ C$$

$$0.890\ g\ H_2O \left(\frac{1\ mol\ H_2O}{18.0\ g\ H_2O}\right)\left(\frac{2\ g\ H}{1\ mol\ H_2O}\right) = 0.099\ g\ H$$

mass of oxygen = 1.55 - 0.49 = 1.056 g O

Dividing each of these masses by the appropriate atomic weight gives us:

moles C = 0.395/12 = 0.033
moles H = 0.099/1 = 0.099
moles O = 1.056/16 = 0.066

Thus the empirical formula is CH_3O_2.

3.37 $$0.608\ g\ BaSO_4 \left(\frac{1\ mol\ BaSO_4}{233\ g\ BaSO_4}\right)\left(\frac{1\ mol\ Ba}{1\ mol\ BaSO_4}\right)\left(\frac{137\ g\ Ba}{1\ mol\ Ba}\right) = 0.357\ g\ Ba$$

wt. percentage Ba = 0.357/0.666 g = 53.6 %

Suppose we have one g of the substance. Of this, 0.536 g is Ba, or (0.536/137) = 3.91×10^{-3} mol Ba. If there is but 1 Ba atom in the empirical formula, this means that 1 g represents also 3.91×10^{-3} mol of the empirical formula. The formula weight is thus

$$1\ g/3.91 \times 10^{-3}\ mol = 256\ g/mol.$$

3.38 $$6.32 \times 10^{-3}\ g\ CO_2 \left(\frac{12\ g\ C}{44\ g\ CO_2}\right) = 1.72\ mg\ C$$

$$2.58 \times 10^{-3}\ g\ H_2O \left(\frac{2.02\ g\ H}{18\ g\ H_2O}\right) = 0.290\ mg\ H$$

mass of O = 2.78 - (1.72 + 0.29) = 0.77 mg O

Dividing by the appropriate atomic masses, then reducing to the lowest ratio of integers gives us C_3H_6O for the empirical formula. The formula weight is 58. Thus, if the molecular weight is between 100 and 150, the correct molecular formula is $C_6H_{12}O_2$.

3.39 (a) $P_4(s) + 5O_2(g) \rightarrow 2P_2O_5(s)$

$$4.00\ g\ P_4 \left(\frac{1\ mol\ P_4}{124\ g\ P_4}\right)\left(\frac{2\ mol\ P_2O_5}{1\ mol\ P_4}\right) = 6.45 \times 10^{-2}\ mol\ P_2O_5$$

(b) From the equation given, we see that 2 moles of H_3PO_4 are formed from 1 mole of P_2O_5. Thus:

$$0.0645 \text{ mol } P_2O_5 \left(\frac{2 \text{ mol } H_3PO_4}{1 \text{ mol } P_2O_5}\right)\left(\frac{98.0 \text{ g } H_3PO_4}{1 \text{ mol } H_3PO_4}\right) = 12.6 \text{ g } H_3PO_4$$

3.40 (a) $$2.55 \text{ g } Li_3N \left(\frac{1 \text{ mol } Li_3N}{34.8 \text{ g } Li_3N}\right)\left(\frac{1 \text{ mol } NH_3}{1 \text{ mol } Li_3N}\right)\left(\frac{17.0 \text{ g } NH_3}{1 \text{ mol } NH_3}\right) = 1.24 \text{ g } NH_3$$

(b) 3.57 g $CaCO_3$; (c) 5.29 g

(d) $$3.46 \text{ g } Cu_2O \left(\frac{1 \text{ mol } Cu_2O}{143 \text{ g } Cu_2O}\right)\left(\frac{2 \text{ mol Mg}}{1 \text{ mol } Cu_2O}\right)\left(\frac{24.3 \text{ g Mg}}{1 \text{ mol Mg}}\right) = 1.18 \text{ g Mg}$$

3.41 (a) $$2.95 \text{ g } Mg_2Si \left(\frac{1 \text{ mol } Mg_2Si}{76.7 \text{ g } Mg_2Si}\right)\left(\frac{1 \text{ mol } SiH_4}{1 \text{ mol } Mg_2Si}\right) = 0.0385 \text{ mol } SiH_4$$

(b) $$2.95 \text{ g } Mg_2Si \left(\frac{1 \text{ mol } Mg_2Si}{76.7 \text{ g } Mg_2Si}\right)\left(\frac{4 \text{ mol } H_2O}{1 \text{ mol } Mg_2Si}\right)\left(\frac{18.0 \text{ g } H_2O}{1 \text{ mol } H_2O}\right) = 2.77 \text{ g } H_2O$$

(c) $$22.5 \times 10^{-3} \text{ g } Mg_2Si \left(\frac{1 \text{ mol } Mg_2Si}{76.7 \text{ g } Mg_2Si}\right)\left(\frac{1 \text{ mol } SiH_4}{1 \text{ mol } Mg_2Si}\right)$$

$$= 2.93 \times 10^{-4} \text{ mol } SiH_4$$

$$14.0 \times 10^{-3} \text{ g } H_2O \left(\frac{1 \text{ mol } H_2O}{18.0 \text{ g } H_2O}\right)\left(\frac{1 \text{ mol } SiH_4}{4 \text{ mol } H_2O}\right) = 1.94 \times 10^{-4} \text{ mol } SiH_4$$

Thus, water is the limiting reagent. A maximum of

$$1.94 \times 10^{-4} \text{ mol } SiH_4 \left(\frac{32.1 \text{ g } SiH_4}{1 \text{ mol } SiH_4}\right) = 6.24 \text{ mg } SiH_4$$

can be formed.

3.42 $5MgO(s) + 2PCl_5(g) \rightarrow 5MgCl_2(s) + P_2O_5(s)$

$$2.5 \times 10^3 \text{ g } P_2O_5 \left(\frac{1 \text{ mol } P_2O_5}{142 \text{ g } P_2O_5}\right)\left(\frac{2 \text{ mol } PCl_5}{1 \text{ mol } P_2O_5}\right)\left(\frac{208 \text{ g } PCl_5}{1 \text{ mol } PCl_5}\right) = 7.32 \text{ kg } PCl_5$$

3.43 $Al_4C_3(s) + 12HCl(aq) \rightarrow 4AlCl_3(aq) + 3CH_4(g)$

$$1.257 \text{ g } CH_4 \left(\frac{1 \text{ mol } CH_4}{16.04 \text{ g } CH_4}\right)\left(\frac{1 \text{ mol } Al_4C_3}{3 \text{ mol } CH_4}\right)\left(\frac{144.0 \text{ g } Al_4C_3}{1 \text{ mol } Al_4C_3}\right) = 3.762 \text{ g } Al_4C_3$$

$$1.257 \text{ g } CH_4 \left(\frac{1 \text{ mol } CH_4}{16.04 \text{ g } CH_4}\right)\left(\frac{4 \text{ mol } AlCl_3}{3 \text{ mol } CH_4}\right)\left(\frac{133.3 \text{ g } AlCl_3}{1 \text{ mol } AlCl_3}\right) = 13.93 \text{ g } AlCl_3$$

3.44 $$1.24 \text{ g NaOH} \left(\frac{1 \text{ mol NaOH}}{40.0 \text{ g NaOH}}\right)\left(\frac{1 \text{ mol } Na_2S}{2 \text{ mol NaOH}}\right) = 0.0155 \text{ mol } Na_2S$$

$$2.05 \text{ g } H_2S \left(\frac{1 \text{ mol } H_2S}{34.1 \text{ g } H_2S}\right)\left(\frac{1 \text{ mol } Na_2S}{1 \text{ mol } H_2S}\right) = 0.0601 \text{ mol } Na_2S$$

NaOH is the limiting reagent. The mass of Na_2S that can be formed is

$$0.0155 \text{ mol} \left(\frac{78.0 \text{ g}}{\text{mol}}\right) = 1.21 \text{ g } Na_2S.$$

3.45 We must determine which is the limiting reagent. The balanced equation is

$$KBr(aq) + AgNO_3(aq) \rightarrow AgBr(s) + KNO_3(aq)$$

$$2.45 \text{ g KBr} \left(\frac{1 \text{ mol KBr}}{119 \text{ g KBr}}\right)\left(\frac{1 \text{ mol AgBr}}{1 \text{ mol KBr}}\right)\left(\frac{188 \text{ g AgBr}}{1 \text{ mol AgBr}}\right) = 3.87 \text{ g AgBr}$$

$$5.86 \text{ g } AgNO_3 \left(\frac{1 \text{ mol } AgNO_3}{170 \text{ g } AgNO_3}\right)\left(\frac{1 \text{ mol AgBr}}{1 \text{ mol } AgNO_3}\right)\left(\frac{188 \text{ g AgBr}}{1 \text{ mol AgBr}}\right) = 6.48 \text{ g AgBr}$$

In this case KBr is the limiting reagent; 3.87 g AgBr is formed in the precipitation reaction.

3.46 The mass of nickel converted to $Ni(CO)_4$ is

$$(2.25 \times 10^{-5} - 1.67 \times 10^{-5}) \text{ g} = 5.8 \times 10^{-6} \text{ g}$$

$$5.8 \times 10^{-6} \text{ g Ni} \left(\frac{1 \text{ mol Ni}}{58.7 \text{ g Ni}}\right)\left(\frac{1 \text{ mol } Ni(CO)_4}{1 \text{ mol Ni}}\right)\left(\frac{171 \text{ g } Ni(CO)_4}{1 \text{ mol } Ni(CO)_4}\right)$$

$$= 1.7 \times 10^{-5} \text{ g } Ni(CO)_4$$

3.47 (a) $\left(\frac{4.55 \text{ g } Na_2CrO_4}{0.250 \text{ L}}\right)\left(\frac{1 \text{ mol } Na_2CrO_4}{162.0 \text{ g } Na_2CrO_4}\right) = 0.112 \text{ M}$

(b) 0.035 M; (c) 4.54×10^{-6} M; (d) 1.2×10^{-3} M (or 1.2 mM)

3.48 $\left(\frac{0.0312 \text{ g } MnC_2O_4 \cdot 2H_2O}{0.100 \text{ L}}\right)\left(\frac{1 \text{ mol } MnC_2O_4 \cdot 2H_2O}{179.0 \text{ g } MnC_2O_4 \cdot 2H_2O}\right) = 1.74 \times 10^{-3} \text{M}$

= 1.74 mM at 25°C

Similarly, at 55°C, solubility is 3.59 mM.

3.49 (a) $\left(\frac{0.112 \text{ mol } CdF_2}{\text{L}}\right)(0.240 \text{ L})\left(\frac{150 \text{ g } CdF_2}{1 \text{ mol } CdF_2}\right) = 4.03 \text{ g } CdF_2$

(b) 0.945 g HNO_3; (c) 6.76 g NH_4CN

(d) $\left(\frac{2.2 \times 10^{-8} \text{M } Cl_4C_{12}H_6}{\text{L}}\right)\left(\frac{1 \text{ L}}{1 \times 10^3 \text{ cm}^3}\right)\left(\frac{100 \text{ cm}}{1 \text{ m}}\right)^3 (5.6 \times 10^{12} \text{m}^3)$

$$\times \left(\frac{292 \text{ g } Cl_4C_{12}H_6}{1 \text{ mol } Cl_4C_{12}H_6}\right) = 3.6 \times 10^{10} \text{ g } Cl_4C_{12}H_6$$

(Tetrachlorobiphenyl is an environmental contaminant, one of the PCB's (polychlorinated biphenyls). The concentration and volume used here are chosen to resemble Lake Michigan. Clearly, Lake Michigan contains a great deal of this and related contaminants!)

3.50 (a) $(4.5 \text{ g } NH_4Cl)\left(\frac{1 \text{ mol } NH_4Cl}{53.5 \text{ g } NH_4Cl}\right)\left(\frac{1 \text{ L Soln}}{0.600 \text{ mol } NH_4Cl}\right) = 0.14 \text{ L Soln}$

(b) $2.35 \text{ mol } H_2SO_4 \left(\frac{1 \text{ L Soln}}{6.55 \text{ mol } H_2SO_4}\right) = 0.359 \text{ L Soln}$

(c) $0.088 \text{ g As} \left(\frac{1 \text{ mol As}}{74.9 \text{ g As}}\right)\left(\frac{1 \text{ mol } H_3AsO_4}{1 \text{ mol As}}\right)\left(\frac{1 \text{ L Soln}}{0.020 \text{ mol } H_3AsO_4}\right)$

$$= 0.059 \text{ L Soln} = 59 \text{ mL Soln}$$

(d) $$1.25 \text{ g Cr} \left(\frac{1 \text{ mol Cr}}{52.0 \text{ g Cr}}\right)\left(\frac{1 \text{ mol } K_2Cr_2O_7}{2 \text{ mol Cr}}\right)\left(\frac{1 \text{ L Soln}}{0.0600 \text{ mol } K_2Cr_2O_7}\right)$$
$$= 0.200 \text{ L Soln}$$

3.51 (a) We first calculate the number of moles of $HClO_4$ in 1 L of 0.100 M perchloric acid solution. This is simply 0.100 moles $HClO_4$. Next, what volume of 2.06 M solution gives us this many moles of $HClO_4$?

$$0.100 \text{ mol } HClO_4 \left(\frac{1 \text{ L Soln}}{2.06 \text{ mol } HClO_4}\right) = 0.0485 \text{ L Soln}$$

We thus must withdraw some of the 2.06 M acid solution with a graduated pipette, and add precisely 48.5 mL to a 1-L volumetric flask containing a few hundred ml of water. Then, with swirling to obtain mixing, we add more water to dilute to precisely the 1-L mark.

(b) $$1.65 \text{ L KOH} \left(\frac{0.880 \text{ mol KOH}}{1 \text{ L Soln}}\right)\left(\frac{1 \text{ mol } HClO_4}{1 \text{ mol KOH}}\right)\left(\frac{1 \text{ L Soln}}{2.06 \text{ mol } HClO_4}\right)$$
$$= 0.705 \text{ L } HClO_4 \text{ Soln}$$

(c) $$35.0 \text{ mL Soln} \left(\frac{1 \text{ L}}{1000 \text{ mL}}\right)\left(\frac{0.125 \text{ mol } Na_2CO_3}{1 \text{ L Soln}}\right)\left(\frac{2 \text{ mol } HClO_4}{1 \text{ mol } Na_2CO_3}\right)$$
$$\times \left(\frac{1 \text{ L } HClO_4 \text{ Soln}}{0.100 \text{ mol } HClO_4}\right) = 0.0875 \text{ L } HClO_4 \text{ Soln}$$

(d) We can employ the same approach outlined in Sample Exercise 3.20:

$$0.0546 \text{ L } HClO_4 \text{ Soln} \left(\frac{0.100 \text{ mol } HClO_4}{1 \text{ L Soln}}\right)\left(\frac{1 \text{ mol NaOH}}{1 \text{ mol } HClO_4}\right)$$
$$= 5.46 \times 10^{-3} \text{ mol NaOH}$$

$$\text{molarity} = \frac{5.46 \times 10^{-3} \text{ mol NaOH}}{0.0340 \text{ L NaOH Soln}} = 0.161 \text{ M}$$

3.52 The balanced equation for the titration is:

$$Sr(NO_3)_2(aq) + Na_2CrO_4(aq) \rightarrow SrCrO_4(s) + 2NaNO_3(aq)$$

Let us begin with the given fact that we have a 0.100 L sample, then do the following conversions:

volume Soln $\rightarrow$ g $Sr(NO_3)_2$ $\rightarrow$ mol $Sr(NO_3)_2$ $\rightarrow$ mol Na_2CrO_4 $\rightarrow$ volume Na_2CrO_4 Soln

$$0.100 \text{ L Soln} \left(\frac{5.55 \text{ g } Sr(NO_3)_2}{0.750 \text{ L Soln}}\right)\left(\frac{1 \text{ mol } Sr(NO_3)_2}{212 \text{ g } Sr(NO_3)_2}\right)\left(\frac{1 \text{ mol } Na_2CrO_4}{1 \text{ mol } Sr(NO_3)_2}\right)$$
$$\times \left(\frac{1 \text{ L Soln}}{0.0460 \text{ mol } Na_2CrO_4}\right) = 0.0759 \text{ L } Na_2CrO_4 \text{ Soln}$$

3.53 $$0.0263 \text{ L HCl Soln} \left(\frac{0.110 \text{ mol HCl}}{1 \text{ L Soln}}\right)\left(\frac{1 \text{ mol RbOH}}{1 \text{ mol HCl}}\right) = 2.89 \times 10^{-3} \text{ mol RbOH}$$

$$\text{molarity} = \frac{2.89 \times 10^{-3} \text{ mol RbOH}}{0.0465 \text{ L RbOH Soln}} = 0.0622 \text{ M}$$

$$0.0622 \text{ mol RbOH} \times \frac{102 \text{ g RbOH}}{1 \text{ mol RbOH}} = 6.34 \text{ g RbOH}$$

3.54 2.94×10^{-3} L $AgNO_3$ Soln $\left(\frac{0.0130 \text{ mol } AgNO_3}{1 \text{ L Soln}}\right)\left(\frac{1 \text{ mol } ZnCl_2}{2 \text{ mol } AgNO_3}\right)$

$$\times \left(\frac{136 \text{ g } ZnCl_2}{1 \text{ mol } ZnCl_2}\right) = 2.60 \times 10^{-3} \text{ g } ZnCl_2$$

The weight percentage is $(2.60 \times 10^{-3}/2.87)\ 1.00 = 9.05 \times 10^{-2}$ %.

3.55 The neutralization reaction here is:

$$2HBr(aq) + Ca(OH)_2(aq) \rightarrow CaBr_2(aq) + 2H_2O(l)$$

$$0.0488 \text{ L HBr Soln} \left(\frac{5.00 \times 10^{-2} \text{ mol HBr}}{1 \text{ L Soln}}\right)\left(\frac{1 \text{ mol } Ca(OH)_2}{2 \text{ mol HBr}}\right)$$

$$= 1.22 \times 10^{-3} \text{ mol } Ca(OH)_2$$

$$\text{molarity of } Ca(OH)_2 \text{ solution is } \frac{1.22 \times 10^{-3} \text{ mol } Ca(OH)_2}{0.100 \text{ L Soln}} = 1.22 \times 10^{-2} \text{ M}$$

Because we have the number of moles of $Ca(OH)_2$ in 0.100 L, we can easily calculate the mass:

$$\left(\frac{1.22 \times 10^{-3} \text{ mol } Ca(OH)_2}{100 \text{ ml Soln}}\right)\left(\frac{74.1 \text{ g } Ca(OH)_2}{1 \text{ mol } Ca(OH)_2}\right) = \frac{0.0906 \text{ g } Ca(OH)_2}{100 \text{ ml Soln}}$$

3.56 (a) $Li_3N(s) + 3H_2O(l) \rightarrow NH_3(g) + 3LiOH(aq)$
(b) $2C_3H_7OH(l) + 9O_2(g) \rightarrow 6CO_2(g) + 8H_2O(g)$
(c) $PBr_3(l) + 3H_2O(l) \rightarrow H_3PO_3(aq) + 3HBr(aq)$
(d) $Mg_3B_2(s) + 6H_2O(l) \rightarrow 3Mg(OH)_2(aq) + B_2H_6(g)$
(e) $2CCl_4(g) + O_2(g) \rightarrow 2CCl_2O(g) + 2Cl_2(g)$
(f) $2La(NO_3)_3(aq) + 3Ba(OH)_2(aq) \rightarrow 2La(OH)_3(s) + 3Ba(NO_3)_2(aq)$

3.57 (a) $HCl(aq) + BaCO_3(s) \rightarrow BaCl_2(aq) + CO_2(g) + H_2O(l)$
(b) $HNO_3(aq) + CaCO_3(s) \rightarrow Ca(NO_3)_2(aq) + CO_2(g) + H_2O(l)$
(c) $K_2O(s) + H_2O(l) \rightarrow 2KOH(aq)$
(d) $Zn(s) + 2HBr(aq) \rightarrow ZnBr_2(aq) + H_2(g)$
(e) $BaO(s) + H_2O(l) \rightarrow Ba(OH)_2(s)$
(Note: $Ba(OH)_2$ is not very soluble in water.)

3.58 The most likely value for the atomic weight of element x is 22 amu. This would mean that there is 1 atom of x in substance 1, four in substance 2 and two in substance 3. A second possible value is any submultiple of 22, for example, 11. However, the number of atoms of x in the substance is doubled from the above numbers if this is the case.

3.59 Avg. mass C = 0.9889(12.0000) + 0.0111(13.003) = 12.011 amu

3.60 The formula weight of H_2SO_3 is 82.08.
wt percentage H is 100(2 x 1.008)/82.08 = 2.46%
wt percentage S is 100(32.06/82.08) = 39.06%
wt percentage O is 100(3 x 16.00)/82.08 = 58.48%

3.61 The formula weight of $K_2Cr_2O_7$ is 294.2.
wt percentage K is 100(2 x 39.10)/294.2 = 26.58%
wt percentage Cr is 100(2 x 52.00)/294.2 = 35.35%
wt percentage O is 100(7 x 16.00)/294.2 = 38.07%

3.62 (a) $12 \text{ g } Na_2CO_3 \left(\frac{1 \text{ mol } Na_2CO_3}{106 \text{ g } Na_2CO_3}\right) = 0.11 \text{ mol } Na_2CO_3$

(b) $44.2 \text{ g Ar} \left(\frac{1 \text{ mol Ar}}{39.9 \text{ g Ar}}\right) = 1.11 \text{ mol Ar}$

(c) $65 \times 10^3 \text{ g } H_2 \left(\frac{1 \text{ mol } H_2}{2.02 \text{ g } H_2}\right) = 3.2 \times 10^4 \text{ mol } H_2$

(d) $2.3 \times 10^{-7} \text{ g } HgCl_2 \left(\frac{1 \text{ mol } HgCl_2}{272 \text{ g } HgCl_2}\right) = 8.5 \times 10^{-10} \text{ mol } HgCl_2$

3.63 Assume we have 1 g of the compound. This sample contains 0.684 g Cr, 0.316 g O.

$$0.684 \text{ g Cr} \left(\frac{1 \text{ mol Cr}}{52.0 \text{ g Cr}}\right) = 1.32 \times 10^{-2} \text{ mol Cr}$$

$$0.316 \text{ g O} \left(\frac{1 \text{ mol O}}{16.0 \text{ g O}}\right) = 1.98 \times 10^{-2} \text{ mol O}$$

These numbers are in the ratio 2:3, since 1.32/1.98 = 0.66. The empirical formula is thus Cr_2O_3.

3.64 The fraction of O in the compound is, by difference equal to 1 - 0.688 - 0.049 = 0.263. Assuming for convenience a 1 g sample,

$$0.688 \text{ g C} \left(\frac{1 \text{ mol C}}{12.0 \text{ g C}}\right) = 0.0573 \text{ mol C}$$

$$0.049 \text{ g H} \left(\frac{1 \text{ mol H}}{1.01 \text{ g H}}\right) = 0.0485 \text{ mol H}$$

$$0.263 \text{ g O} \left(\frac{1 \text{ mol O}}{16.0 \text{ g O}}\right) = 0.0164 \text{ mol O}$$

Dividing by the smallest of these, we obtain 3.5 mol C: 3 mol H: 1 mol O. Doubling these to obtain whole numbers we obtain the empirical formula, $C_7H_6O_2$. The formula weight of this empirical formula is 122, so this is also the molecular weight.

3.65 $3.14 \text{ g } CO_2 \left(\frac{1 \text{ mol } CO_2}{44.0 \text{ g } CO_2}\right)\left(\frac{1 \text{ mol C}}{1 \text{ mol } CO_2}\right) = 0.0714 \text{ mol C}$

$$1.29 \text{ g } H_2O \left(\frac{1 \text{ mol } H_2O}{18.02 \text{ g } H_2O}\right)\left(\frac{2 \text{ mol H}}{1 \text{ mol } H_2O}\right) = 0.143 \text{ mol H}$$

We see that the molar ratio of H to C is 2.0. The simplest formula is CH_2.

3.66 $\left(\frac{0.230 \text{ g F}}{1 \text{ g Compd}}\right)\left(\frac{1 \text{ mol F}}{19.0 \text{ g F}}\right)\left(\frac{1 \text{ mol Compd}}{2 \text{ mol F}}\right) = \frac{6.05 \times 10^{-3} \text{ mol Compd}}{1 \text{ g Compd}}$

The reciprocal of this quantity is the molecular wt: 165 g/mol.

3.67 We proceed as usual, assuming we have 1 g of the substance of interest.

$$0.122 \text{ g C} \left(\frac{1 \text{ mol C}}{12.0 \text{ g C}}\right) = 0.0102 \text{ mol C}$$

$$0.0051 \text{ g H} \left(\frac{1 \text{ mol H}}{1.0 \text{ g H}}\right) = 5.1 \times 10^{-3} \text{ mol H}$$

$$0.289 \text{ g F} \left(\frac{1 \text{ mol F}}{19.0 \text{ g F}}\right) = 0.0152 \text{ mol F}$$

$$0.180 \text{ g Cl} \left(\frac{1 \text{ mol Cl}}{35.4 \text{ g Cl}}\right) = 5.08 \times 10^{-3} \text{ mol Cl}$$

$$0.404 \text{ g Br} \left(\frac{1 \text{ mol Br}}{79.9 \text{ g Br}}\right) = 5.06 \times 10^{-3} \text{ mol Br}$$

Dividing through by 5.1×10^{-3}, we obtain the whole number ratio of atoms: C_2HF_3ClBr.

3.68 $$80 \times 10^3 \text{ g } ZnSO_3 \left(\frac{1 \text{ mol } ZnSO_3}{145 \text{ g } ZnSO_3}\right)\left(\frac{1 \text{ mol ZnO}}{1 \text{ mol } ZnSO_3}\right)\left(\frac{81.4 \text{ g ZnO}}{1 \text{ mol ZnO}}\right)$$

$$= 4.5 \times 10^4 \text{ g ZnO} = 45 \text{ kg ZnO}$$

$$(2 \times 10^3 \text{ g } SO_2)\left(\frac{1 \text{ mol } SO_2}{64 \text{ g } SO_2}\right)\left(\frac{1 \text{ mol } ZnSO_3}{1 \text{ mol } SO_2}\right)\left(\frac{145 \text{ g } ZnSO_3}{1 \text{ mol } ZnSO_3}\right) = 4.5 \times 10^3 \text{ g } ZnSO_3$$

3.69 The balanced equation for reaction is:

$$2H_2(g) + O_2(g) \rightarrow 2H_2O(g)$$

We calculate first how many moles of water are formed from the amounts of H_2 and O_2 given:

$$3.50 \; H_2 \left(\frac{1 \text{ mol } H_2}{2.02 \text{ g } H_2}\right)\left(\frac{1 \text{ mol } H_2O}{1 \text{ mol } H_2}\right) = 1.73 \text{ mol } H_2O$$

$$26.0 \text{ g } O_2 \left(\frac{1 \text{ mol } O_2}{32.0 \text{ g } O_2}\right)\left(\frac{2 \text{ mol } H_2O}{1 \text{ mol } O_2}\right) = 1.62 \text{ mol } H_2O$$

We see that O_2 is the limiting reagent. There will be essentially no O_2 present when reaction is complete. There will be $1.73 - 1.62 = 0.11$ mol H_2 and 1.62 mol H_2O.

$$0.11 \text{ mol } H_2 \left(\frac{2.0 \text{ g } H_2}{1 \text{ mol } H_2}\right) = 0.22 \text{ g } H_2$$

$$1.62 \text{ mol } H_2O \left(\frac{18.0 \text{ g } H_2O}{1 \text{ mol } H_2O}\right) = 29.2 \text{ g } H_2O$$

3.70 $S(s) + O_2(g) \rightarrow SO_2(g)$; $SO_2(g) + CaO(s) \rightarrow CaSO_3(s)$

$$2000 \text{ tons coal} \left(\frac{2000 \text{ lb}}{1 \text{ ton}}\right)\left(\frac{1 \text{ kg}}{2.20 \text{ lb}}\right)\left(\frac{1000 \text{ g}}{1 \text{ kg}}\right)\left(\frac{0.028 \text{ g S}}{1 \text{ g coal}}\right)\left(\frac{1 \text{ mol S}}{32 \text{ g S}}\right)$$

$$\times \left(\frac{1 \text{ mol } SO_2}{1 \text{ mol S}}\right)\left(\frac{1 \text{ mol } CaSO_3}{1 \text{ mol } SO_2}\right)\left(\frac{120 \text{ g } CaSO_3}{1 \text{ mol } CaSO_3}\right)\left(\frac{1 \text{ kg } CaSO_3}{1000 \text{ g } CaSO_3}\right)$$

$$= 1.9 \times 10^5 \text{ kg } CaSO_3$$

Incidentally, this converts to over 200 tons of $CaSO_3$ per day as a waste byproduct.

3.71 Missing iron has a mass of $(8.00 - 5.92) \times 10^{-6}$ g $= 2.08 \times 10^{-6}$ g.

$$Fe + 5CO \rightarrow Fe(CO)_5$$

$$2.08 \times 10^{-6} \text{ g Fe} \left(\frac{1 \text{ mol Fe}}{55.8 \text{ g Fe}}\right)\left(\frac{1 \text{ mol Fe(CO)}_5}{1 \text{ mol Fe}}\right)\left(\frac{195 \text{ g Fe(CO)}_5}{1 \text{ mol Fe(CO)}_5}\right)$$

$$= 7.26 \times 10^{-6} \text{ g Fe(CO)}_5$$

3.72 $$4.60 \text{ kg } NH_4HCO_3 \left(\frac{1000 \text{ g}}{1 \text{ kg}}\right)\left(\frac{1 \text{ mol } NH_4HCO_3}{79.1 \text{ g } NH_4HCO_3}\right)\left(\frac{1 \text{ mol } NaHCO_3}{1 \text{ mol } NH_4HCO_3}\right)$$

$$\left(\frac{84.0 \text{ g } NaHCO_3}{1 \text{ mol } NaHCO_3}\right)\left(\frac{1 \text{ kg}}{1000 \text{ g}}\right) = 4.88 \text{ kg } NaHCO_3$$

3.73 We proceed in all these cases by assuming that we have 1 g of sample, calculate the number of moles of each element present in that 1 g, then obtain the ratios of the numbers of moles as smallest whole numbers.

(a) $$0.590 \text{ g C} \left(\frac{1 \text{ mol C}}{12.0 \text{ g C}}\right) = 0.0492 \text{ mol C}$$

$$0.071 \text{ g H} \left(\frac{1 \text{ mol H}}{1.0 \text{ g H}}\right) = 0.071 \text{ mol H}$$

$$0.262 \text{ g O} \left(\frac{1 \text{ mol O}}{16.0 \text{ g O}}\right) = .0164 \text{ mol O}$$

$$0.077 \text{ g N} \left(\frac{1 \text{ mol N}}{14.0 \text{ g N}}\right) = 0.0055 \text{ mol N}$$

Dividing through by the smallest of these, 0.0055, and rounding to the nearest whole number, we obtain $C_9H_{13}O_3N$. This corresponds to a formula weight of 183, so it is therefore also the molecular formula.

(b) C_5H_7N is the empirical formula. This has a formula weight of 81. The molecular formula must then be twice the empirical formula, $C_{10}H_{14}N_2$.

(c) CH_3O is the empirical formula, with formula weight 30. The molecular formula is thus $C_2H_6O_2$.

(d) Similarly, we find that $C_4H_5N_2O$ is the empirical formula, with formula weight of 97. The molecular formula must be twice this, $C_8H_{10}N_4O_2$.

3.74 $$\left(\frac{0.0129 \text{ g AgCl}}{1.03 \text{ g ointment}}\right)\left(\frac{1 \text{ mol AgCl}}{143 \text{ g AgCl}}\right)\left(\frac{1 \text{ mol } C_{11}H_{12}O_5N_2Cl_2}{2 \text{ mol AgCl}}\right)\left(\frac{323.1 \text{ g } C_{11}H_{12}O_5N_2Cl_2}{1 \text{ mol } C_{11}H_{12}O_5N_2Cl_2}\right)$$

$$= \frac{0.0141 \text{ g chloromycetin}}{1 \text{ g ointment}} = 1.41 \text{ wt percent}$$

3.75 $$0.0250 \text{ L Soln} \left(\frac{0.102 \text{ mol } Ag^+}{1 \text{ L Soln}}\right)\left(\frac{1 \text{ mol } Ag_3AsO_4}{3 \text{ mol } Ag^+}\right)\left(\frac{1 \text{ mol As}}{1 \text{ mol } Ag_3AsO_4}\right)$$

$$\times \left(\frac{74.9 \text{ g As}}{1 \text{ mol As}}\right) = 0.0637 \text{ g As}$$

$$\text{wt percentage} = 100 \left(\frac{0.0637 \text{ g As}}{1.22 \text{ g sample}}\right) = 5.22 \text{ wt percent}$$

3.76 $$58.8 \text{ g CoO} \left(\frac{1 \text{ mol CoO}}{74.9 \text{ g CoO}}\right)\left(\frac{1 \text{ mol } O_2}{4 \text{ mol CoO}}\right)\left(\frac{32.0 \text{ g } O_2}{1 \text{ mol } O_2}\right)\left(\frac{1 \text{ g CO}}{85 \times 10^{-6} \text{ g } O_2}\right)$$

$$= 7.4 \times 10^4 \text{ g CO}$$

The last conversion factor used here derives from what we are told in the problem, that there are 85 x 10^{-6} g of O_2 in each g of total gas mixture. Because 85 x 10^{-6} is small in comparison with 1, we can state as a reasonable approximation that there are 85 x 10^{-6} g O_2 per g of CO.

3.77 (a) $$\frac{35.0 \text{ g } H_2SO_4}{0.600 \text{ L Soln}}\left(\frac{1 \text{ mol } H_2SO_4}{98.0 \text{ g } H_2SO_4}\right) = 0.595 \text{ M}$$

(b) $$0.0420 \text{ L } HNO_3 \text{ Soln}\left(\frac{0.550 \text{ mol } HNO_3}{1 \text{ L Soln}}\right) = 0.0231 \text{ mol } HNO_3$$

$$\text{molarity} = \frac{0.0231 \text{ mol } HNO_3}{0.500 \text{ L Soln}} = 0.0462 \text{ M}$$

(c) $$\frac{2.56 \text{ g NaBr}}{0.0650 \text{ L Soln}}\left(\frac{1 \text{ mol NaBr}}{103 \text{ g NaBr}}\right) = 0.382 \text{ M}$$

(d) $$(0.035 \text{ L KBr Soln})\left(\frac{0.50 \text{ mol KBr}}{1 \text{ L}}\right) = 0.018 \text{ mol KBr}$$

$$(0.065 \text{ L KBr Soln})\left(\frac{0.36 \text{ mol KBr}}{1 \text{ L}}\right) = 0.023 \text{ mol KBr}$$

$$\text{molarity} = \frac{(0.018 + 0.023) \text{ mol KBr}}{0.100 \text{ L Soln}} = 0.41 \text{ M}$$

3.78 $$\left(\frac{63 \text{ g } Na_3PO_4}{1 \text{ L Soln}}\right)\left(\frac{1 \text{ mol } Na_3PO_4}{164 \text{ g } Na_3PO_4}\right) = 0.38 \text{ M}$$

3.79 (a) Calculate the number of moles of KOH required, then the mass of KOH to which this corresponds. Weigh out this mass of KOH. Dissolve in water, then add more water with stirring to bring the total volume to precisely 300 ml in a 300 ml volumetric flask, or in a graduated cylinder (accurate enough for two-place precision).

$$0.300 \text{ L Soln}\left(\frac{0.35 \text{ mol KOH}}{1 \text{ L Soln}}\right)\left(\frac{56 \text{ g KOH}}{1 \text{ mol KOH}}\right) = 5.9 \text{ g KOH}$$

(b) $$1.00 \text{ L Soln}\left(\frac{0.500 \text{ mol } AgNO_3}{1 \text{ L Soln}}\right)\left(\frac{170 \text{ g } AgNO_3}{1 \text{ mol } AgNO_3}\right) = 85.0 \text{ g } AgNO_3$$

Weigh out this mass of $AgNO_3$, then dissolve in a few hundred mL water. Add to a 1 L volumetric flask, rinsing with a couple of small additional volumes of water. Then add water with stirring to bring volume to accurate 1 L mark on flask.

(c) You first need to determine the concentration of the H_2SO_4 solution. Let us assume it is on the order of perhaps 6 M; we want to take a volume such that it will not require a huge volume of NaOH solution to neutralize it. Let us say we accurately withdraw 2.00 mL from a beaker containing the concentrated solution. We titrate this with the 0.150 M NaOH solution. From the volume required to reach the equivalence point, say y liters, we can calculate the molarity of the concentrated H_2SO_4 solution:

$$\left(\frac{\text{y L NaOH Soln}}{2.00 \times 10^{-3} \text{ L } H_2SO_4 \text{ Soln}}\right)\left(\frac{0.150 \text{ mol NaOH}}{1 \text{ L NaOH Soln}}\right)\left(\frac{1 \text{ mol } H_2SO_4}{2 \text{ mol NaOH}}\right)$$

$$= \text{? M } H_2SO_4 \text{ Soln}$$

We know that in the final result we want a total of 1 L x 0.200 mol/L = 0.200 mol H_2SO_4. From a knowledge of the concentration of the stock solution we can calculate the volume of stock solution that yields 0.200 mol H_2SO_4. This volume is then added to a volumetric flask and diluted to 1 L total volume.

3.80 $C(s) + 2Cl_2(g) + 2H_2O(l) \rightarrow CO_2(g) + 4HCl(aq)$

$$6.0 \times 10^3 m^3 \text{ Soln} \left(\frac{100 \text{ cm}}{\text{m}}\right)^3 \left(\frac{1 \text{ L}}{1000 \text{ cm}^3}\right) \left(\frac{3.2 \times 10^{-7} \text{ mol } Cl_2}{1 \text{ L Soln}}\right)$$
$$\times \left(\frac{1 \text{ mol } CO_2}{2 \text{ mol } Cl_2}\right) \left(\frac{44 \text{ g } CO_2}{1 \text{ mol } CO_2}\right) = 42 \text{ g } CO_2$$

3.81 $$\frac{3.41 \times 10^{-5} \text{ g P}}{1 \text{ L Soln}} \left(\frac{1 \text{ mol P}}{31.0 \text{ g P}}\right) = 1.10 \times 10^{-6} \text{ mol P/L}$$

$$\left(\frac{7.98 \times 10^{-4} \text{ g N}}{1 \text{ L Soln}}\right) \left(\frac{1 \text{ mol N}}{14.0 \text{ g N}}\right) = 5.70 \times 10^{-5} \text{ mol N/L}$$

3.82 $$\left(\frac{0.0640 \text{ L NaOH Soln}}{0.0100 \text{ L } H_2SO_4 \text{ Soln}}\right) \left(\frac{1.06 \text{ mol NaOH}}{1 \text{ L NaOH Soln}}\right) \left(\frac{1 \text{ mol } H_2SO_4}{2 \text{ mol NaOH}}\right) = 3.39 \text{ M } H_2SO_4 \text{ Soln}$$

3.83 (a) $$1.5 \times 10^5 \text{ g } C_9H_8O_4 \left(\frac{1 \text{ mol } C_9H_8O_4}{180 \text{ g } C_9H_8O_4}\right) \left(\frac{1 \text{ mol } C_7H_6O_3}{1 \text{ mol } C_9H_8O_4}\right) \left(\frac{138 \text{ g } C_7H_6O_3}{1 \text{ mol } C_7H_6O_3}\right)$$
$$= 1.2 \times 10^2 \text{ kg } C_7H_6O_3$$

(b) If only 80 percent of the acid reacts, then we need 1/0.80 = 1.25 times as much to obtain the same mass of product: 1.25 x 1.2 x 10^2 kg = 1.5 x 10^2 kg.

3.84 $$18.0 \text{ g Zn} \left(\frac{1 \text{ mol Zn}}{65.4 \text{ g Zn}}\right) \left(\frac{2 \text{ mol HBr}}{1 \text{ mol Zn}}\right) \left(\frac{1 \text{ L Soln}}{0.20 \text{ M}}\right) = 2.75 \text{ L Soln required}$$

3.85 First calculate the number of moles of HBr to begin with:

$$0.400 \text{ ml Soln} \left(\frac{0.550 \text{ mol HBr}}{1 \text{ L}}\right) = 0.220 \text{ mol HBr}$$

Next calculate the number of moles of HBr that remain after reaction with $Zn(OH)_2$. We get this from the NaOH titration data:

$$\left(\frac{0.500 \text{ mol NaOH}}{1 \text{ L Soln}}\right) (0.165 \text{ L Soln}) \left(\frac{1 \text{ mol HBr}}{1 \text{ mol NaOH}}\right) = 0.0825 \text{ mol HBr}$$

The difference in the two values, 0.138 moles HBr, is the quantity that reacted with $Zn(OH)_2$ in the reaction:

$$Zn(OH)_2(s) + 2HBr(aq) \rightarrow ZnBr_2(aq) + 2H_2O(l)$$

$$0.138 \text{ mol HBr} \left(\frac{1 \text{ mol } Zn(OH)_2}{2 \text{ mol HBr}}\right) \left(\frac{99.4 \text{ g } Zn(OH)_2}{1 \text{ mol } Zn(OH)_2}\right) = 6.86 \text{ g } Zn(OH)_2$$

4

Energy relationships in chemical systems

4.1 A portable radio is an example of a dry cell use in which work is done. Some of the work is electrical, as in moving electrons through transistors; some is mechanical, as in moving the cone of the speaker back and forward to create sound impulses. Many children's toys operate on batteries; examples are toy autos, toy shooting galleries, and so forth. A flashlight represents an example in which radiant energy, including the infrared rays that we experience as heat, results from conversion of the batteries' chemical energy. The fresh dry cell represents a form of chemical potential energy.

4.2 Kinetic energy is energy a body possesses by virtue of its motion relative to another body. Potential energy is energy that a body possesses by virtue of the position of one part of matter with respect to another.

When a rubber band is stretched, the chemical energy stored in our bodies is converted in part into mechanical potential energy. When the stretched rubber band is released, the potential energy is converted into kinetic energy and then into heat. A candle represents chemical potential energy. When the candle burns, this energy is in part released as radiant energy. Some of the radiant energy is in the visible portion of the spectrum, some of it is in the infrared region; we refer to this as heat (this aspect of energy is discussed more in Section 5.1). A brick possesses potential energy by virture of its location. If it is picked up its mechanical potential energy increases. When it is dropped that potential energy is converted into kinetic energy. When the brick strikes the ground the kinetic energy it has gained in falling is converted into heat energy.

4.3 Kinetic energy $= 1/2\ mv^2$

$$= 1/2(7.3 \times 10^{22}\ \text{kg})(1.0 \times 10^{5}\ \text{cm/sec})^2 \left(\frac{1\ \text{m}}{100\ \text{cm}}\right)^2$$

$$= 3.6 \times 10^{28}\ \text{kg-m}^2/\text{sec}^2 = 3.6 \times 10^{28}\ \text{J}$$

4.4 Kinetic energy = 1/2 mv^2

$$= 0.5(6.00\ \text{kg})(3.20\ \text{km/sec})^2\left(\frac{1000\ \text{m}}{1\ \text{km}}\right)^2$$

$$= 3.07 \times 10^7\ \text{J}\left(\frac{9.48 \times 10^{-4}\ \text{Btu}}{1\ \text{J}}\right) = 2.91 \times 10^4\ \text{Btu}$$

4.5 At time of entry, $KE = 0.5(32\ \text{g})\left(\frac{22\ \text{km}}{\text{sec}}\right)^2\left(\frac{10^3\ \text{m}}{1\ \text{km}}\right)^2\left(\frac{1\ \text{kg}}{10^3\ \text{g}}\right) = 7.7 \times 10^6\ \text{J}$

At time of striking earth,

$$KE = 0.5(32\ \text{g})\left(\frac{0.3\ \text{km}}{\text{sec}}\right)^2\left(\frac{10^3\ \text{m}}{1\ \text{km}}\right)^2\left(\frac{1\ \text{kg}}{10^3\ \text{g}}\right) = 1.4 \times 10^3\ \text{J}$$

The change in kinetic energy is essentially all of the 7.7×10^6 J. The loss in kinetic energy comes from contact of the meteorite with earth's atmosphere. In effect, the kinetic energy of the meteorite has been converted into heat by collisions with molecules of the atmosphere. This conversion of kinetic energy to heat energy causes the meteorite to heat up greatly during entry, to the extent that it may become white hot, and be visible as a "falling star."

4.6 The change in potential energy of the ball at the instant it is at the top of its rise is

$$E_2 - E_1 = 55\ \text{g}\ (9.81\ \text{m/sec}^2)(2.00\ \text{m} - 1.80\ \text{m}) = \left(\frac{99\ \text{g m}^2}{\text{sec}^2}\right)\left(\frac{1\ \text{kg}}{1000\ \text{g}}\right)$$

$$= \frac{9.9 \times 10^{-2}\ \text{kg-m}^2}{\text{sec}^2} = 9.9 \times 10^{-2}\ \text{J}$$

This much energy has been converted to some other form. It cannot be kinetic energy, because at the instant the ball comes to the top of its rise it is motionless. The lost potential energy is due to imperfect elasticity; it has appeared as heat.

4.7 work = f x d = m x g x h

$$= 165\ \text{lb}\left(\frac{1\ \text{kg}}{2.20\ \text{lb}}\right)\left(\frac{9.81\ \text{m}}{\text{sec}^2}\right)(192\ \text{ft})\left(\frac{12\ \text{in}}{1\ \text{ft}}\right)\left(\frac{2.54\ \text{cm}}{1\ \text{in}}\right)$$

$$\times\left(\frac{1\ \text{m}}{1 \times 10^2\ \text{cm}}\right) = 4.30 \times 10^4\ \text{kg-m}^2/\text{sec}^2 = 4.30 \times 10^4\ \text{J}$$

4.8 Combustion of the fuel represents conversion of chemical potential energy into mechanical potential energy (the airplane rises against gravitational force), kinetic energy (motion of the airplane) and heat. The potential energy gained by the craft in climbing is converted in part to kinetic energy as it descends, and in part to heat through frictional contact with the air. Eventually, as the plane lands, it will have lost all of the potential energy gained. As it comes to a stop, all the kinetic energy gained will have been lost. Both kinetic and potential energy will have been converted to heat, distributed throughout the environment. As you think about it, you will realize that this is the eventual result of all our energy conversions. The first law tells us that the total amount of heat generated must equal the chemical energy released in combustion of the plane's fuel supply.

4.9 (a) exothermic; (b) endothermic; (c) endothermic; (d) exothermic; (e) exothermic.

4.10 (a) ΔH positive. Heat must be added to the system to evaporate the liquid. (b) ΔH negative. Combustion of methyl alcohol releases heat to the surroundings. (c) ΔH positive. If the temperature decreases when

$NH_4Cl(s)$ dissolves, this tells us that heat energy is required by the system for the process. This heat is taken from the surroundings, or from the water itself, so temperature decreases. ΔH is positive because heat is being added to the system from surroundings. (d) ΔH positive. Heat must be added to break up the stable HCl molecules. We know that ΔH is negative (heat is evolved) for the reverse reaction, and we know that when a reaction is reversed the sign of ΔH changes.

4.11 The balanced equation for combustion of $C_6H_6(l)$ is

$$2C_6H_6(l) + 15O_2(g) \rightarrow 12CO_2(g) + 6H_2O(l) \qquad \Delta H = 2(-3273\ kJ)$$

We must add to this the equation

$$6H_2O(l) \rightarrow 6H_2O(g) \qquad \Delta H = 6[(-241.8) - (-285.9)] = 264.6$$

$$2C_6H_6(l) + 15O_2(g) \rightarrow 12CO_2(g) + 6H_2O(g) \qquad \Delta H = -6281\ kJ$$

$$\left(\frac{-6281\ kJ}{2\ mol\ C_6H_6}\right)\left(\frac{1\ mol\ C_6H_6}{78.1\ g\ C_6H_6}\right) = \frac{-40.2\ kJ}{1\ g\ C_6H_6}$$

$$\left(\frac{-40.2\ kJ}{1\ g\ C_6H_6}\right)\left(\frac{0.878\ g\ C_6H_6}{1\ cm^3\ C_6H_6}\right) = -35.3\ kJ/cm^3\ C_6H_6$$

Similarly,

$$2CH_3OH(l) + 3O_2(g) \rightarrow 2CO_2(g) + 4H_2O(l) \qquad \Delta H = 2(-728\ kJ)$$

$$4H_2O(l) \rightarrow 4H_2O(g) \qquad \Delta H = 4[(-241.8) - (-285.9)] = 176.4\ kJ$$

$$2CH_3OH(l) + 3O_2(g) \rightarrow 2CO_2(g) + 4H_2O(g) \qquad \Delta H = -1280\ kJ$$

$$\left(\frac{-1280\ kJ}{2\ mol\ CH_3OH}\right)\left(\frac{1\ mol\ CH_3OH}{32.0\ g\ CH_3OH}\right) = \frac{-20.0\ kJ}{1\ g\ CH_3OH}$$

$$\left(\frac{-20.0\ kJ}{1\ g\ CH_3OH}\right)\left(\frac{0.790\ g\ CH_3OH}{1\ cm^3\ CH_3OH}\right) = -15.8\ kJ/cm^3\ CH_3OH$$

The benzene thus produces 35.3/15.8 = 2.23 times as much heat per unit volume. One could expect to get roughly 2.2 times as much mileage from a gallon of benzene as compared with a gallon of methyl alcohol.

4.12 $$3.5\ g\ Na\left(\frac{1\ mol\ Na}{23\ g\ Na}\right)\left(\frac{-368\ kJ}{2\ mol\ Na}\right) = -56\ kJ$$

4.13 We have that $2C(s) + H_2(g) \rightarrow C_2H_2(g)$ $\Delta H = +227$ kJ. The heat of the reaction

$$2C_2H_2(g) + 5O_2(g) \rightarrow 4CO_2(g) + 2H_2O(g)$$

$$\Delta H_{rxn} = 4\Delta H_f(CO_2(g)) + 2\Delta H_f(H_2O(l)) - 2\Delta H_f(C_2H_2(g))$$
$$= 4(-393.5) + 2(-241.8) - 2(227) = -2512\ kJ$$

Similarly,

$$C_2H_4(g) + 3O_2(g) \rightarrow 2CO_2(g) + 2H_2O(g)$$

$$\Delta H_{rxn} = 2(-393.5) + 2(-241.8) - 52.3 = -1323\ kJ$$

$$2C_2H_6(g) + 7O_2(g) \rightarrow 4CO_2(g) + 6H_2O(g)$$

$$\Delta H_{rxn} = 4(-393.5) + 6(-241.8) - 2(-84.7) = -2855\ kJ$$

We now calculate the heat of combustion per kg of each substance.

acetylene: $\left(\frac{-2512\ kJ}{2\ mol\ C_2H_2}\right)\left(\frac{1\ mol\ C_2H_2}{26.04\ g\ C_2H_2}\right)\left(\frac{1000\ g}{1\ kg}\right) = \frac{-4.82 \times 10^4\ kJ}{1\ kg\ C_2H_2}$

ethylene: $\left(\frac{-1323\ kJ}{1\ mol\ C_2H_4}\right)\left(\frac{1\ mol\ C_2H_4}{28.05\ g\ C_2H_4}\right)\left(\frac{1000\ g}{1\ kg}\right) = \frac{-4.72 \times 10^4\ kJ}{kg\ C_2H_4}$

ethane: $\left(\frac{-2855\ kJ}{2\ mol\ C_2H_6}\right)\left(\frac{1\ mol\ C_2H_6}{30.07\ g\ C_2H_6}\right)\left(\frac{1000\ g}{1\ kg}\right) = \frac{-4.75 \times 10^4\ kJ}{1\ kg\ C_2H_6}$

Acetylene is the most efficient fuel in terms of heat evolved per unit mass.

4.14 We want the enthalpy change in the reaction

$$C_2H_4(g) + 6F_2(g) \rightarrow 2CF_4(g) + 4HF(g)$$

The process of putting the three equations together to obtain this one is somewhat trial and error in nature. We see that we need twice the first and second equations to obtain 4HF and $2CF_4$ as products:

$$2H_2(g) + 2F_2(g) \rightarrow 4HF(g) \qquad \Delta H = 2(-537\ kJ)$$
$$2C(s) + 4F_2(g) \rightarrow 2CF_4(g) \qquad \Delta H = 2(-680\ kJ)$$

reversing the third equation and adding gives us what we need:

$$C_2H_4(g) \rightarrow 2C(s) + 2H_2(g) \qquad \Delta H = -52.3\ kJ$$
$$\Delta H = 2(-537) + 2(-680) + (-52) = -2486\ kJ$$

4.15 To obtain the reaction $S(s) + O_2(g) \rightarrow SO_2(g)$, we must reverse the first reaction given, add it to the second equation, then divide through by 2:

$$2SO_3(g) \rightarrow 2SO_2(g) + O_2(g) \qquad \Delta H = +196\ kJ$$
$$2S(s) + 3O_2(g) \rightarrow 2SO_3(g) \qquad \Delta H = -790\ kJ$$
$$2S(s) + 2O_2(g) \rightarrow 2SO_2(g) \qquad \Delta H = -594\ kJ$$
$$S(s) + O_2(g) \rightarrow SO_2(g) \qquad \Delta H = 1/2(-594\ kJ) = -297\ kJ$$

4.16 (a) $\Delta H_{rxn} = \Delta H_f(NH_4CN) + \Delta H_f(H_2O) - \Delta H_f(CO) - 2\Delta H_f(NH_3)$. We must be sure to choose the ΔH_f value appropriate to the physical state given in the equation.

$$\Delta H_{rxn} = 0 + (-241.8) - (-110.5) - 2(-46.2) = -38.9\ kJ$$

(b) $\Delta H_{rxn} = 81.6 + 2(-241.8) - (-365.6) = -36.4\ kJ$

(c) $\Delta H_{rxn} = -3012.5 - (-1640.1) = -1372.4\ kJ$

(d) $\Delta H_{rxn} = 3(-592.0) + (-435.9) - (-391.2) - 3(-338.9)$
$= -804.0\ kJ$

4.17 $HO(g) + Cl_2(g) \rightarrow HOCl(g) + Cl(g)$. To obtain the desired products, we need to add the first and third reactions. We then reverse the second reaction and add it:

$$Cl_2(g) \rightarrow 2Cl(g) \qquad \Delta H = 242\ kJ$$
$$1/2(H_2O_2(g) + 2Cl(g) \rightarrow 2HOCl(g)) \qquad \Delta H = 1/2(-209\ kJ)$$
$$1/2H_2O_2(g) + Cl_2(g) \rightarrow Cl(g) + HOCl(g) \qquad \Delta H = 137.5\ kJ$$

We now add 1/2 times the reverse of the second reaction:

$$\begin{array}{lr} 1/2H_2O_2(g) + Cl_2(g) \rightarrow Cl(g) + HOCl(g) & \Delta H = 137.5 \text{ kJ} \\ 1/2[2HO(g) \rightarrow H_2O_2(g)] & \Delta H = 1/2(-134 \text{ kJ}) \\ \hline HO(g) + Cl_2(g) \rightarrow Cl(g) + HOCl(g) & \Delta H = 70.5 \text{ kJ} \end{array}$$

Note that the reaction is endothermic.

4.18 $C_{14}H_{10}(s) + H_2(g) \rightarrow C_{14}H_{12}(s)$

$$\Delta H_{rxn} = 66.4 \text{ kJ} - 121.3 \text{ kJ} = -54.9 \text{ kJ}$$

4.19 For nitroethane, we have:

$$\left(\frac{1348 \text{ kJ}}{\text{mol } C_2H_5NO_2}\right)\left(\frac{1 \text{ mol } C_2H_5NO_2}{75.1 \text{ g } C_2H_5NO_2}\right)\left(\frac{1.052 \text{ g } C_2H_5NO_2}{1 \text{ cm}^3 \ C_2H_5NO_2}\right) = 18.9 \text{ kJ/cm}^3$$

Similarly, we obtain for ethanol 23.4 kJ/cm^3, and for diethyl ether, 26.3 kJ/cm^3. Thus, diethyl ether has the highest heat value per unit volume.

4.20 (a) $\Delta H_{rxn} = \Delta H_f(Ca(OH)_2(s)) - \Delta H_f(H_2O(l)) - \Delta H_f(CaO(s))$
$= -986.2 - (-285.8) - (-635.5) = -64.9$ kJ

(b) $\Delta H_{rxn} = -245.6 - 241.8 - (-296.9) - 2(-92.3) = -5.9$ kJ

(c) $\Delta H_{rxn} = 2(-446.2) - 277.4 - (-421.3) - 2(-360.6) = -27.3$ kJ

(d) $\Delta H_{rxn} = 2(-241.8) - 296.9 - 2(-133.2) = -514.1$ kJ

4.21 $$240 \text{ m}^3 \ H_2O \left(\frac{100 \text{ cm}}{1 \text{ m}}\right)^3\left(\frac{1 \text{ g } H_2O}{1 \text{ cm}^3 \ H_2O}\right)\left(\frac{4.184 \text{ J}}{°\text{C-g}}\right)(8°\text{C}) = 8 \times 10^6 \text{ kJ}$$

4.22 The total heat gained by the water is given by:

$$150 \text{ g } H_2O \left(\frac{4.184 \text{ J}}{1°\text{C-g}}\right)(3.4°\text{C}) = 2.13 \times 10^3 \text{ J}$$

The metal object changed temperature from 100°C to 28.0°C, so $\Delta T = 72.0°C$. The heat lost by the molybdenum is precisely that gained by the water. Thus, the heat capacity is

$$\frac{\text{J}}{°\text{C-g}} = \frac{2.13 \times 10^3 \text{ J}}{(72.0°\text{C})(110 \text{ g})} = 0.27 \text{ J/°C-g}$$

4.23 We can describe what we are given in the problem in terms of the mass of benzoic acid that gives a certain temperature change:

$$\frac{0.235 \text{ g benzoic acid}}{3.62°\text{C change observed}}\left(\frac{26.38 \text{ kJ}}{1 \text{ g benzoic acid}}\right) = \frac{1.71 \text{ kJ}}{1°\text{C}}$$

Now we employ the observed temperature change for the given mass of citric acid:

$$\left(\frac{1.83°\text{C rise}}{0.305 \text{ g citric acid}}\right)\left(\frac{1.71 \text{ kJ}}{1°\text{C-g}}\right)\left(\frac{192 \text{ g citric acid}}{1 \text{ mol citric acid}}\right)$$
$$= 1.97 \times 10^3 \text{ kJ/mol citric acid}$$

4.24 We will use the same approach outlined in Sample Exercise 4.9, with the assumptions mentioned in the exercise.

$$100 \text{ g}\left(\frac{4.2 \text{ J}}{°\text{C-g}}\right)(0.55°\text{C}) = 2.3 \times 10^2 \text{ J}$$

The number of moles of solutes that react is $\left(\frac{0.050 \text{ mol}}{1 \text{ L}}\right)(0.050 \text{ L})$
$= 2.5 \times 10^{-3}$ mol. Thus, the heat evolved per mol is
$2.3 \times 10^2 \text{ J}/2.5 \times 10^{-3} \text{ mol} = 92$ kJ/mol.

4.25 The total heat capacity of the solution is given by 4.18 J/°C-g times the total mass, 100 + 1.38 = 101.4 g.

$$\left(\frac{4.18\ \text{J}}{°\text{C} - \text{g}}\right)(101\ \text{g}) = 422\ \text{J/°C}$$

The temperature increase is 32.00 - 25.55 = 6.45°C. This corresponds to

$$\left(\frac{422\ \text{J}}{°\text{C}}\right)(6.45°\text{C}) = -2.72\ \text{kJ}$$

(The negative sign denotes that heat is evolved.) This is the heat of solution of 1.38 g PCl_3 in water. We calculate the molar heat of solution as follows:

$$\left(\frac{-272\ \text{kJ}}{1.38\ \text{g PCl}_3}\right)\left(\frac{137\ \text{g PCl}_3}{1\ \text{mol PCl}_3}\right) = -271\ \text{kJ/mol PCl}_3$$

4.26 We multiply the quantity of each type of food by the caloric value per g, then add.

$$6\ \text{g protein}\left(\frac{17\ \text{kJ}}{1\ \text{g protein}}\right) + 19\ \text{g carb.}\left(\frac{17\ \text{kJ}}{1\ \text{g carb}}\right) + 1\ \text{g fat}\left(\frac{38\ \text{kJ}}{1\ \text{g fat}}\right)$$

$$= 463\ \text{kJ}$$

$$463\ \text{kJ}\left(\frac{1000\ \text{calories}}{4.184\ \text{kJ}}\right)\left(\frac{1\ \text{Calorie}}{1000\ \text{calories}}\right) = 111\ \text{Calories}$$

(That is, this is the number of nutritional calories, as ordinarily reported on food containers.)

4.27 $$62\ \text{Calories}\left(\frac{1000\ \text{cal}}{1\ \text{Calorie}}\right)\left(\frac{4.184\ \text{kJ}}{1000\ \text{cal}}\right)\left(\frac{1\ \text{g carb.}}{17\ \text{kJ}}\right) = 15\ \text{g carb.}$$

The percentage weight of the apple that is carbohydrate is then

$$100\left(\frac{15}{120}\right) = 13\%.$$ Thus, 87% by weight is water.

4.28 $$214\ \text{g carb.}\left(\frac{17\ \text{kJ}}{1\ \text{g carb.}}\right)\left(\frac{1\ \text{Cal}}{4.184\ \text{kJ}}\right) = 870\ \text{Cal}$$

$$146\ \text{g fat}\left(\frac{38\ \text{kJ}}{1\ \text{g fat}}\right)\left(\frac{1\ \text{Cal}}{4.184\ \text{kJ}}\right) = 1.33 \times 10^3\ \text{Cal}$$

$$79\ \text{g protein}\left(\frac{17\ \text{kJ}}{1\ \text{g protein}}\right)\left(\frac{1\ \text{Cal}}{4.184\ \text{kJ}}\right) = 322\ \text{Cal}$$

Total fuel value, measured in nutritional calories, is thus 2500 Cal. In a 50 g bar, there would be 50/454 times as much, or about 275 Cal.

$$2600\ \text{Cal}\left(\frac{454\ \text{g}}{2500\ \text{Cal}}\right) = 470\ \text{g required per day}$$

(We have used two significant figures here, because the precision of the quantities discussed does not call for more.

4.29 $$355\ \text{mL}\left(\frac{1\ \text{g beer}}{1\ \text{mL}}\right)\left(\frac{.037\ \text{g ethanol}}{1\ \text{g beer}}\right)\left(\frac{1\ \text{mol ethanol}}{46\ \text{g ethanol}}\right)\left(\frac{1371\ \text{kJ}}{1\ \text{mol ethanol}}\right)$$

$$\times\left(\frac{1\ \text{Cal}}{4.18\ \text{kJ}}\right) = 94\ \text{Cal}$$

Notice that this is more than half the calorie content of ordinary beers (about 150 Cal/12 oz bottle), and nearly all the calorie content of so-called "light" beers.)

4.30 The reaction of interest is:

$$2(CH_2OH)_2CHOH(l) + 7O_2(g) \rightarrow 6CO_2(g) + 8H_2O(l)$$

$$\Delta H_{rxn} = 6(-393.5) + 8(-285.8) - 2(-666) = -3315 \text{ kJ}$$

$$\left(\frac{-3315 \text{ kJ}}{2 \text{ mol glycerin}}\right)\left(\frac{1 \text{ mol glycerin}}{92.1 \text{ g } C_3H_8O_3}\right) = \frac{18.0 \text{ kJ}}{\text{g } C_3H_8O_3}$$

Thus, complete metabolism of 10 g of $C_3H_8O_3$ would yield 180 kJ. In terms of nutritional calories, this amounts to

$$180 \text{ kJ}\left(\frac{1 \text{ Cal}}{4.18 \text{ kJ}}\right) = 43 \text{ Cal}$$

4.31 $$\frac{2.556°\text{C}}{3.556 \text{ g sample}}\left(\frac{13.62 \text{ kJ}}{1°\text{C}}\right) = \frac{9.790 \text{ kJ}}{1 \text{ g sample}}$$

Let us calculate the heat that would be evolved from a weight of sample equal to the molecular weight of pure salicylic acid, 138.1 g/mol.

$$9.790 \frac{\text{kJ}}{\text{g}}\left(\frac{258.2 \text{ g } C_7O_3H_6}{1 \text{ mol}}\right) = \frac{1352 \text{ kJ}}{1 \text{ mol}}$$

This is less than the heat expected for pure acid, 3.00×10^3 kJ/mol. The percentage salicylic acid in the sample is thus

$$100\left(\frac{1.35 \times 10^3}{3.00 \times 10^3}\right) = 45.0\%$$

The boric oxide contaminant is thus 55.0% by weight.

4.32 $$19 \text{ quad}\left(\frac{10^{15} \text{ Btu}}{1 \text{ quad}}\right)\left(\frac{1.05 \text{ kJ}}{1 \text{ Btu}}\right) = 2.0 \times 10^{16} \text{ kJ}$$

If half this is to come from anthracite coal, we have

$$1 \times 10^{16} \text{ kJ}\left(\frac{1 \text{ g}}{31 \text{ kJ}}\right)\left(\frac{1 \text{ kg}}{1000 \text{ g}}\right) = 3.2 \times 10^{11} \text{ kg}$$

$$3.2 \times 10^{11} \text{ kg}\left(\frac{2.2 \text{ lb}}{1 \text{ kg}}\right)\left(\frac{1 \text{ ton}}{2 \times 10^3 \text{ lb}}\right) = 3.2 \times 10^8 \text{ tons anthracite}$$

The other half comes from bituminous coal; we calculate in the same way that this yields 3.1×10^{11} kg, or 3.4×10^8 tons. Thus, the total coal requirement is 6.9×10^8 tons/yr.

4.33 $$2.09 \times 10^{11} \text{ ft}^3\left(\frac{980 \text{ kJ}}{1 \text{ ft}^3}\right) = 2.05 \times 10^{14} \text{ kJ}$$

$$2.05 \times 10^{14} \text{ kJ}\left(\frac{1 \text{ g coal}}{31 \text{ kJ}}\right) = 6.6 \times 10^{12} \text{ g coal}\left(\frac{1 \text{ lb}}{454 \text{ g}}\right)\left(\frac{1 \text{ ton}}{2000 \text{ lb}}\right)$$
$$= 7.3 \times 10^6 \text{ tons}$$

4.34 (a) Heat of combustion is the heat evolved when a substance is burned to yield carbon dioxide and water. Heat of formation refers to the heat change in a reaction in which a substance is formed from the elements in their standard states. (b) A calorie is a unit of heat; it equals (1/4.184) = 0.2390 Joules. The Calorie is a unit of heat used in nutrition; it equals 1000 calories, or 1 kcal. (c) Hess's law is a statement that the heat changes that occur in a sequence of reactions at constant pressure total to the heat change that occurs in a single reaction that is the net sum of adding all those reactions together. It is a consequence of the first law of thermodynamics, which is a statement of our experience that energy is conserved in any process. The first law is a completely general statement; Hess's law is a particular application of it. (d) The heat capacity of a system is the heat required to cause a given temperature change in that system. For example, it requires a certain number of calories to raise the temperature of a calorimeter and its contents by 1°C. That number of calories is the heat capacity of that system. Fuel value is the heat evolved upon combustion of some substance, given in units of heat evolved per unit mass of the substance. (e) Work is a form of energy, or an expression of energy. Work done represents energy expended. Work represents a conversion of energy from one form to another. Heat is a form of energy; it is a measure of the total energies of motion of all the particles that make up a substance. Heat can be converted partially into work, as in an auto engine, or steam engine. (f) A state function is some property of a system that can be specified by the condition of the system at a given time (that is, by specifying such things as temperature, pressure, location and so forth). The heat content, or enthalpy, is a state function. For any given system, when we specify the temperature, pressure, quantity of matter, and so forth, we have determined the enthalpy.

4.35 The three reactions involved are

a) $C(s) + H_2O(g) \rightarrow CO(g) + H_2(g)$
b) $C(s) + 2H_2(g) \rightarrow CH_4(g)$
c) $CO(g) + H_2O(g) \rightarrow CO_2(g) + H_2(g)$

We choose the gaseous state for all components except carbon, because the reactions are run at high temperatures. However, carbon is non-volatile even under reaction conditions. We assume that ΔH_f of C(s) is zero.

a) $\Delta H_{rxn} = -110.5 - (-241.8) = +131.3$ kJ
b) $\Delta H_{rxn} = -74.8$ kJ
c) $\Delta H_{rxn} = -393.5 - (-241.8) - (-110.5) = -41.2$ kJ

Notice that the first reaction is endothermic. As we will see in more detail later (Chapter 18), this indicates that the reaction may not proceed very far to the right. However, we must be careful in interpreting these ΔH_{rxn} values. They are calculated as though all the substances were at 25°C, which is in fact not the case. The reactions are actually seen at much higher temperatures, and ΔH values may be much different.

4.36 $$2 \times 10^8 \text{ gal} \left(\frac{3.8 \text{ L}}{1 \text{ gal}}\right)\left(\frac{1 \times 10^3 \text{ cm}^3}{1 \text{ L}}\right)\left(\frac{0.90 \text{ g}}{1 \text{ cm}^3}\right)\left(\frac{46 \text{ kJ}}{\text{g}}\right) = 3.1 \times 10^{13} \text{ kJ}$$

4.37 $$1.5 \text{ quad} \left(\frac{1.05 \times 10^{15} \text{ kJ}}{1 \text{ quad}}\right)\left(\frac{1 \text{ g oil}}{45 \text{ kJ}}\right)\left(\frac{1 \text{ kg}}{1 \times 10^3 \text{ g}}\right) = 3.5 \times 10^{10} \text{ kg}$$

Incidentally, assuming a density of 0.9 g/cm^3, and that 1 barrel equals 42 gallons, the unit generally used with oil, this quantity represents 243 million barrels per year. At $30 per barrel, this represents $7.3 billion dollars. It is evident why we must learn to conserve in oil use.

4.38 $$\text{Kinetic energy} = 1/2\ mv^2$$
$$= 0.5(1500 \text{ kg})(85 \text{ km/hr})^2 = 5.4 \times 10^6 \text{ kg-km}^2/\text{hr}^2$$
or
$$= 0.5(2200 \text{ kg})(60 \text{ km/hr})^2 = 4.0 \times 10^6 \text{ kg-km}^2/\text{hr}^2$$

The lighter car traveling more rapidly possesses the larger kinetic energy. Now, let us convert units:

$$\left(\frac{5.4 \times 10^6 \text{ kg km}^2}{\text{hr}^2}\right)\left(\frac{1 \times 10^3 \text{ m}}{1 \text{ km}}\right)\left(\frac{1 \text{ hr}}{3.6 \times 10^3 \text{ sec}}\right)^2 = 4.2 \times 10^5 \text{ kg-m}^2/\text{sec}^2$$

(These are the units of the Joule.) We divide by 4.18 to convert to calories, which gives 1.0×10^5 cal.

$$\left(\frac{4.0 \times 10^6 \text{ kg km}^2}{\text{hr}^2}\right)\left(\frac{1 \times 10^3 \text{ m}}{1 \text{ km}}\right)\left(\frac{1 \text{ hr}}{3.6 \times 10^3 \text{ sec}}\right)^2 = 3.1 \times 10^5 \text{ J}$$

$$= 7.4 \times 10^4 \text{ cal.}$$

4.39 The chemical energy stored in gunpowder is converted into heat as a rifle shell is caused to ignite. This, in turn, through gas expansion, is converted into the kinetic energy of the bullet. The chemical energy in foods is converted into heat and various forms of work as an animal metabolizes the food. The chemical energy in alcohol can be used to propel a car if the alcohol is added to gasoline.

4.40 We need to reverse the second and third reactions, then add them to the first:

$$1/2H_2(g) + 1/2Br_2(g) \rightarrow HBr(g) \qquad \Delta H = -36 \text{ kJ}$$
$$H(g) \rightarrow 1/2H_2(g) \qquad \Delta H = -218 \text{ kJ}$$
$$Br(g) \rightarrow 1/2Br_2(g) \qquad \Delta H = -112 \text{ kJ}$$
$$H(g) + Br(g) \rightarrow HBr(g) \qquad \Delta H = -366 \text{ kJ}$$

(This quantity of heat represents the H-Br bond energy; more on this in Chapter 7.)

4.41 In (a) and (b), ΔH_{rxn} is just ΔH_f° for the product, since the other reactants are elements, for which ΔH_f° is zero.

(a) $\Delta H_{rxn} = 4(9.2) = +36.8$ kJ (an endothermic process)

(b) $\Delta H_{rxn} = -1117$ kJ (an exothermic process)

(c) $\Delta H_{rxn} = 2(-36.2) - 2(-268.6) = +464.8$ kJ (an endothermic process)

(d) $\Delta H_{rxn} = -1207.1 - (-393.5) - (-635.5)$

$= -178.1$ kJ (an exothermic process)

4.42 The reaction for which we wish to know ΔH is

$$CH_4(g) + 2O_2(g) \rightarrow CO_2(g) + 2H_2O(g)$$

Adding the first and fourth reaction:

$$CH_4(g) \rightarrow C(g) + 4H(g) \qquad \Delta H = 1660 \text{ kJ}$$
$$C(g) + 2O(g) \rightarrow CO_2(g) \qquad \Delta H = -1610 \text{ kJ}$$
$$CH_4(g) + 2O(g) \rightarrow 4H(g) + CO_2(g) \qquad \Delta H = +50 \text{ kJ}$$

Double and add the third reaction:

$$4H(g) + 2O(g) \rightarrow 2H_2O(g) \qquad \Delta H = -1860 \text{ kJ}$$
$$CH_4(g) + 4O(g) \rightarrow CO_2(g) + 2H_2O(g) \qquad \Delta H = -1810 \text{ kJ}$$

And now double and add the second reaction:

$$2O_2(g) \rightarrow 4O(g) \qquad \Delta H = 980 \text{ kJ}$$
$$CH_4(g) + 2O_2(g) \rightarrow CO_2(g) + 2H_2O(g) \qquad \Delta H = -830 \text{ kJ}$$

(Because of experimental uncertainties the value of ΔH calculated in this way differs a little from that given in Eq. [4.8] in the text.)

4.43

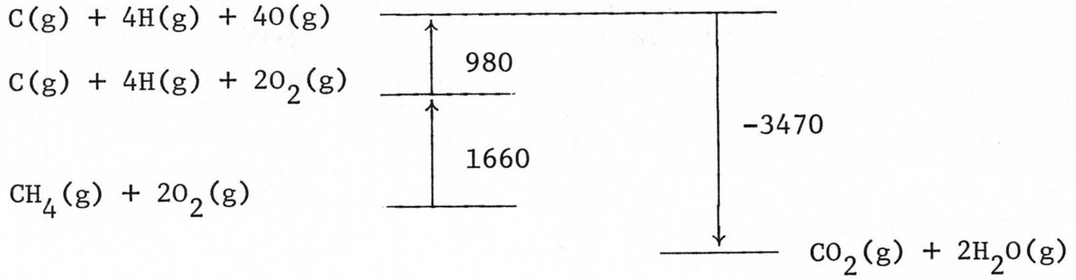

4.44 We can write the cooling-off reaction as

$$Na_2SO_4(aq) + 10H_2O(l) \rightarrow Na_2SO_4 \cdot 10H_2O(g)$$

$$\Delta H = -4324 - 10(-286) - (-1387) = -77 \text{ kJ/mol } Na_2SO_4{\cdot}10H_2O$$

$$\times \left(\frac{1 \text{ mol } Na_2SO_4{\cdot}10H_2O}{322 \text{ g } Na_2SO_4{\cdot}10H_2O}\right)\left(\frac{1000 \text{ g}}{1 \text{ kg}}\right) = 239 \text{ kJ/kg } Na_2SO_4{\cdot}10H_2O$$

$$6 \times 10^4 \text{ Btu} \left(\frac{1.05 \text{ kJ}}{1 \text{ Btu}}\right)\left(\frac{1 \text{ kg } Na_2SO_4{\cdot}10H_2O}{239 \text{ kJ}}\right) = 2.6 \times 10^2 \text{ kg } Na_2SO_4{\cdot}10H_2O$$

4.45 $2C(s) + 4H_2(g) + O_2(g) \rightarrow 2CH_3OH(g)$ $\quad \Delta H = -402.6$ kJ

$CH_3OH(l) \rightarrow CH_3OH(g)$ $\quad \Delta H = 37.4$ kJ

Reverse the second reaction and double it:

$2CH_3OH(g) \rightarrow 2CH_3OH(l)$ $\quad \Delta H = -74.8$ kJ

Add this to the first to obtain the desired reaction:

$2C(s) + 4H_2(g) + O_2(g) \rightarrow 2CH_3OH(l)$ $\quad \Delta H = -477.4$ kJ

4.46 The reaction for which we want ΔH is:

$$4NH_3(l) + 3O_2(g) \rightarrow 2N_2(g) + 6H_2O(g) \qquad \text{(A)}$$

Before we can calculate ΔH for this, we must calculate ΔH_f for $NH_3(l)$. We have that ΔH_f for $NH_3(g)$ is -46.2 kJ, and that for

$NH_3(l) \rightarrow NH_3(g)$, $\quad \Delta H = 4.6$ kJ

Thus, $\Delta H = \Delta H_f(NH_3(g)) - \Delta H_f(NH_3(l))$

$4.6 = -46.2 - \Delta H_f(NH_3(l))$

$\Delta H_f(NH_3(l)) = -50.8$ kJ/mol

Then for reaction (A), the enthalpy change is:

$$\Delta H = 6(-241.8) - 4(-50.8) = -1248 \text{ kJ}$$

$$\left(\frac{-1248 \text{ kJ}}{4 \text{ mol } NH_3}\right)\left(\frac{1 \text{ mol } NH_3}{17.0 \text{ g } NH_3}\right)\left(\frac{0.81 \text{ g } NH_3}{1 \text{ cm}^3}\right)\left(\frac{1000 \text{ cm}^3}{1 \text{ L}}\right) = 1.49 \times 10^4 \text{ kJ/L } NH_3$$

$$2CH_3OH(l) + 3O_2(g) \rightarrow 2CO_2(g) + 4H_2O(g)$$

$$\Delta H = 2(-393.5) + 4(-241.8) - 2(-239) = -1276 \text{ kJ}$$

$$\left(\frac{-1276 \text{ kJ}}{2 \text{ mol } CH_3OH}\right)\left(\frac{1 \text{ mol } CH_3OH}{32.0 \text{ g } CH_3OH}\right)\left(\frac{0.792 \text{ g } CH_3OH}{1 \text{ cm}^3}\right)\left(\frac{1000 \text{ cm}^3}{1 \text{ L}}\right)$$

$$= 1.58 \times 10^4 \text{ kJ/liter } CH_3OH$$

In terms of heat obtained per unit volume of fuel, liquid ammonia is a better fuel than methanol. It is not a practical fuel for several reasons, among them its low boiling point, and the fact that nitrogen oxides rather than N_2 are products.

<u>4.47</u> $$1.80 \text{ g } C_5H_8O_4 \left(\frac{1 \text{ mol } C_5H_8O_4}{132 \text{ g } C_5H_8O_4}\right)\left(\frac{2166 \text{ kJ}}{1 \text{ mol } C_5H_8O_4}\right)\left(\frac{1°\text{C}}{7.74 \text{ kJ}}\right) = 3.82°\text{C rise}$$

<u>4.48</u> $$8 \text{ g protein} \left(\frac{17 \text{ kJ}}{1 \text{ g protein}}\right) + 12 \text{ g carb.} \left(\frac{17 \text{ kJ}}{1 \text{ g carb.}}\right) + 4 \text{ g fat} \left(\frac{38 \text{ kJ}}{1 \text{ g fat}}\right)$$

$$= 4.9 \times 10^2 \text{ kJ}$$

$$4.9 \times 10^2 \text{ kJ} \left(\frac{1 \text{ Cal}}{4.18 \text{ kJ}}\right) = 1.2 \times 10^2 \text{ Cal}$$

<u>4.49</u> $$\frac{1000 \text{ tons}}{\text{hr}}\left(\frac{2 \times 10^3 \text{ lb}}{1 \text{ ton}}\right)\left(\frac{1.05 \times 10^4 \text{ Btu}}{1 \text{ lb}}\right)\left(\frac{1.05 \text{ kJ}}{1 \text{ Btu}}\right) = 2.2 \times 10^{10} \text{ kJ/hr}$$

$$\left(\frac{1000 \text{ tons}}{\text{hr}}\right)\left(\frac{2 \times 10^3 \text{ lb}}{1 \text{ ton}}\right)\left(\frac{1 \text{ kg}}{2.2 \text{ lb}}\right)\left(\frac{10^3 \text{g}}{1 \text{ kg}}\right)\left(\frac{0.92 \text{ g S}}{100 \text{ g coal}}\right) = 8.4 \times 10^6 \text{ g S/hr}$$

The reaction we assume is $S(s) + O_2(g) \rightarrow SO_2(g)$, for which ΔH is just ΔH_f° for $SO_2(g)$, -297 kJ. Thus,

$$\left(\frac{8.4 \times 10^6 \text{ g S}}{1 \text{ hr}}\right)\left(\frac{1 \text{ mol S}}{32 \text{ g}}\right)\left(\frac{-297 \text{ kJ}}{1 \text{ mol S}}\right) = 7.8 \times 10^7 \text{ kJ/hr}$$

<u>4.50</u> $$2.5 \times 10^{11} \text{ tons} \left(\frac{2 \times 10^3 \text{ lb}}{1 \text{ ton}}\right)\left(\frac{1 \text{ kg}}{2.2 \text{ lb}}\right)\left(\frac{10^3 \text{ g}}{1 \text{ kg}}\right)\left(\frac{31 \text{ kJ}}{1 \text{ g}}\right) = 7.0 \times 10^{18} \text{ kJ}$$

$$7.0 \times 10^{18} \text{ kJ} \left(\frac{1 \text{ quad}}{1.05 \times 10^{15} \text{ kJ}}\right)\left(\frac{1 \text{ yr}}{105 \text{ quad}}\right) = 64 \text{ yr}$$

We first calculate the weight fraction of carbon in the coal as 0.85. Then,

$$2.5 \times 10^{11} \text{ ton} \left(\frac{2 \times 10^3 \text{ lb}}{1 \text{ ton}}\right)\left(\frac{1 \text{ kg}}{2.2 \text{ lb}}\right)\left(\frac{10^3 \text{ g}}{1 \text{ kg}}\right)\left(\frac{85 \text{ g C}}{100 \text{ g coal}}\right)\left(\frac{44 \text{ g } CO_2}{12 \text{ g C}}\right)$$

$$= 7.1 \times 10^{17} \text{ g } CO_2$$

5

The electronic structures of atoms

5.1 In order of increasing wavelength; x-rays, green light, red light, microwaves, FM radio signal.

5.2 (a) $\lambda = c/\nu = (3.00 \times 10^8 \text{ m/sec})/(5.00 \times 10^{13}/\text{sec})$

$= 6.00 \times 10^{-6} \text{ m} = 6.00 \text{ }\mu\text{m}$

(b) $\nu = c/\lambda = (3.00 \times 10^8 \text{m/sec})/(1.50 \times 10^{-17} \text{ m}) = 2.00 \times 10^{25}/\text{sec}$

(c) $\left(\frac{3.00 \times 10^8 \text{ m}}{\text{sec}}\right)(1 \text{ min})\left(\frac{60 \text{ sec}}{1 \text{ min}}\right) = 1.80 \times 10^{10} \text{ m}$

5.3 $\nu = (3.00 \times 10^8 \text{ m/sec})/(6.16 \times 10^{-7} \text{ m}) = 4.87 \times 10^{14}/\text{sec}$
The color of this radiation would be between yellow and red; that is, an orange shade.

5.4 (a) $(3.00 \times 10^8 \text{ m/sec})/(96.3 \times 10^6/\text{sec}) = 3.11 \text{ m}$

(b) $(3.00 \times 10^8 \text{ m/sec})/(2.10 \times 10^8/\text{sec}) = 1.43 \text{ m}$

Half this length is 0.71 m, the optimal antenna length.

5.5 (a) $\left(\frac{3.00 \times 10^8 \text{ m}}{\text{sec}}\right)(1 \text{ yr})\left(\frac{365.25 \text{ day}}{1 \text{ yr}}\right)\left(\frac{24 \text{ hr}}{1 \text{ day}}\right)\left(\frac{3600 \text{ sec}}{1 \text{ hr}}\right) = 9.47 \times 10^{15} \text{ m}$

$9.47 \times 10^{15} \text{ m}\left(\frac{1.094 \text{ yd}}{1 \text{ m}}\right)\left(\frac{1 \text{ mi}}{1760 \text{ yd}}\right) = 5.89 \times 10^{12} \text{ mi}$

(b) $3.26 \text{ yr}\left(\frac{9.47 \times 10^{15} \text{ m}}{\text{yr}}\right)\left(\frac{1 \text{ km}}{10^3 \text{ m}}\right) = 3.09 \times 10^{13} \text{ km}$

5.6 $2.39 \times 10^5 \text{ mi}\left(\frac{1760 \text{ yd}}{1 \text{ mi}}\right)\left(\frac{1 \text{ m}}{1.094 \text{ yd}}\right)\left(\frac{1 \text{ sec}}{3.00 \times 10^8 \text{ m}}\right) = 1.28 \text{ sec}$

5.7 (a) $E = h\nu = (6.625 \times 10^{34} \text{ J-sec})(5.00 \times 10^{13} \text{ sec}) = 3.31 \times 10^{-20} \text{ J}$

(b) $E = h\nu = hc/\lambda = (6.625 \times 10^{-34}\ \text{J-sec})\left(\frac{3.00 \times 10^{8}\ \text{m/sec}}{2.00 \times 10^{-7}\ \text{m}}\right)$

$= 9.94 \times 10^{-19}\ \text{J}$

(c) $\nu = E/h = 1.30 \times 10^{-19}\ \text{J}/(6.62 \times 10^{-34}\ \text{J-sec}) = 1.96 \times 10^{14}/\text{sec}$

5.8 $E = hc/\lambda = (6.625 \times 10^{-34}\ \text{J-sec})(3.00 \times 10^{8}\ \text{m/sec})/\lambda$

$= 1.99 \times 10^{-25}\ \text{J-m}/\lambda$

When $\lambda = 0.15\ \text{nm} = 1.5 \times 10^{-10}\ \text{m}$, $E = 1.99 \times 10^{-25}\ \text{J-m}/1.5 \times 10^{-10}\ \text{m}$

$= 1.33 \times 10^{-15}\ \text{J}$

When $\lambda = 1.0\ \text{mm} = 10^{-3}\ \text{m}$, $E = 1.99 \times 10^{-25}\ \text{J-m}/1.00 \times 10^{-3}\ \text{m}$

$= 1.99 \times 10^{-22}\ \text{J}$

5.9 $E = hc/\lambda = \frac{(6.625 \times 10^{-34}\ \text{J-sec})(3.00 \times 10^{8}\ \text{m/sec})}{6.71 \times 10^{-7}\ \text{m}}$

$= 2.96 \times 10^{-19}\ \text{J} = 2.96 \times 10^{-22}\ \text{kJ}$

$$\left(\frac{2.96 \times 10^{-22}\ \text{kJ}}{\text{photon}}\right)\left(\frac{6.023 \times 10^{23}\ \text{photons}}{1\ \text{mol}}\right) = 178\ \text{kJ/mol}$$

5.10 First calculate the energy per photon of radiant energy of wavelength 550 nm:

$E = hc/\lambda = \frac{(6.625 \times 10^{-34}\ \text{J-sec})(3.00 \times 10^{8}\ \text{m/sec})}{5.50 \times 10^{-7}\ \text{m}}$

$= 3.61 \times 10^{-19}\ \text{J/photon}$

The number of photons is then $1.45 \times 10^{-17}\ \text{J}\left(\frac{1\ \text{photon}}{3.61 \times 10^{-19}\ \text{J}}\right)$

$= 40\ \text{photons}$

5.11 $E_1 = \frac{(6.625 \times 10^{-34}\ \text{J-sec})(3.00 \times 10^{8}\ \text{m/sec})}{4.60 \times 10^{-7}\ \text{m}} = 4.32 \times 10^{-19}\ \text{J}$

$E_2 = \frac{(6.625 \times 10^{-34}\ \text{J-sec})(3.00 \times 10^{8}\ \text{m/sec})}{6.60 \times 10^{-7}\ \text{m}} = 3.01 \times 10^{-19}\ \text{J}$

$\Delta E = E_1 - E_2 = 1.31 \times 10^{-19}\ \text{J}$

5.12 $\lambda = hc/E = \frac{(6.625 \times 10^{-34}\ \text{J-sec})(3.00 \times 10^{8}\ \text{m/sec})}{6.69 \times 10^{-19}\ \text{J}} = 2.97 \times 10^{-7}\text{m}$

$= 297\ \text{nm}$

When photons of this energy strike the copper surface an electron may be ejected. When the intensity of the radiation doubles, the number of photons per unit time doubles, so the number of electrons ejected per unit time doubles. However, there is no change in the energy of the ejected electrons.

5.13 (a) $E = h\nu = (6.625 \times 10^{-34}\ \text{J-sec})(4.60 \times 10^{14}/\text{sec})$

$= 3.05 \times 10^{-19}\ \text{J}$

is the minimum energy photon that can produce photoemission.

(b) wavelength $= c/\nu = (3.00 \times 10^{8}\ \text{m/sec})/4.6 \times 10^{14}/\text{sec})$

$= 6.52 \times 10^{-7}\ \text{m}$

(c) $E = hc/\lambda = \frac{(6.625 \times 10^{-34}\ \text{J-sec})(3.00 \times 10^{8}\ \text{m/sec})}{5.40 \times 10^{-7}\ \text{m}} = 3.68 \times 10^{-19}\text{J}$

The difference between this energy and the minimum energy required, 3.05×10^{-19} J, is 6.3×10^{-20} J. This energy appears as kinetic energy of the electron.

5.14 $$\frac{(6.625 \times 10^{-34} \text{ J-sec})(3.00 \times 10^{8} \text{ m/sec})(6.02 \times 10^{23})}{4.50 \times 10^{-7} \text{ m}}$$

$= 2.66 \times 10^{5}$ J/mol $= 2.66 \times 10^{5}$ J/einstein

5.15 (a) Energy is emitted, because the electron moves to a lower energy (more stable) orbital; (b) absorbed; (c) absorbed

5.16 $\Delta E = R_H(1/n_1^2 - 1/n_2^2) = 2.179 \times 10^{-18}$ J $(1/1 - 1/9) = 1.94 \times 10^{-18}$ J

5.17 $\Delta E = 2.179 \times 10^{-18}$ J$(1/25 - 1/4) = 4.57 \times 10^{-19}$ J

$E = h\nu = hc/\lambda$

$$\lambda = hc/E = \left[\frac{(6.625 \times 10^{-34} \text{ J-sec})(3.00 \times 10^{8} \text{ m/sec})}{4.57 \times 10^{-19} \text{ J}}\right]\left(\frac{10^{9} \text{ nm}}{1 \text{ m}}\right) = 435 \text{ nm}$$

Note that there is a blue line at 434.2 nm that corresponds to this transition.

5.18 $\Delta E = 2.179 \times 10^{-18}$ J$(1/1 - 1/25) = 2.09 \times 10^{-18}$ J

$\Delta E = hc/\lambda$

$$\lambda = hc/\Delta E = \frac{(6.625 \times 10^{-34} \text{ J-sec})(3.00 \times 10^{8} \text{ m/sec})}{2.09 \times 10^{-18} \text{ J}}$$

$= 9.52 \times 10^{-8}$ m $= 95.2$ nm

Light of this wavelength lies in the ultraviolet region of the spectrum.

5.19 The ionization potential for Li^{2+} is much larger than that for H, because of the larger nuclear charge. Li^{2+} and H both have just one electron, but the nuclear charge is +3 in Li^{2+}, whereas it is +1 in H.

5.20 The energy for ionization is $\Delta E = (2)^2(2.179 \times 10^{-18}$ J$)(1 - 0)$

$= 8.716 \times 10^{-18}$ J

The larger energy requirement arises because the electron is attracted more strongly by the higher nuclear charge in He^+.

5.21 (a) $r = (3)^2(0.53 \times 10^{-8}$ cm$) = 4.8 \times 10^{-8}$ cm

(b) $r = (1)^2(0.53 \times 10^{-8}$ cm$)/3 = 0.18 \times 10^{-8}$ cm

5.22 Electrons are diffracted by the atoms or ions that make up a crystal lattice; electron diffraction patterns are the result. Neutrons can also be diffracted in a similar manner, and neutron diffraction patterns have been observed.

5.23 (a) momentum $= mv = (1.67 \times 10^{-24}$ g$)(3.78 \times 10^{3}$ m/sec$)$

$= 6.31 \times 10^{-21}$ g-m/sec $= 6.31 \times 10^{-24}$ kg-m/sec

(b) $\lambda = h/mv$

$$= \frac{6.625 \times 10^{-34} \text{ kg-m}^2\text{/sec}}{6.31 \times 10^{-24} \text{ kg-m/sec}} = 1.05 \times 10^{-10} \text{ m}$$

(c) $mv = h/\lambda$, $v = h/m\lambda$

$v = (6.625 \times 10^{-34}$ kg-m^2/sec$)/(1.67 \times 10^{-27}$ kg$)(2.00 \times 10^{-10}$ m$)$

$= 1.99 \times 10^{3}$ m/sec

5.24 momentum = mv

$$mv = (82.5)\left(\frac{255\ \text{km}}{\text{hr}}\right)\left(\frac{1\ \text{kg}}{10^3\ \text{g}}\right)\left(\frac{10^3\ \text{m}}{1\ \text{km}}\right)\left(\frac{1\ \text{hr}}{3600\ \text{sec}}\right) = 5.84\ \text{kg-m/sec}$$

$$\lambda = h/mv = \frac{(6.625 \times 10^{-34}\ \text{kg-m}^2/\text{sec})}{(5.84\ \text{kg-m/sec})} = 1.13 \times 10^{-34}\ \text{m}$$

5.25 The momentum of the electron is mv,

$$= (9.11 \times 10^{-31}\ \text{kg})(3.00 \times 10^{7}\ \text{m/sec}) = 2.73 \times 10^{-23}\ \text{kg-m/sec}$$

$$\lambda = h/mv = \frac{(6.625 \times 10^{-34}\ \text{kg-m}^2/\text{sec})}{(2.73 \times 10^{-23}\ \text{kg-m/sec})} = 2.43 \times 10^{-11}\ \text{m}$$

5.26

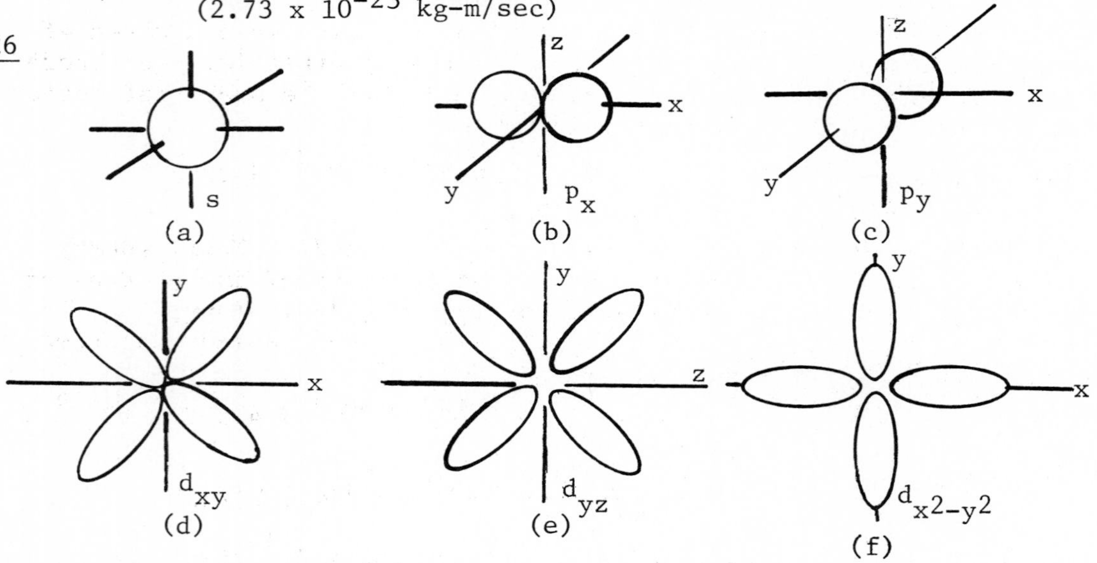

5.27 The incorrect designations are:
2d − ℓ cannot equal 2 when n equals 2.
1p − ℓ cannot equal 1 when n equals 1.
3f − ℓ cannot equal 3 when n equals 3.

5.28 The values of n, ℓ and m_ℓ for all the 4f orbitals are: 4,3,−3; 4,3,−2; 4,3,−1; 4,3,0; 4,3,+1; 4,3,+2; 4,3,+3. For the orbitals of the n = 2 shell: 2,0,0; 2,1,−1; 2,1,0; 2,1,+1.

5.29 The 2p and 3p orbitals differ in the value of the major quantum number n. This means that there are differences in the shapes of the orbitals near the nucleus, but the overall shape of both 2p and 3p orbitals is like that shown in Figure 5.18. The 3p orbitals extend farther from the nucleus than the 2p.

5.30 1s < 2s = 2p < 3s = 3d < 4s = 4f.

5.31 A wave function is a mathematical expression that describes the distribution in space of an orbital. The square of the wave function has the physical significance of probability. The value of the wave function squared over any small volume in space measures the relative probability that the electron will at any given instant be found in that volume.

5.32 Permissible sets are: (a) 3p and (d) 9s. (b) is incorrect because m_ℓ may not exceed ℓ. (c) is incorrect because ℓ may not exceed n − 1. (e) is incorrect because ℓ may not have negative values.

5.33 A node is the locus of points in space at which the wave function has zero value. The locus of points may be a plane or a spherical surface. The square of the wave function measures the probability that the electron will be found at a certain place; thus, the probability of the electron's being found at a node is zero.

5.34 (a) The principal quantum number determines the radial extent of the orbital; that is, the average distance from the nucleus. It also determines the energy of the orbital. (b) The azimuthal quantum number ℓ determines the angular shape of the orbital. (c) The magnetic quantum number determines the direction in which an orbital is oriented. For example, there are three 2p orbitals all with the same shape (Figure 5.18). The three are, however, differently oriented in space according to their different values of m_ℓ.

5.35 (a) For n = 4, we may have one 4s, three 4p, five 4d and seven 4f orbitals for a total of 16. (b) There are three 4p orbitals. (c) There are no 2d orbitals. (d) There is just one $3d_{xy}$ orbital; the subscript letter designation specifies a particular value of m_ℓ.

5.36 (a) $E = h\nu$, $\nu = 1.45 \times 10^{-19}$ J/6.625 x 10^{-34} J-sec

$= 2.19 \times 10^{14}$/sec

(b) We have that $E = h\nu$. Since $\nu\lambda = c$, $E = hc/\lambda$. Thus, energy is inversely proportional to wavelength. The energies of two photons of wavelengths λ_1 and λ_2 thus are related as $E_1/E_2 = \lambda_2/\lambda_1$. Thus, E(535 nm)/E(0.15 nm) = 0.15 nm/535 = 2.80 x 10^{-4} times as many calories from the longer wavelength photon.

5.37 (a) $$E = \frac{(6.625 \times 10^{-34}\ \text{J-sec})(2.998 \times 10^{8}\ \text{m/sec})}{4.358 \times 10^{-7}\ \text{m}} = 4.558 \times 10^{-19}\ \text{J}$$

(b) $$\left(\frac{4.558 \times 10^{-19}\ \text{J}}{\text{photon}}\right)\left(\frac{6.022 \times 10^{-23}\ \text{photons}}{\text{mol}}\right) = 274.4\ \text{kJ/mol}$$

5.38 $\nu = (3.00 \times 10^{8}$ m/sec)/(4.00 x 10^{-7} m) = 7.50 x 10^{14}/sec

$E = h\nu = (6.625 \times 10^{-34}$ J-sec)(7.50 x 10^{14}/sec) = 4.97 x 10^{-19} J

5.39 (a) First calculate the energy per photon:

$$E = hc/\lambda = \frac{(6.625 \times 10^{-34}\ \text{J-sec})(3.00 \times 10^{8}\ \text{m/sec})}{6.50 \times 10^{-7}\ \text{m}} = 3.06 \times 10^{-19}\ \text{J}$$

$$1\ \text{J}\left(\frac{1\ \text{photon}}{3.06 \times 10^{-19}\ \text{J}}\right) = 3.27 \times 10^{18}\ \text{photons}$$

(b) Similarly, $$E = \frac{(6.625 \times 10^{-34}\ \text{J-sec})(3.00 \times 10^{8}\ \text{m/sec})}{4.00 \times 10^{-7}\ \text{m}}$$

$= 4.97 \times 10^{-19}$ J

$$1\ \text{J}\left(\frac{1\ \text{photon}}{4.97 \times 10^{-19}\ \text{J}}\right) = 2.01 \times 10^{18}\ \text{photons}$$

5.40 The 1s to 2s transition in a hydrogen atom is calculated using the Bohr model to have an energy change:

$\Delta E = 2.179 \times 10^{-18}$ J(1/1 - 1/4) = 1.63 x 10^{-18} J

5.41 (a) $$\lambda = \frac{344\ \text{m/sec}}{30/\text{sec}} = 11.5\ \text{m}$$

(b) $$\lambda = \frac{344\ \text{m/sec}}{1.5 \times 10^{4}/\text{sec}} = 2.3 \times 10^{-2}\ \text{m} = 2.3\ \text{cm}$$

5.42 (a) $$E = \frac{hc}{\lambda} = \frac{(6.625 \times 10^{-34}\ \text{J-sec})(3.00 \times 10^{8}\ \text{m/sec})}{5.20 \times 10^{-7}\ \text{m}}$$

$= 3.82 \times 10^{-19}$ J. On a molar basis, this is

$$\frac{3.82 \times 10^{-19}\ \text{J}}{\text{photon}}\left(\frac{6.02 \times 10^{23}\ \text{photons}}{1\ \text{mol}}\right) = 230\ \text{kJ/mol}$$

This energy represents the work function for lithium.

$$\text{(b) } E = \frac{(6.625 \times 10^{-34}\ \text{J-sec})(3.00 \times 10^{8}\ \text{m/sec})}{3.60 \times 10^{-7}\ \text{m}} = 5.52 \times 10^{-19}\ \text{J}$$

The kinetic energy imparted to the electron is this energy less the work function calculated in part (a):

$$KE = 5.52 \times 10^{-19}\ \text{J} - 3.82 \times 10^{-19}\ \text{J} = 1.70 \times 10^{-19}\ \text{J per electron.}$$

5.43 (a) An orbit is an allowed circular path of the electron in Bohr's model of the hydrogen atom. An orbital is an allowed state of the electron in the quantum-mechanical model of the atom. (b) The wavelength of radiant energy represents the distance between corresponding points in each cycle as the radiation moves through space. The frequency is the number of cycles that pass a point in one second as the wave passes by. (c) An s orbital has a value of 0 for the azimuthal quantum number; a p orbital has a value of 1. (d) The ground state of an atom is the stable, lowest energy state. When the atom absorbs energy an electron may be raised in energy, to occupy a higher energy orbital. The atom is then in an excited state. (e) A continuous spectrum is obtained from radiant energy that contains photons of all wavelengths over a given region such as the visible. A line spectrum, on the other hand, is obtained from radiation that contains photons of only certain wavelengths. (f) The principal quantum number, n, determines the radial extent of an orbital. The azimuthal quantum number, ℓ, determines the shape of the orbital.

5.44 radius $= 0.53 \times 10^{-10}\ \text{m}(n^2) = 0.53 \times 10^{-10}\ \text{m} \times (4) = 2.12 \times 10^{-10}\ \text{m}$

circumference $= 2\pi r = 13.3 \times 10^{-10}\ \text{m}$ $\lambda = h/mv$. Rearranging,

$$v = h/m\lambda = \frac{(6.625 \times 10^{-34}\ \text{kg-m}^2/\text{sec})}{(9.11 \times 10^{-31}\ \text{kg})(13.3 \times 10^{-10}\ \text{m})} = 5.47 \times 10^{5}\ \text{m/sec}$$

5.45 $4.2 \times 10^{8}\ \text{mi}\left(\frac{1760\ \text{yd}}{1\ \text{mi}}\right)\left(\frac{1\ \text{m}}{1.094\ \text{yd}}\right)\left(\frac{1\ \text{sec}}{3.00 \times 10^{8}\ \text{m}}\right) = 2.2 \times 10^{3}\ \text{sec}$

5.46 $\lambda = \frac{3.00 \times 10^{8}\ \text{m/sec}}{1.12 \times 10^{6}/\text{sec}} = 2.68 \times 10^{2}\ \text{m}$

$$E = (6.625 \times 10^{-34}\ \text{J-sec})(1.12 \times 10^{6}/\text{sec}) = 7.42 \times 10^{-28}\ \text{J}$$

5.47 $\nu = \frac{3.00 \times 10^{8}\ \text{m/sec}}{0.21\ \text{m}} = 1.4 \times 10^{9}/\text{sec}$

$$E = (6.625 \times 10^{-34}\ \text{J-sec})(1.4 \times 10^{9}/\text{sec}) = 9.5 \times 10^{-25}\ \text{J}$$

5.48 $E = h\nu = hc/\lambda$. Rearranging, $\lambda = hc/E$

$$= \frac{(6.625 \times 10^{-34}\ \text{J-sec})(3.00 \times 10^{8}\ \text{m/sec})}{7.00 \times 10^{-19}\ \text{J}} = 2.84 \times 10^{-7}\ \text{m} = 284\ \text{nm}$$

We see that this is a much shorter wavelength than the lamp photons, 540 nm. Thus, the lamp photons cannot cause photoemission of electrons from chromium.

5.49 The emissions from the neon lamp arise from transitions of the electrons in neon atoms that have been excited by an electric discharge. The electrons may exist only in allowed states of definite energy. When they undergo a transition from one allowed state to another of lower energy, a photon of energy that corresponds to the difference in energies of the two allowed states is emitted.

5.50 (a) The number of allowed subshells is determined by the number of different values that ℓ may have. When n = 1, ℓ has only one value, 0. When n = 2, ℓ has 2 possible values, 0 and 1, and so on. Thus, the number of different subshells is just n. (b) The number of allowed orbitals varies with n as follows:

n	orbitals
1	1 (1s)
2	4 (2s, three 2p)
3	9 (3s, three 3p, five 3d)
4	16 (4s, three 4p, five 4d, seven 4f)

In general, the number of orbitals with a given n varies as n^2.

5.51 See Figures 5.16, 5.18 and 5.19.

5.52 (a) one; (b) five; (c) one; (d) three

5.53 $\lambda = h/mv$. Rearranging, $v = \frac{h}{m\lambda} = \frac{6.625 \times 10^{-34}\ \text{kg-m}^2/\text{sec}}{(9.11 \times 10^{-31}\ \text{kg})(3.32 \times 10^{-10}\ \text{m})}$

$= 2.19 \times 10^6$ m/sec

5.54 $E = 1/2\ mv^2$. Thus,

$$v = \left(\frac{2(8.0 \times 10^{-21}\ \text{kg-m}^2/\text{sec}^2)}{1.67 \times 10^{-27}\ \text{kg}}\right)^{1/2} = 3.09 \times 10^3\ \text{m/sec}$$

$$\lambda = h/mv = \frac{6.63 \times 10^{-34}\ \text{kg-m}^2/\text{sec}}{(1.67 \times 10^{-27}\ \text{kg})(3.09 \times 10^3\ \text{m/sec})} = 1.28 \times 10^{-10}\text{m} = 1.28\ \text{Å}$$

$$v = \left(\frac{2(1.60 \times 10^{-20}\ \text{kg-m}^2/\text{sec}^2)}{1.67 \times 10^{-27}\ \text{kg}}\right)^{1/2} = 4.38 \times 10^3\ \text{m/sec}$$

$$\lambda = \frac{6.63 \times 10^{-34}\ \text{kg-m}^2/\text{sec}}{(1.67 \times 10^{-27}\ \text{kg})(4.38 \times 10^3\ \text{m/sec})} = 0.907 \times 10^{-10}\ \text{m} = 0.907\ \text{Å}$$

5.55 The change in major quantum number is the only factor that determines the energy change. The change is determined by $(1/n_1^2 - 1/n_2^2)$. This factor is smallest for $n_1 = 2$, $n_2 = 3$. Thus, the 2p → 3s transition requires the lowest energy photon.

5.56 They would be similar, but with a stronger dependence on r. That is, ψ^2 would drop off more rapidly as r increases for the He^+ ion than for H. This occurs because the electron is more strongly attracted to the higher nuclear charge, and thus is, on the average, closer to the nucleus. This is true for any value of n.

5.57 (a) 1s : n = 1, ℓ = 0, m_ℓ = 0

(b) 2p : n = 2, ℓ = 1, m_ℓ = 1,0,-1

(c) 3d : n = 3, ℓ = 2, m_ℓ = 2,1,0,-1,-2

6

Periodic relationships among the elements

6.1 (a) 3p; (b) 2s; (c) 3s; (d) 3d

6.2 The beryllium nucleus has a charge of +4, whereas Li has a nuclear charge of +3. The electron configurations are $1s^22s^2$ and $1s^22s^1$, respectively. Each of the 2s electrons in beryllium only partially shields the other 2s electron from the higher nuclear charge.

6.3 In a hydrogen atom, the orbital energy depends only on n. In a many-electron atom, the energy depends on both n and ℓ.

6.4 The ionization energy is given by the negative of the energy of the lowest energy orbital, with n = 1. This is given by $I = R_HZ^2$. If I is 1312 kJ/mol for H, it is $1312(3)^2$ for Li^{2+}, for which Z = 3. Thus, $I = 1.181 \times 10^4$ kJ/mol.

6.5 In a many-electron atom, the $3p_x$ and $3p_y$ orbitals are degenerate.

6.6 In fluorine, the 2s and 2p electrons experience different shielding effects. The 2s electrons penetrate on the average closer to the nucleus, and thus are influenced by a higher effective nuclear charge than the 2p electrons. In a one-electron atom, no shielding of one electron by another occurs, so the 2s and 2p orbitals remain degenerate.

6.7 (a) For n = 3, $\ell = 1$, m_ℓ may have values +1,0,−1, and m_s may have values +1/2 or −1/2. Hence, a total of six electrons. (b) ten; (c) two; (d) 32; (e) nine. These nine are n = 3, $\ell = 2$ with $m_\ell = 2,1,0,-1,-2$; $\ell = 1$ with $m_\ell = 1,0,-1$; $\ell = 0$ with $m_\ell = 0$. There is just one electron in each orbital with $m_s = +1/2$.

6.8 (a) ten; (b) six; (c) 14; (d) two

6.9 n = 1, $\ell = 0$, $m_\ell = 0$, $m_s = +1/2$ or $-1/2$; n = 2, $\ell = 0$, $m_\ell = 0$, $m_s = +1/2$ or $-1/2$

6.10 (a) [Ar]$4s^1$; (b) [Ne]$3s^2 3p^2$; (c) [Ar]$4s^2 3d^{10} 4p^4$; (d) [Ar]$4s^2 3d^5$; (e) [Xe]$6s^2 5d^1$

6.11 (a) $1s^2 2s^2 2p^6 3s^2 3p^6 4s^2 3d^{10} 4p^6 5s^2 4d^{10}$

(b) $1s^2 2s^2 2p^6 3s^2 3p^4$

(c) $1s^2 2s^2 2p^6 3s^2 3p^6 4s^2 3d^{10} 4p^6 5s^2 4d^{10} 5p^6 6s^2 4f^6$

(d) $1s^2 2s^2 2p^6 3s^2 3p^6 4s^2 3d^{10} 4p^6 5s^2 4d^{10} 5p^6 6s^2 4f^{14} 5d^{10} 6p^2$

(e) $1s^2 2s^2 2p^6 3s^2 3p^6 4s^2 3d^{10} 4p^6$

6.12 (a) [Ne] 3s [↑↓] 3p [↑|↑|↑]

(b) [Ar] 4s [↑↓] 3d [↑|↑|↑| |]

(c) [Kr] 5s [↑↓]

(d) [Xe] 6s [↑↓] 4f [↑↓|↑↓|↑↓|↑↓|↑↓|↑↓|↑↓] 5d [↑| | | |]

(e) [Kr] 5s [↑↓] 4d [↑↓|↑↓|↑↓|↑↓|↑↓] 5p [↑↓|↑↓|↑]

6.13 (a) Be; (b) O; (c) Ti; (d) Cl; (e) In; (f) Tc

6.14 (a) ground state of Be; (b) excited state of Li (ground state would be $1s^2 2s^1$; (c) excited state of Al (ground state would be [Ne]$3s^2 3p^1$; (d) ground state of Ti; (e) excited state of Na (ground state would be $1s^2 2s^2 2p^6 3s^1$).

6.15 (a) Ca - active metal; (b) Fe - transition metal; (c) Nd - inner transition metal; (d) Rb - active metal; (e) Se - non-metal; a representative element; (f) Xe - non-metal; one of the rare gases; (g) Ag - a transition metal; (h) U - an inner transition metal of the actinide series; (i) Pb - a metal from the representative elements

6.16 (a) alkaline earths = ns^2 (n = 2,...,7); (b) group 1B - $ns^1(n-1)d^{10}$ (n = 4,5,6); (c) group 5A - ns^2np^3 (n = 2....6); (d) halogens - ns^2np^5 (n = 2....6); (e) noble gases - ns^2np^6(n = 2...6); (f) group 5B - $ns^2(n-1)d^3$ (n = 4,5,6)

6.17 The electron responsible for the line spectra in Li is most likely to be the 2s electron. The transitions of this electron from higher energy orbitals back to the 2s level should be similar in many ways to transitions of the hydrogen electron from higher energy orbitals to the 2s orbital.

6.18 (a) Atomic size decreases from left to right in a horizontal row of the table. (b) Ionization energy increases. (c) increases. As one moves down a family in the table, atomic size generally increases, ionization energy decreases, and electron affinity remains relatively constant.

6.19 (a) Na; (b) Li; (c) P; (d) Si

6.20 (a) F; (b) Mg; (c) O; (d) C; (e) N (because it has a stable, half-filled shell)

6.21 (a) larger size, O; larger ionization energy, F; more negative electron affinity, F. (b) larger size, P; larger ionization energy, N; more negative electron affinity, P. (c) larger size, Ca; larger ionization energy, Sc; more negative electron affinity, Sc.

6.22 The stability of the closed shell octet of electrons makes it very difficult to remove an electron from a noble gas atom. Thus, chemical activity that involves removal or partial removal of an electron from the atom is energetically non-favored. The noble gas atom has almost no tendency whatever to add an electron, because the closed shell already provides a very stable, spherical electron distribution. Thus, the avenues to forming chemical bonds are not really open to the noble gases. However, the heavier noble gases, particularly Xe, with lower ionization energy, do exhibit some chemical activity (Chapter 21).

6.23 The third electron removed from Mg comes from the 2p subshell. These 2p electrons experience a much larger effective nuclear charge because they do not completely shield one another from the nucleus. The 3s electrons, on the other hand, were well shielded by the 2p electrons, and thus experienced a much smaller effective nuclear charge.

6.24 As the effective nuclear charge increases in going from potassium through krypton, the electrons are drawn in more closely, so that atomic size decreases. At the same time, the energy required to remove an electron from the attractive field of the effective nuclear charge increases.

6.25 (a) $2K(s) + H_2O(\ell) \rightarrow 2KOH(aq) + H_2(g)$
(b) $Li_2O(s) + H_2O(\ell) \rightarrow 2LiOH(aq)$
(c) $4Li(s) + O_2(g) \rightarrow 2Li_2O(s)$
(d) $2Na(\ell) + H_2(g) \rightarrow 2NaH(s)$
(e) $2Li(s) + I_2(g) \rightarrow 2LiI(s)$

6.26 (a) Li is more metallic; (b) Li is larger; (c) Li has one 2s electron; Be has two; (d) Be has the higher ionization energy; (e) Li has a more negative electron affinity; (f) LiCl; $BeCl_2$

6.27 The nuclear charge in Zn is ten larger than in Ca. The ten corresponding electrons fill the 3d orbitals. However, the 4s electrons of Zn are not completely shielded from the nucleus by these 3d electrons. As a result, they experience a larger effective nuclear charge than the 4s electrons of Ca.

6.28 The reversals of relative atomic weights occur at Ar 39.948, K 39.098; Co 58.9332; Ni 58.71; Te 127.60, I 126.9045; Th 232.04, Pa 231.04; U 238.03, Np 237.048. We know that within certain limits the number of neutrons in the nuclei of a given element may vary. These "violations" occur because the average number of neutrons in the nuclei of the element with one less proton is sufficient to outweigh the smaller mass of protons. Thus, for example, there are, on the average, about 22 neutrons in the nuclei of the naturally occurring isotopes of Ar, and about 20 neutrons in the nuclei of the naturally occurring isotopes of K.

The modern version of the periodic law is that the chemical and physical properties of the elements are periodic functions of their atomic number.

6.29 H_2S; H_2Se; H_2Te; H_2Po

6.30 The average distance of a 2p electron is greater, as evidenced by the fact that it requires less energy to remove a 2p electron from a neon atom than to remove a 2s electron. The 2s electron has a smaller average distance from the nucleus because of the form of the 2s wave function. As a result, it experiences a higher effective nuclear charge.

6.31 (a) 2s; (b) 2p; (c) 3d

6.32 $1s^2 2s^2 2p^6 3s^1$. The 1s electrons experience the greatest effective nuclear charge; the 3s electron experiences the smallest.

6.33 We use $E = -R_H(Z)^2/n^2$, because all these species are one-electron in nature. On a molar basis for H,

$$E = -2.179 \times 10^{-19}\ \text{J}\ \frac{(1)^2}{(1)^2}(6.02 \times 10^{23}) = -1312\ \text{kJ}$$

for He^+, with Z = 2, E is 4 x (-1312) = -5248 kJ

for Li^{2+}, with Z = 3, E is 9 x (-1312) = -1.181×10^4 kJ

These 1s orbital energies differ because of the increasing electron-nuclear attraction as nuclear charge increases.

6.34 A separation of atoms into two beams can occur only when there is an unpaired electron that can be oriented up or down with respect to the magnetic field. This could occur with H or Li, but not with He or Be. These results would show that a pairing of electrons occurs when the second electron is added to the 1s orbital. This is consistent with the idea that m_s must have different values for the two electrons. When a third electron is added, there is again an unpaired spin present, suggesting that this third electron has gone into a new orbital. The fourth electron again demonstrates pairing.

6.35 [Kr] 5s [↑] 4d [↑↓|↑↓|↑↓|↑↓|↑↓]

One unpaired spin; see Table 6.2.

6.36 The quantum-mechanical treatment of the atom leads to the prediction of the major quantum numbers n, ℓ and m_ℓ. Thus, we know that there is just one orbital (1s) with n = 1, four (2s, $2p_x$, $2p_y$, $2p_z$) with n = 2, and so on. But we need the Pauli principle to tell us that there can be but two electrons in each orbital. The recognition that the orbitals can contain at most two electrons accounts for the octet of electrons that completes each set of ns and np orbitals, and that characterizes each noble gas element.

Thus, 2 electrons fill all the orbitals with principal quantum number equal to 1; eight more fill the orbitals of principal quantum number equal to 2; another eight fill the s and p orbitals of quantum number equal to 3. Before another set of s and p orbitals can be filled, it is necessary to fill the 3d orbitals (10 electrons), so when the 4s and 4p orbitals are filled a total of 18 electrons has been added. This brings us to atomic number 36. Another 18 electrons brings us to atomic number 54.

6.37 If there were three possible values for m_s, this would suggest that it would require three electrons to fully occupy each orbital. Thus, using an orbital diagram, we would have for carbon,

1s [↑ → ↓] 2s [↑ → ↓]

We conclude that carbon would have no unpaired electrons. Its electron configuration would be $1s^3 2s^3$.

6.38 The g orbitals correspond to ℓ = 4. Thus, the lowest value which n could have would be 5. There are 2ℓ + 1 orbitals of a given value of ℓ and n. Thus, there are 9 g orbitals. There are no known elements with electrons occupying g orbitals in the ground state.

6.39 (a) $1s^2 2s^2 2p^6 3s^2 3p^5$; (b) $1s^2 2s^2 2p^6 3s^2 3p^6 4s^2$;

(c) $1s^2 2s^2 2p^6 3s^2 3p^6 4s^2 3d^1$; (d) $1s^2 2s^2 2p^6 3s^2 3p^6 4s^2 3d^{10} 4p^6 5s^2 4d^{10} 5p^3$;

(e) $1s^2 2s^2 2p^6 3s^2 3p^6 4s^2 3d^{10} 4p^2$; (f) $1s^2 2s^2 2p^6 3s^2 3p^6 4s^2 3d^{10} 4p^6 5s^2 4d^{10} 5p^6 6s^2 4f^{14} 5d^{10} 6p^6 7s^2 5f^3 6d^1$ (see Table 6.2)

6.40 (a) [Ar] 4s [↑↓] 3d [↑↓|↑↓|↑↓|↑↓|↑↓] 4p [↑|↑|↑]

(b) [Ar] 4s [↑] 3d [↑|↑|↑|↑|↑]

(c) [Kr] 5s [↑↓] 4d [↑| | | |]

(d) [Ar] 4s [↑]

(e) [Ne] 3s [↑↓] 3p [↑| |]

6.41 (a) C possesses two unpaired electrons; (b) none; (c) five; (d) two

6.42 (a) incorrect; the 2s orbital can contain at most two electrons; (b) correct; an excited state configuration; (c) correct; (d) correct; an excited state configuration; (e) incorrect; the 3d orbitals, of which there are five, can contain at most ten electrons; (f) incorrect; there are no f orbitals of major quantum number n = 3.

6.43 (a) chlorine; (b) alkaline earth metals (Mg, Ca, etc.); (c) Co, Rh (actually, Rh is not exactly this, but $5s^14d^8$ (see Table 6.2). Note that Ir, a member of the same family, is not included. It possesses 4f electrons in addition to the s and d (see Table 6.2).

6.44 (a) K; (b) He; (c) N; (d) He; (e) F or B; (f) chlorine

6.45 H is unique in that it has one electron in its valence orbital, in analogy to the alkali metals, and at the same time is just one electron short of attaining the noble gas configuration, in analogy to the halogens. Thus, it displays a valence of one; that is, a tendency to lose, gain or share an electron with another atom. Depending on what it reacts with, it may thus appear similar to an alkali metal such as Li, or to a halogen such as F.

6.46 In Be the two 2s electrons incompletely shield one another from the nuclear charge. When one additional charge is added to the nucleus, the next electron added, in B, goes into the 2p orbital. Because the 2p orbital extends further from the nucleus than the 2s, this 2p electron is quite well shielded from the nucleus by the 2s electrons as well as the 1s. Its ionization energy is lowered because it is, on the average, more distant from the nucleus than the 2s electron of Be. The lowered ionization energy of O as compared with N arises from the fact that the last electron added to the O atom must go into a 2p orbital that is already occupied. This gives rise to greater electron-electron repulsions than are present in N, in which the three 2p electrons singly occupy each of the three 2p orbitals.

6.47

Element	Density (g/cm^3)	At. Wt.	Chloride	Oxide	I_1 (kJ/mol)	I_2	I_3	Melting Point (°C)
Be	1.86	9.01	$BeCl_2$	BeO	899	1757	14849	1277
Mg	1.74	24.3	$MgCl_2$	MgO	735	1445	7730	650
Ca	1.55	40.1	$CaCl_2$	CaO	590	1145	4912	850
Sr	2.54	87.6	$SrCl_2$	SrO	550	1064	4207	769
Ba	3.51	137.3	$BaCl_2$	BaO	503	965	--	725
Ra	5.5	226.0	$RaCl_2$	--	509	979	--	700

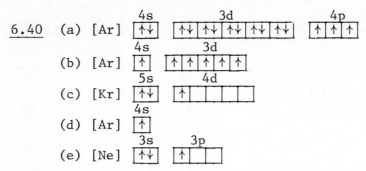

The similarities among these elements are most evident in the sequence of ionization energies, especially in the large jump at I_3, and in the formulas of the chlorides and oxides. The density and melting point reflect the effects of increasing atomic mass and radius as we go downward in the family.

6.48 We expect for Tc properties similar to those for its lighter neighbor, Mn, in the same way as Mo is similar to Cr. Let us look at some properties as a basis for predicting the properties of Tc.

Element	Density (g/cm^3)	I_1	I_2	Melting Point (°C)	Oxides	Electron Config.
Cr	7.20	653	1496	1857	CrO, Cr_2O_3, CrO_3	$4s^13d^5$
Mo	10.2	685	1558	2610	Mo_2O_3, MO_2, Mo_2O_5, MoO_3	$5s^14d^5$
Mn	7.47	717	1509	1244	MnO, Mn_2O_3, MnO_2, Mn_3O_4, MnO_3, Mn_2O_7	$4s^23d^5$

On the basis of these comparisons we expect that Tc will have a density of about 10.3 g/cm^3; that it will have I_1 and I_2 values a bit higher than those of Mn, say 750 and 1570 kJ/mol; that it will melt about 750°C higher than Mn, say 2000°C; that its electron configuration will be $5s^24d^5$. As to the oxides, manganese seems to possess nearly every possible oxide formula! We might guess that Tc will not be as stable in the lowest oxidation states, just as Mo is not in comparison with Cr. Thus, oxides of the form Tc_2O_3, TcO_2, Tc_3O_4, TcO_3 and Tc_2O_7 might be expected.

It is known only that Tc has a melting point over 1300°C; that I_1 and I_2 are 702 and 1472 kJ/mol, respectively; oxides TcO_2 and Tc_2O_7 have been isolated.

6.49 I_3 for Mg represents removal of a 2p electron. For all the other elements of Table 6.3, I_3 represents removal of an electron from an orbital with n = 3. The effective nuclear charge experienced by a 2p electron in Mg is much greater than that felt by any of the 3s or 3p electrons of the other elements involved.

The electron configuration of P^{2+} is $[Ne]3s^23p^1$. Removal of the single remaining 3p electron empties the 3p level. There is no effect of partial shielding of the nuclear charge by other p electrons. Hence, in comparison with the ionizations of Si^{2+} or S^{2+}, in which partial screening by equivalent electrons is involved, the effective nuclear charge is lower for P^{2+}. (Notice that a similar dip occurs in I_1 at Al in comparison with Mg and Si. Again, the outer electron configuration ns^2np^1 is involved.)

6.50 The first ionization energy for Ga corresponds to removal of the 4p electron. In Sc it corresponds to removal of a single 3d electron. The ionization energy is lower in Ga because the 4p orbital in Ga extends out rather far in space, farther than the 3d orbital in Sc. Thus, the energy required to remove this electron is a bit lower in Ga. Once the first electron is gone, however, the second and third electrons come from the 4s orbital in both cases. The electron configurations for Sc^+ and Ga^+ are:

Sc^+ $[Ar]\ 3d^14s^1$ Ga^+ $[Ar]\ 3d^{10}4s^2$

At this point the 4s electron of Ga^+ experiences a higher effective nuclear charge because the 4s electrons are incompletely screened by the 3d electrons. As a result, I_2 and I_3 are higher for Ga. Note that the comparison of Ga^+ with Sc^+ is very similar to the comparison of I_1 and I_2 for Zn and Ca (Table 6.6).

6.51 When we move one place to the right in a horizontal row of the table, for example, from Li to Be, there is an increase in ionization energy. When we move downward in a given family, for example from Be to Mg, there is usually a decrease in ionization energy. Similarly, atomic size decreases in moving one place to the right, increases in moving downward. Thus, two elements such as Li and Mg that are diagonally related tend to have similar ionization energies and atomic sizes. This in turn gives rise to some similarities in chemical behavior. Note, however, that the valences expected for the elements are not the same. That is, lithium still appears as Li^{+}, magnesium as Mg^{2+}.

6.52 Ground state:

One possible excited state:
(promote the 2s electron
to the vacant 2p)

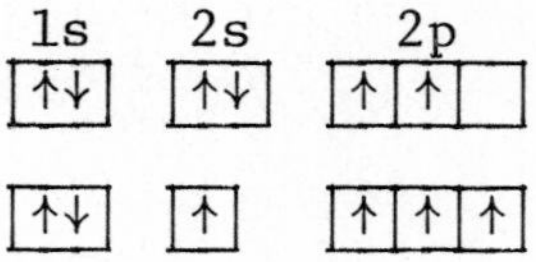

7

Chemical bonding

7.1 (a) •Ca• ; (b) •S̈e̤• (c) :B̈r̤• (d) •Ḃ•

7.2 Atoms or ions with octets are (b) O^{2-}; (c) Ne; (d) Ca^{2+}.

7.3 (a) $[Ar]4s^1$, K• ; (b) $[Ne]3s^2 3p^2$, •Ṩi•

(c) $[Ar]4s^2 3d^{10} 4p^3$:Ȧṣ• (We do not write the 3d electrons in the Lewis symbol because they comprise a completed lower level subshell that is not chemically active in arsenic.)

(d) $[Ne]3s^2 3p^1$, •Ȧl•

7.4 (a) Ca^{2+} (b) :Är̤: (c) $[:\ddot{S}:]^{2-}$ (d) Na (e) :Ï̤•

7.5 (a) $SrBr_2$; (b) BeO; (c) Al_2S_3 (aluminum has 3 valence electrons, sulfur has six. Al tends to form the Al^{3+} ion, sulfur tends to form the S^{2-} ion. Thus, it requires three sulfur atoms to balance the electrons lost from two Al atoms. (d) ZnF_2; (e) K_2Se; (f) ZnO

7.6 (a) Mn outer electron configuration is $4s^2 3d^5$; Mn^{4+} is $3d^3$; (b) $3s^2 3p^6$; (c) $3d^{10}$; (d) none; (e) $3d^3$; (f) $6s^2 4f^{14} 5d^{10}$

7.7 Noble gas configurations are possessed by (a) Te^{2-}, (d) Mg^{2+}, and (e) Sc^{3+}. The electron configurations for the others are (b) Ga^+, $4s^2 3d^{10}$; (c) In^{3+}, $3d^{10}$; (f) Fe^{2+}, $3d^6$.

7.8 (a) Each ion in MgO has a higher charge, +2 and -2, as compared with +1 and -1 in KCl. (b) The smaller oxide ion allows the cation and anion to approach more closely. The lattice energy grows more negative (more stable) as distance between oppositely charged ions decreases (1/r dependence). (c) Chloride ion is smaller than bromide ion; see the discussion in part (b).

7.9 (a) $Ba^{2+}\left[:\overset{\displaystyle\cdot\cdot}{\underset{\displaystyle\cdot\cdot}{O}}:\right]^{2-}$; (b) $[Na^+]_2\left[:\overset{\displaystyle\cdot\cdot}{\underset{\displaystyle\cdot\cdot}{O}}:\right]^{2-}$; (c) $Mg^{2+}\left[:\overset{\displaystyle\cdot\cdot}{\underset{\displaystyle\cdot\cdot}{Br}}:\right]_2^{-}$; (d) $Al^{3+}\left[:\overset{\displaystyle\cdot\cdot}{\underset{\displaystyle\cdot\cdot}{F}}:\right]_3^{-}$

7.10 (a) $Al(NO_3)_3$; (b) Fe_2O_3; (c) $CrCl_3$; (d) $(NH_4)_2SO_4$; (e) CaO

7.11 (a) $CaBr_2$; (b) $Al_2(SO_4)_3$; (c) $(NH_4)_3PO_4$; (d) K_3N; (e) TiF_4

7.12 (a) Cs^+; (b) Cl^-; (c) Fe^{2+}; (d) S^{2-} (S^{2-} and Cl^- are isoelectronic; that is, they contain the same number of electrons. The nuclear charge in S^{2-} is one less than in Cl^-. Thus, the electrons in S^{2-} are less tightly pulled in toward the nucleus. (e) Zn; (f) K^+

7.13 (a) Na^+, Mg^{2+}; (b) Cl^-, Ar, K^+; (c) Se^{2-}, Rb^+; (d) none

7.14 (a) $Mg^{2+} < Ca^{2+} < Sr^{2+}$; (b) $Ca^{2+} < Na^+ < K^+ < Cl^-$ (Here the higher charge on Ca^{2+} compensates for the fact that it is in the next horizontal row relative to Na^+.)   (c) $F < O < O^{2-}$

7.15 Anions contain a larger number of electrons in relation to the nuclear charge than do cations. The electron-electron repulsions cause the electrons to spread out, to occupy a larger volume, so as to decrease the repulsive energies.

7.16 Ionic substances are generally rather hard, brittle solids that cleave easily along regular planes. They generally have rather high melting points. When they dissolve in water or can be melted, their solutions or the melt conduct electricity.

Covalent substances are often gases or liquids with low boiling points. Many covalent solids are soft and low melting. Many covalent compounds tend to be soluble in organic liquids, and neither their melts nor solutions conduct electricity very well. On the other hand, some covalent substances, such as diamond, are hard and high melting.

7.17 (a) covalent; (b) ionic; (c) covalent; (d) covalent; (e) ionic; (f) ionic.

7.18 (a)
$$
\begin{array}{c}
H \\
| \\
H - Si - H \\
| \\
H
\end{array}
$$
  (b) $H - \overset{\displaystyle\cdot\cdot}{\underset{\displaystyle\cdot\cdot}{S}} - H$ (c) $:C \equiv O:$ (d) $:N \equiv N:$

(e) $H - \overset{\displaystyle\cdot\cdot}{\underset{\displaystyle\cdot\cdot}{O}} - \overset{\displaystyle\cdot\cdot}{\underset{\displaystyle\cdot\cdot}{O}} - H$ (f) $\overset{\displaystyle\cdot\cdot}{\underset{\displaystyle\cdot\cdot}{S}} = C = \overset{\displaystyle\cdot\cdot}{\underset{\displaystyle\cdot\cdot}{S}}$

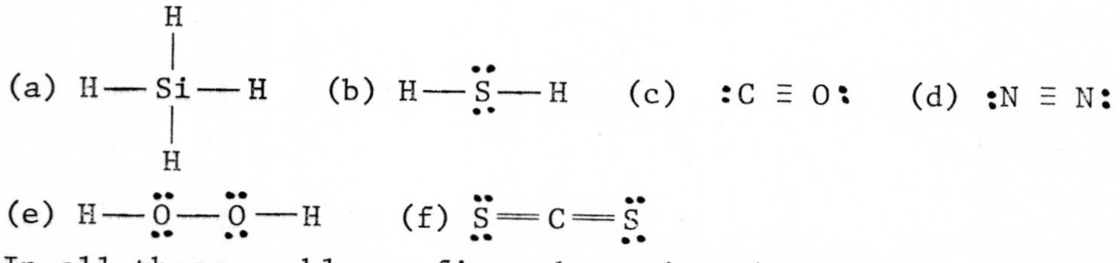

7.19 In all these problems, first determine the number of valence-shell electrons to be placed. Be sure to consider the charge on the ion.

(a) $\left[:\overset{\cdot\cdot}{\underset{\cdot\cdot}{Cl}} - \overset{\cdot\cdot}{\underset{\cdot\cdot}{O}}:\right]^{-}$ (b) $\left[:\overset{\cdot\cdot}{\underset{\cdot\cdot}{O}} - \overset{\displaystyle|}{\underset{\displaystyle|}{S}} - \overset{\cdot\cdot}{\underset{\cdot\cdot}{O}}:\,\underset{:O:}{}\right]^{2-}$

(c) $\left[\begin{array}{c} :\overset{\cdot\cdot}{O}: \\ | \\ :\overset{\cdot\cdot}{\underset{\cdot\cdot}{O}} - P - \overset{\cdot\cdot}{\underset{\cdot\cdot}{O}}: \\ | \\ :\overset{}{\underset{\cdot\cdot}{O}}: \end{array}\right]^{3-}$ (d) $\left[:C \equiv N:\right]^{-}$ (e) $\left[:N \equiv O:\right]^{+}$

(f) $\left[\begin{array}{c} :\overset{\cdot\cdot}{F}: \\ | \\ :\overset{\cdot\cdot}{\underset{\cdot\cdot}{F}} - B - \overset{\cdot\cdot}{\underset{\cdot\cdot}{F}}: \\ | \\ :\overset{}{\underset{\cdot\cdot}{F}}: \end{array}\right]^{-}$

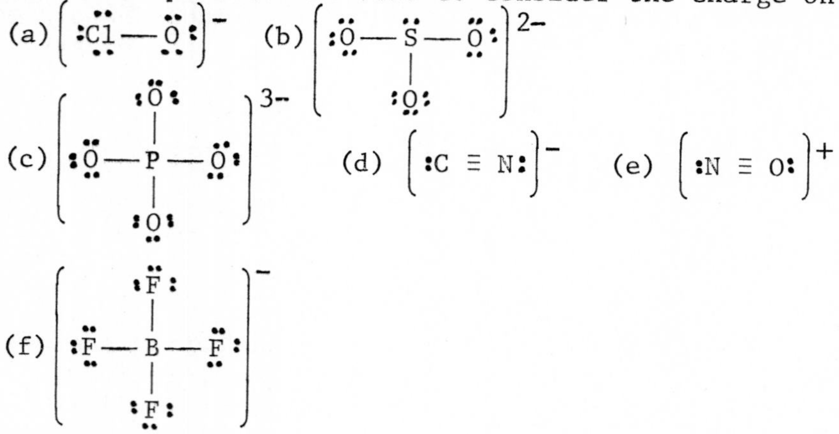

7.20 (a) To attain an octet of electrons about each atom in this compound is possible only when Cl is bonded to N, not O. We have a total of

7 + 6 + 5 = 18 valence electrons. 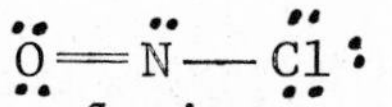 The manner in which the formula is written in this case is confusing, and not an indication of the bonding arrangement.

(b) H—C ≡ C—H (c) 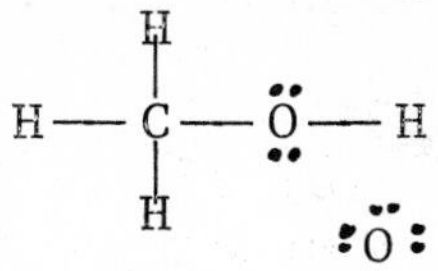 (d) 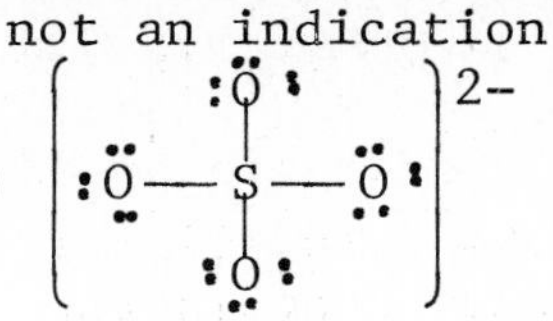

7.21 (a) 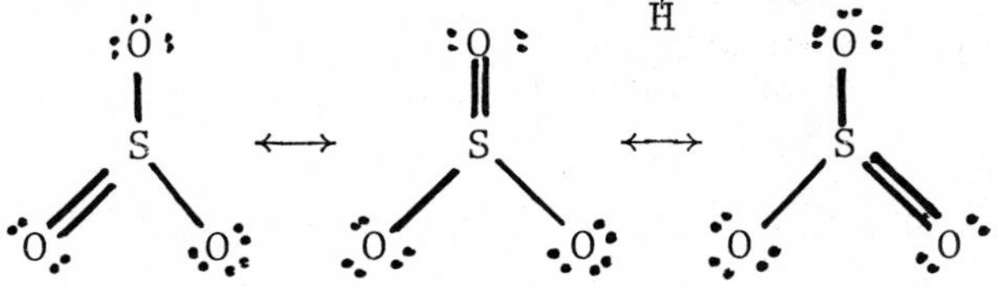

(b) 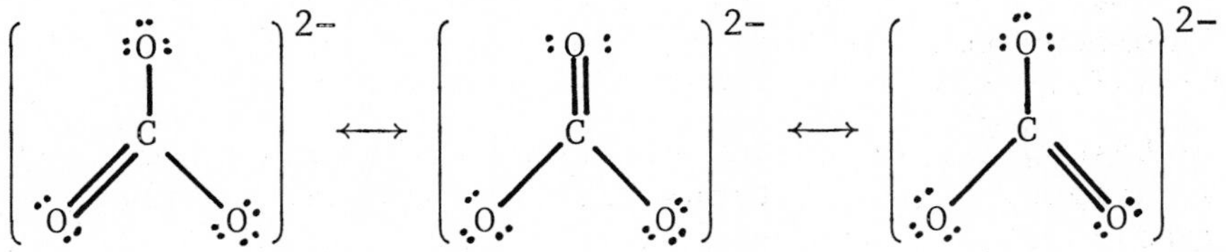

(c) 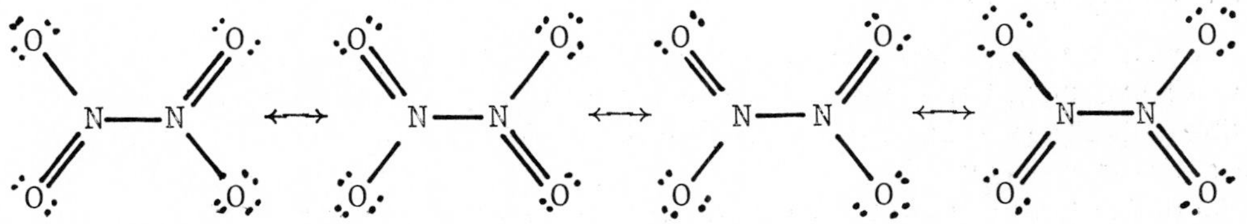

(d) 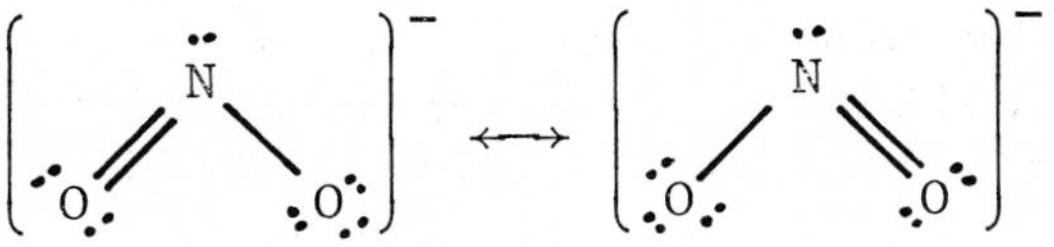

7.22 In CO_2 the carbon-oxygen bond is a full double bond. In CO_3^{2-} it has 2/3 the character of a single bond, 1/3 the character of a double bond (see 7.21(b) above). Thus we can say that the C–O bond order in CO_3^{2-} is 1.33. This means that the C–O bond length in CO_3^{2-} should be closer to a single bond length than a double bond length. We thus expect the C–O bond length to be longer in CO_3^{2-} than in CO_2.

7.23 The experimental facts are that the NO_3^- ion is planar, triangular in shape, with three equal N–O bond lengths. A single Lewis structure is shown on the right. We cannot account for the symmetrical character of the NO_3^- ion in terms of this one structure. Instead, we need to write three such structures, in which the N=O double bond is alternately placed between N and each of the three oxygens. We say that the real NO_3^- structure is a kind of average of the three equivalent resonance forms.

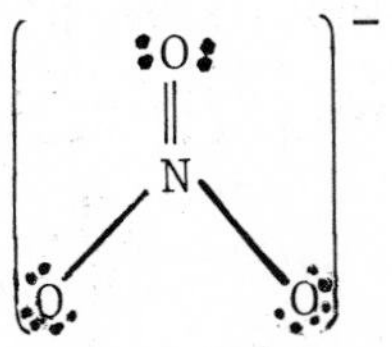

7.24 (a) In SF_6 sulfur has 12 electrons in valence shell orbitals. (c) In BCl_3 boron has only six valence shell electrons. (e) In PCl_5 phosphorus possesses 10 valence shell electrons. (g) In XeF_2 the Xe atom possesses 10 valence shell electrons.

(a) 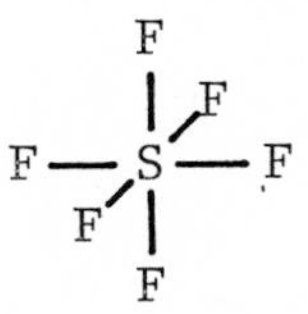 (c) 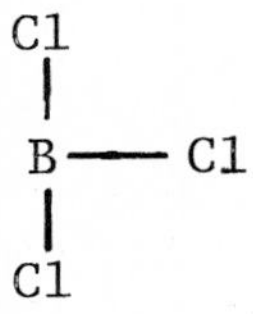 (e) 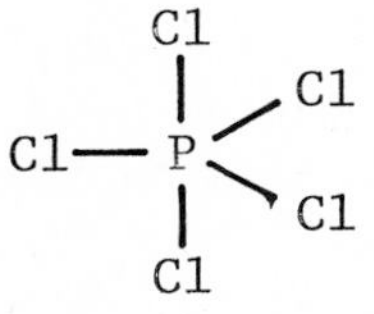 (g) 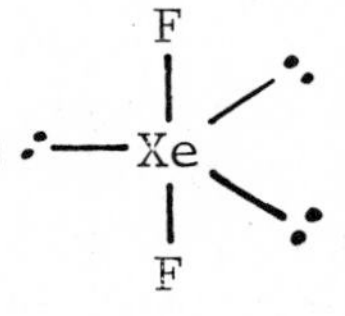

7.25 (a) Si - it has available 3d orbitals; C does not. (b) P - it has available 3d orbitals; N does not. (c) B - B has only three valence electrons, and may thus end up with six electrons in the valence shell by forming three single bonds, as in BF_3.

7.26 To stabilize a compound in which there is in excess of an octet of electrons about the central atom, two things are needed. The surrounding atoms must be small (F is smaller than Cl), and the surrounding atoms should be able to form strong covalent bonds. Usually this means that a more electronegative atom is better. Thus fluorine, more electronegative than chlorine, forms stronger bonds to S (Table 7.3).

7.27 (a) C-F is stronger, because fluorine is closer in size to carbon; (b) C=O is stronger, because bond energies increase as the bond order increases. (That is, triple bonds are stronger than double bonds, which in turn are stronger than single bonds.) (c) The S-S bond is stronger; the second-row nonmetallic elements (except for carbon) do not form as strong single bonds as the heavier members of their respective families; (d) H-O. The electronegativity difference is greater.

7.28 (a) $\Delta H = -D(O\text{-}H) - D(C\text{-}I) + D(C\text{-}O) + D(H\text{-}I) = -463 - 240 + 358 + 299 = -46$ kJ

(b) $\Delta H = -D(C{=}O) - D(H\text{-}Cl) + D(C\text{-}H) + D(C\text{-}Cl) = -799 - 431 + 413 + 328 = -489$ kJ

(c) $\Delta H = -D(H\text{-}H) - 2D(C{=}O) + 2D(O\text{-}H) + D(C{\equiv}O) = -436 - 2(799) + 2(463) + 1072 = -36$ kJ

7.29 The average O-H bond strength in H_2O is just the average of the two successive bond dissociation energies: 1/2(494 + 427) = 460 kJ.

7.30 (a) P < S < O; (b) Mg < Al < Si; (c) S < Br < Cl; (d) Si < C < N

7.31 (a) Cl-I is polar; Cl is the more electronegative atom; (b) P-P is non-polar; (c) C-N is polar; N is the more electronegative atom; (d) F-F is non-polar; (e) O-H is polar; O is the more electronegative atom.

7.32 (a) HH < HC < HF; (b) PS < SiCl < AlCl

7.33 Polarities increase in the order BrCl < ICl < ClF < BrF.

7.34 Let us consider this question in light of the definition of electronegativity as the average of the first ionization energy and the electron affinity. We know that the ionization energy decreases steadily as we go downward in the halogen family (Figure 6.6). We also know that electron affinities do not change much among the halogens (Table 6.4). This means that the ease with which electrons are transferred away from the halogen increases as we move downward in group 7A; thus, electronegativity decreases.

7.35 In a horizontal row of the periodic table, the effective nuclear charge increases from left to right. This variation arises because electrons of the same valence level only partially shield one another from the nuclear charge. The increasing effective nuclear charge results in an increase in the extent to which electrons are attracted toward the atom.

7.36 (a) N,+4; O,-2 (b) Cu,+1; Cl,-1 (c) H,+1; S,+4; O,-2 (d) O,-2; H,+1 (e) Na,+1; H,+1; O,-2; P,+5 (f) S,0

7.37 (a) N may range in oxidation state from -3(Na_3N) to +5(HNO_3). (b) Cl may range from -1(Cl^-) to +7(ClO_4^-). (c) F may range from -1(HF) to 0(F_2). (d) Na may range from 0(Na) to +1(Na^+) (Actually, someone recently prepared a compound containing Na^-, but this is a highly unusual situation).

(e) Se may range from $-2(Na_2Se)$ to $+6(H_2SeO_4)$.

7.38 (a) C is oxidized, O is reduced. (b) S is oxidized, N is reduced. (c) Mg is oxidized, H is reduced. (d) Cl is both oxidized and reduced; one Cl goes from oxidation zero to −1, the other from zero to +1.

7.39 (a) yes; (b) no; (c) yes; (d) no.

7.40 (a) nitrogen trifluoride or nitrogen(III) fluoride; (b) dichromium hexaoxide or chromium(VI) oxide; (c) niobium pentafluoride or niobium(V) fluoride; (d) selenium dioxide or selenium(IV) oxide; (e) diphosphorus tetrachloride or phosphorus(II) chloride; (f) titanium trifluoride or titanium(III) fluoride.

7.41 (a) titanium(II) ion; (b) chromium(II) (or chromous) ion; (c) gallium(III) ion; (d) calcium ion (no need to add (II); (e) iodite ion; (f) sulfide ion; (g) permanganate ion; (h) sulfite ion

7.42 (a) oxygen difluoride; (b) cadmium oxide (or cadmium(II) oxide); (c) lead(IV) oxide (or lead dioxide) (It could also be lead(II) peroxide!); (d) aluminum phosphate; (e) iron(II) chlorite or ferrous chlorite; (f) lithium bromate

7.43 (a) XeF_4; (b) Cr_2O_3; (c) CuBr; (d) N_2O; (e) Ga_2S_3.

7.44 (a) $SrO(s) + H_2O(\ell) \rightarrow Sr(OH)_2(s)$, or we could write $Sr(OH)_2(aq)$ as product. The solubility of $Sr(OH)_2$ in water is about 18 g/L at room temperature. (b) $SO_3(g) + 2KOH(aq) \rightarrow K_2SO_4(aq) + H_2O(\ell)$ or $SO_3(g) + KOH(aq) \rightarrow KHSO_4(aq)$; (c) $P_4O_{10}(s) + 6H_2O(\ell) \rightarrow 4H_3PO_4(aq)$; (d) $2HCl(aq) + CoO(s) \rightarrow CoCl_2(aq) + H_2O(\ell)$; (e) $Ga_2O_3(s) + 6NaOH(aq) \rightarrow 2Na_3GaO_3(aq) + 3H_2O(\ell)$

7.45 The acidity of the oxides increases with the nonmetallic character of the element forming the oxide. Thus, we expect that CO_2 should be the most acidic oxide of the group 4A elements, and that PbO_2 should be the least acidic, or most basic.

7.46 Since silicon is a nonmetallic or metalloid element, the oxide of this element might be expected to react with strongly basic solution. The reaction would be of the form:

$$SiO_2(s) + 2NaOH(aq) \rightarrow Na_2SiO_3(aq) + H_2O(\ell)$$

Thus glass is slowly etched by strongly basic solutions. However, glass generally holds up well in the presence of more dilute base solutions, over relatively short periods of time.

7.47 (a) MgO. This compound has an ionic structure in the solid state. The substantial lattice energy of the solid must be overcome to form an aqueous solution. (b) SiO_2. This oxide forms a network structure that resists disruption. In addition, SiO_2 is neither strongly acidic nor basic in character. By contrast, Na_2O has a relatively much lower lattice energy and dissolves easily to form a strongly basic solution.

7.48 Lattice energy is related to the charges on the ions and inversely to the distances separating oppositely charged ions. In NaF the distances between cation and surrounding anions is shorter than in KCl, hence the interionic attractive forces are greater.

7.49 Solid NaCl is a very stable substance because of the electrostatic attractions between anions and cations throughout the three dimensional solid. We call the energy of the ionic solid the lattice energy. Gaseous Na^+Cl^- ion pairs at room temperature would spontaneously come together to form the more stable solid ionic lattice. Thus, the vapor pressures of ionic solids at room temperature are very low.

7.50 (a) Na_2O is ionic; (b) P + S would form covalent compounds together; (c) C and O would form covalent compounds (CO, CO_2); (d) Ba and I would be expected to form an ionic material, BaI_2; (e) Sc and F should form ionic ScF_3. In all these cases, we look first at the electronegativity difference, then at the usual oxidation states assumed by the elements.

7.51 (a) should double energy of ionic interaction; (b) should double energy of ionic interaction; (c) should quadruple the energy of ionic interaction; (d) should reduce the energy of interaction to one-half

7.52 $E = \frac{k\, Q_1 Q_2}{d}$

$$\text{(a) } E = \left(\frac{8.99 \times 10^{9} \text{ J-m}}{\text{coul}^2}\right)\left(\frac{(-1.60 \times 10^{-19})(1.60 \times 10^{-19}) \text{ coul}^2}{2.13 \times 10^{-10} \text{ m}}\right)$$

$$= -1.08 \times 10^{-18} \text{ J}$$

$$\text{(b) } E = \left(\frac{8.99 \times 10^{9} \text{ J-m}}{\text{coul}^2}\right)\left(\frac{(-1.60 \times 10^{-19})(1.60 \times 10^{-19}) \text{ coul}^2}{4.62 \times 10^{-10} \text{ m}}\right)$$

$$= -4.98 \times 10^{-19} \text{ J}$$

On a molar basis (multiply by Avogadro's number), these energies are -650 kJ and -300 kJ/mol, respectively.

7.53 $$E = \left(\frac{8.99 \times 10^{9} \text{ J-m}}{\text{coul}^2}\right)\left(\frac{3(1.60 \times 10^{-19})(-1.60 \times 10^{-19}) \text{ coul}^2}{1.78 \times 10^{-10} \text{ m}}\right)$$

$$= -3.88 \times 10^{-18} \text{ J/ion-pair}$$

$$\frac{3.88 \times 10^{-18} \text{ J}}{\text{ion-pair}}\left(\frac{6.02 \times 10^{23} \text{ ion-pair}}{\text{mol}}\right) = 2.33 \times 10^{6} \text{ J/mol}$$

7.54 Rearranging Equation 7.2,

$$d = \frac{kQ_1Q_2}{E} = \left(\frac{8.99 \times 10^{9} \text{ J-m}}{\text{coul}^2}\right)\left(\frac{(-1.60 \times 10^{-19})^2 \text{ coul}^2}{1.00 \times 10^{-19} \text{ J}}\right)$$

$$= 2.30 \times 10^{-9} \text{ m} = 23.0 \text{ Å}$$

7.55 It is surely KCl, an ionic solid which should be colorless, crystalline and high melting. NO_2 should be a gas or liquid with low boiling point. Al should have the characteristics we associate with metals; that is, metallic lustre, malleability, and so forth.

7.56 The effective nuclear charge increases in moving from left to right, and ionic radii decrease. These quantities vary in opposite senses, because as the effective nuclear charge increases, the electrons are drawn more tightly toward the nucleus, thus reducing the atomic or ionic radius.

7.57 As^{3-}, Se^{2-}, Br^{-}, Rb^{+}, Sr^{2+}, Y^{3+}, and Zr^{4+} are all isoelectronic with Kr.

7.58 (a) covalent; (b) metallic; (c) covalent; (d) ionic; (e) covalent; (f) ionic

7.59

	Melting Point °C	Boiling Point °C	Solubility in Liquid CS_2	Solubility in H_2O
KBr	734	1398	insoluble	5.5 M
Br_2	-7.3	59	soluble	0.22 M

The most important indications are the high melting and boiling point for KBr, and the low corresponding values for Br_2. The solubilities are not definitive indications, but they support our view of KBr as ionic, Br_2 as covalent. Note that in the polar solvent water, KBr is quite soluble, whereas Br_2 is not very soluble. In carbon disulfide, a non-polar liquid, KBr is insoluble, Br_2 is soluble. However, it should be emphasized that solubilities can only provide supporting evidence, and are not definitive. Many ionic substances have very low solubilities in water (for example, MgO), whereas many covalent compounds are very soluble (for example, ethyl alcohol, or glycerin, or sugar).

7.60 Mn^{2+}: $[Ar]3d^5$; Mn^{4+}: $[Ar]3d^3$; Mn^{7+}: [Ar] Oxides are MnO, MnO_2 and Mn_2O_7.

7.61 (a) 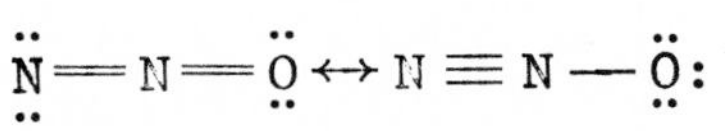 (b) 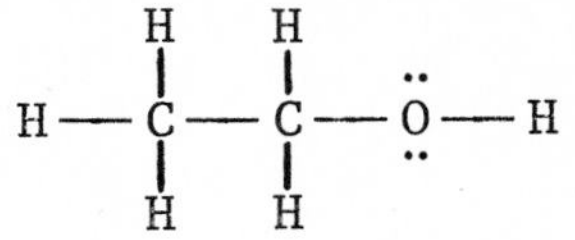

(c) 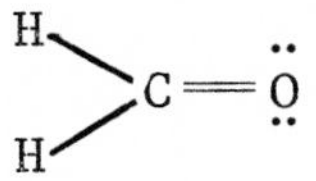 (d) 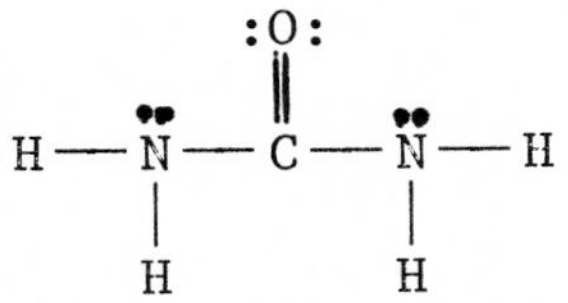

7.62 (a) [Ar]; (b) [Xe]; (c) $[Xe]6s^2 4f^{14} 5d^{10}$; (d) $[Ar]3d^5$; (e) $[Ar]3d^{10}$

7.63 The lattice energy of MgO is about four times larger than that for NaF, because the ions in MgO carry twice the charge, and are of comparable size. As a result, the melting point of MgO is much higher.

7.64 The Lewis structures are as follows:

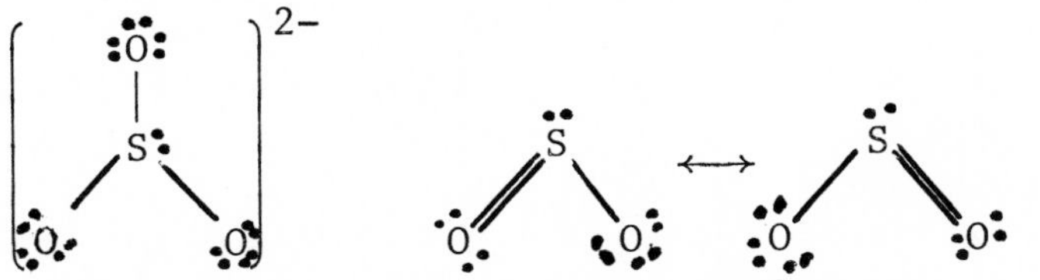

We see that the S-O bond order in SO_3^{2-} is 1, whereas it is 1.5 in SO_2. Thus, the S-O bond distance should be shorter and S-O bond energy higher in SO_2 than in SO_3^{2-}.

7.65 (a) SnF_2; (b) $BaSO_4$; (c) AgBr; (d) KI

7.66 (a) lead(II) bromide; (b) germanium tetrachloride or germanium(IV) chloride; (c) selenium dioxide or selenium(IV) oxide; (d) cobalt(II) chloride; (e) potassium nitrate; (f) barium peroxide

7.67 (a) $NiCl_2$; (b) Cr_2O_3; (c) $FeBr_2$; (d) NO_2; (e) Na_2SO_4; (f) NH_4Cl; (g) $Al(OH)_3$; (h) Na_2O_2

7.68 (a) Mn,+4; O,-2; (b) N,+3; O,-2; (c) N,-2; H,+1

7.69 (a) $CH_4(g) + 2O_2(g) \rightarrow CO_2(g) + 2H_2O(g)$
(b) $H_2S(g) + 2O_2(g) \rightarrow SO_3(g) + H_2O(g)$
(c) $2PH_3(g) + 4O_2(g) \rightarrow P_2O_5(s) + 3H_2O(g)$

7.70 (a) Mn is reduced from oxidation state +7 to +2. (b) There is no change in oxidation number. (c) Cu is reduced from +2 to 0 oxidation state. (d) Oxygen is reduced from the -1 to -2 oxidation states.

7.71 (a) $\Delta H = -4D(C-H) - D(C-C) + 2D(H-H) + D(C\equiv C)$
$= -4(413) - (348) + 2(436) + 839 = -289$ kJ
(b) $\Delta H = -D(Cl-Cl) - 2D(H-F) + D(F-F) + 2D(H-Cl)$
$= -242 - 2(567) + 155 + 2(431) = -359$ kJ
(c) $\Delta H = -4D(C-H) - 2D(O-H) + 3D(H-H) + D(C\equiv O)$
$= -4(413) - 2(463) + 3(436) + 1072 = -198$ kJ

7.72 Sulfur dioxide dissolves in water to form weakly acidic H_2SO_3 solutions. (In addition, SO_2 is oxidized in air to form SO_3, which forms strongly acidic H_2SO_4 in solution.) The acid solution attacks metals, causing oxidation of the metal, e.,g $Fe(s) + H_2SO_3(aq) \rightarrow FeSO_3(s) + H_2(g)$ It reacts with marble ($CaCO_3$) to produce $CO_2(g)$:

$$CaCO_3(s) + H_2SO_3(aq) \rightarrow CaSO_3(s) + H_2O(\ell) + CO_2(g).$$

7.73 The average Ti-Cl bond dissociation energy is just the average of the four values listed, 430 kJ/mol.

7.74 The bond distance vs. bond energy data are graphed in the accompanying figure. Note that most of the data fall on a rather smooth curve which shows that bond energy increases with decreasing bond distance. There is a group of exceptions; these are the bonds between the second row elements oxygen, nitrogen and fluorine. In these cases, the bond energies are much smaller than would be predicted from the data for the other cases shown. Notice that in all three instances, both of the bonded atoms, of rather small radius, have at least one unshared pair on them. The same phenomenon is not seen when one of the two second row atoms is carbon, which has no unshared pairs in its covalent bonding arrangements. We will have more on electron pair repulsions in the next chapter.

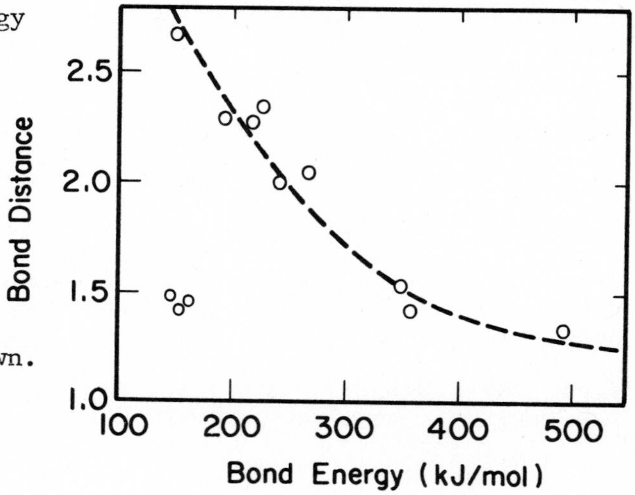

8

Geometries of molecules; molecular orbitals

8.1 (a) 180°; (b) 120°; (c) 109.5°; (d) 90°

8.2 (a) In CF_4 there are four shared electron pairs about carbon; the geometry is tetrahedral. (b) There are four electron pairs about S. One is an unshared pair, the other three are in the S-O bonds. The four electron pairs are arranged in a tetrahedral arrangement. The structure we "see" is that of a trigonal pyramid, as in NH_3. (c) planar about C, 120° bond angles; (d) see-saw, as in SF_4, Table 8.3; (e) linear, 180° bond angles; (f) planar trigonal, 120° bond angles

8.3 (a) tetrahedral about carbon; (b) tetrahedral about P; (c) bent, or nonlinear; (d) trigonal pyramid (Note that H_3O^+ is isoelectronic with NH_3.) (e) linear; (f) tetrahedral

8.4 The geometry about carbon is approximately trigonal planar. Using VSEPR theory, we predict that the bond angles about nitrogen will be approximately a trigonal pyramid. However, we can write resonance structures such as

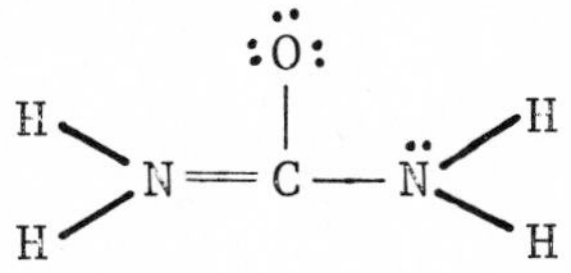

in which N is double-bonded to C. For this structure, we would again predict a trigonal plane about carbon, but now we would also predict that the bonds about the doubly bonded nitrogen would be trigonal planar. Experimentally, a nearly planar bond arrangement is found about both carbon and the two nitrogens.

8.5 There are four electron pairs about the central atom in each of the three compounds. In CH_4 all four are equivalent, and exact tetrahedral geometry is found. In NH_3 one of the four electron pairs is unshared, and it exerts a greater repulsive effect on the other three pairs, thus pushing the bond angles in. In H_2O there are two such unshared pairs so the effect is even more evident.

8.6 (a) BF_3 (3 electrons pairs vs. 4 in NF_3); (b) NH_3 (smaller lone pair repulsions than in H_2O); (c) C_2H_2 (two bonds to adjacent atoms vs. three in C_2H_4); (d) H_2O Since F is more electronegative than H, the electrons in the O-F bond are constricted to a smaller volume than the electron pairs in the O-H bonds of H_2O. Thus they can be pushed back together to a greater extent by the lone-pair electrons.

8.7 Write the Lewis structures:

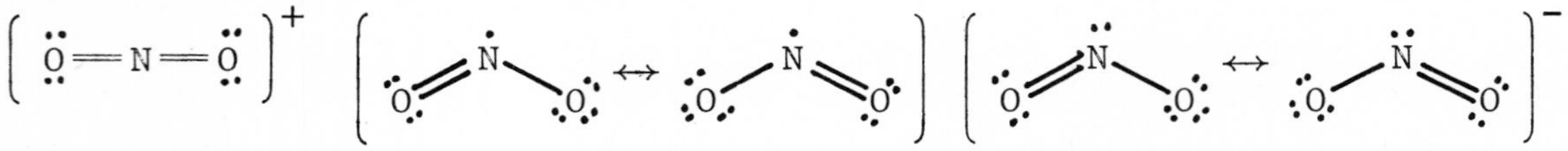

Note that in NO_2^+ (isoelectronic with CO_2) there are two multiple bonds, no unshared pairs, so we predict 180°. In NO_2 there are two bonds and an orbital with one electron. The repulsive effect of this one electron is less than that of an electron pair, so the O-N-O angle is larger than the 120° we would predict for 3 equivalent electron pairs. In NO_2^-, the repulsion from the unshared pair is greater than that from the shared pairs, so the O-N-O angle is less than the idealized 120°.

8.8 Dipole moment measures the separation of centers of positive and negative charge. We expect that this separation will be larger as the electronegativity difference between A and B increases, inasmuch as the A-B bond polarity increases.

8.9 (a) $CHCl_3$ is polar; (b) CF_4 is nonpolar (the four C-F bonds are each polar, but their polarities cancel because of the symmetry around carbon. (c) non-polar; (d) polar; (e) polar (however, CO has a surprisingly small dipole moment of only 0.12 debye); (f) polar

8.10 CS_2 must have zero dipole moment because the two C-S bond dipoles are oppositely directed and thus cancel. However, the C-S and C-O bonds are not equivalent. There is thus only a partial cancellation and a net dipole moment in SCO.

8.11 If the two P-Cl bonds are oppositely directed along the vertical axis, the dipole moment must be zero. If the two P-Cl bonds are in the equatorial positions, they would make an angle of 120°, and thus not completely cancel. The molecule would have an overall dipole moment, because the P-F and P-Cl bonds have different polarities.

8.12 (a) $\mu = Qr = (1.60 \times 10^{-19} \text{ coul})(1.41 \times 10^{-10} \text{ m})$

$$= 2.26 \times 10^{-29} \text{ coul-m} \left(\frac{1 \text{ D}}{3.33 \times 10^{-30} \text{ m}}\right) = 6.78 \text{ D}$$

(b) The observed dipole moment of HBr, 0.79 D, is much smaller than that corresponding to a full separation of one electronic charge. This is so because the H-Br bond is not ionic; to a large extent the bonding electron pair is shared between H and Br. The dipole moment is a reflection of the shift of the electron pair partly toward the more electronegative Br atom.

8.13 (a) 180°; (b) 120°; (c) 109.5°; (d) 90°; (e) 90° angles between axial and equatorial; 120° angles in equatorial plane; (f) 90°; (g) 90°

8.14 (a) sp^3; no π bonds; (b) sp^2; one π bond; (c) sp; two π bonds; (d) sp^2 about carbon; one π bond

8.15 (a) sp^3; (b) approximately sp^3; (c) dsp^3; (d) dsp^3

8.16 (a) The bond angles about the oxygen-bearing carbon atom are 120°C; about the second carbon 109°; (b) about the oxygen-bearing carbon, approximately sp^2; about the second carbon atom, approximately sp^3; (c) seven σ bonds; (d) one π bond

8.17 The overlap of atomic orbitals represents a region of space in which the wave functions for two different orbitals both have appreciable values. Thus, an electron in either atomic orbital has an appreciable probability of being located in the overlap region.

8.18 The resonance structures for a molecule or ion represent two different ways of locating certain electrons. The electrons in question may be placed in two or more different locations, which is another way of saying that those electrons are delocalized over the region represented in the resonance structures. For example, the π electrons in the CO_3^{2-} ion may be represented as being between carbon and any of the three oxygen atoms (see answer to 7.21(b)). We can say that this pair of π electrons is delocalized over these three bonds.

8.19

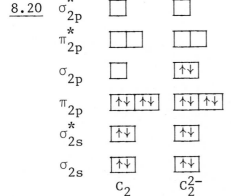

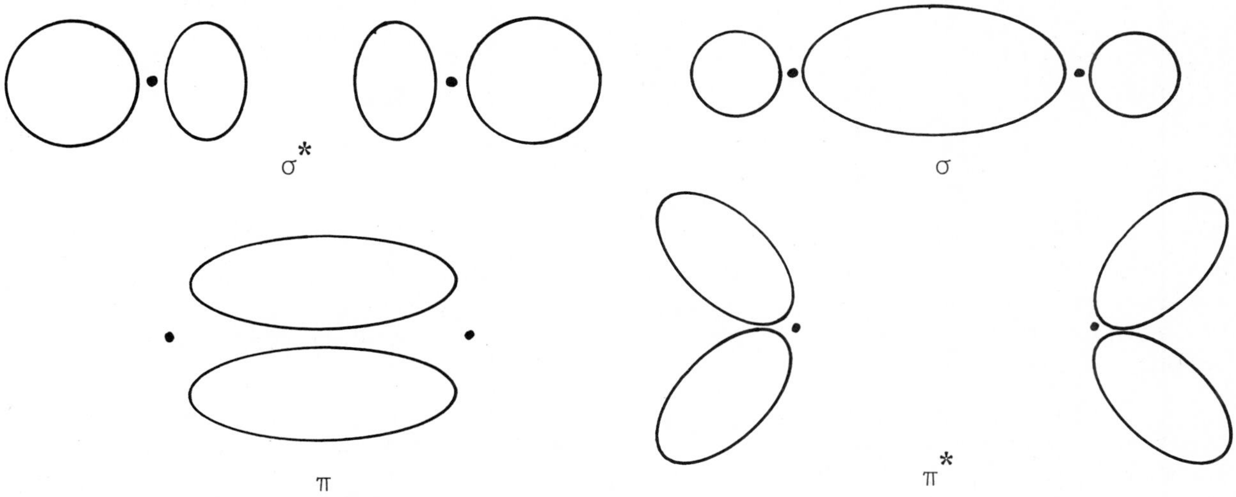

8.20

	C_2	C_2^{2-}
σ^*_{2p}	□	□
π^*_{2p}	□□	□□
σ_{2p}	□	↑↓
π_{2p}	↑↓ ↑↓	↑↓ ↑↓
σ^*_{2s}	↑↓	↑↓
σ_{2s}	↑↓	↑↓

The C_2^{2-} ion contains two electrons more than C_2; these are located in the σ_{2p} orbital, a bonding orbital. The bond order in C_2 is 2; in C_2^{2-} it is 3. We thus expect the C-C bond energy in C_2^{2-} to be greater and the C-C distance to be shorter than in C_2. Both species are expected to be diamagnetic. Note that C_2^{2-} is isoelectronic with N_2.

8.21 In SO we have the same number of valence-shell electrons that are found in O_2. We predict that the molecule, like O_2, is paramagnetic with two unpaired electrons (see Table 8.7). SO is relatively unstable because sulfur does not form such stable π bonds as oxygen. Furthermore, if SO decomposes to elemental sulfur and oxygen, the S-S single-bond energy is recovered. This is a relatively strong bond, and in the S_8 molecule each sulfur forms two such bonds.

8.22 A major factor in the stability of the bonding molecular orbital is that electrons can reside in a region of lower potential energy than in the separated atomic orbitals. The bonding molecular orbital is concentrated between two nuclear centers, thus affording electron-nuclear attractive interactions with two positive charges. This more than overcomes any additional electron-electron repulsions.

8.23 The highest energy electron in O_2 is in an antibonding orbital. Thus its removal leads to a net increase in bond strength. On the other hand, the highest energy electron in N_2 is in a bonding orbital; its removal thus weakens the bond.

8.24 There are no unpaired electrons in N_2, so all electrons in the molecule are paired up in some way. It is reasonable to suppose that the pairing results from bond formation. The N-N distance in N_2 is remarkably short (Table 8.7) and the N-N bond energy is especially large. These observations suggest a very strong bonding interaction.

8.25 Ozone and dioxygen, O_2, are the two allotropic forms of the element. The structure for O_2 is described in Table 8.7. Note that the O-O bond order is two. Ozone is described in terms of two resonance structures, as described in Section 7.6. Inspection of these leads us to conclude that the average O-O bond order in ozone is 1.5. Thus, the O-O bond energy in O_2 should be larger than the average O-O bond energy in O_3, and this is indeed found to be the case.

8.26 The nonmetallic elements of the third row of the periodic table and beyond do not form strong homonuclear π bonds. On the other hand, their σ bonds are relatively strong. Thus phosphorus is more stable as an extended sheet-like structure involving P-P σ bonds between phosphorus and three neighboring phosphorus atoms.

8.27 Silicon does not form strong π bonds with other silicon atoms. An extended π bond structure is involved in graphite. On the other hand, silicon does form stable σ bonds to other Si atoms. Thus, it is relatively more stable in the diamond-type structure, in which there are only σ bonds between silicon atoms.

8.28 The three electron pairs that determine the geometry about the central oxygen in O_3 consist of one unshared electron pair and two bonding pairs in the O-O bonds. The volume requirement of the unshared pair is greater than that of the two shared pairs, so the O-O bonds are pushed together to make the O-O-O bond angle less than 120°.

8.29 The two major resonance forms for O_3 (see Section 7.6) represent delocalization of a pair of π electrons over the two O-O bonds.

8.30 (a) Graphite consists of planar layers of carbon atoms in which each carbon is bonded to three others with 120° C-C-C bond angles. A p orbital on each carbon, perpendicular to the plane of atoms, overlaps with p orbitals on adjacent atoms to form a delocalized orbital extending over the sheet. The electrons in this extended orbital are free to move over the entire layer; when an electrical potential is applied they move under its influence. (b) Diamond consists of a three-dimensional network structure of tetrahedral carbon atoms bonded to other carbon atoms through strong C-C single bonds. This provides a very rigid structure in which bonds are not easily broken. (c) In white phosphorus, P_4, each phosphorus atom is bonded to three other P atoms with 60° P-P-P bond angles. There is a great deal of strain in the molecule because of the required close approach of bonding electron pairs. Thus the P_4 tetrahedron readily comes apart.

8.31 (a) tetrahedral; (b) octahedral; (c) In $SnCl_2$ there are four electrons from Sn, one from each Cl for a total of six. One Lewis structure is thus

$$:\ddot{\underset{..}{Cl}}-\ddot{Sn}-\ddot{\underset{..}{Cl}}:$$

We can also write Lewis structures like $\ddot{\underset{..}{Cl}}=\ddot{Sn}-\ddot{\underset{..}{Cl}}:$. We predict a bent molecule with Cl-Sn-Cl bond angle less than 120° because of the greater repulsion produced by the unshared pair. (d) tetrahedral; (e) a bent T; the Lewis structure is shown below:

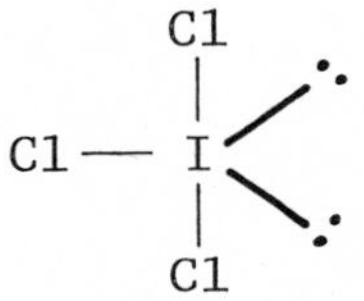

The VSEPR model predicts that the unshared pairs will go into equatorial positions. Since they exert a larger repulsive effect than shared pairs, the Cl-I-Cl angle should be less than 90°.

(f) $\left[\ddot{\underset{..}{O}}=C=\ddot{\underset{..}{N}}\right]^-$ linear (isoelectronic with CO_2).

8.32 In BF_3 there are three shared electron pairs in the B valence orbitals. The molecule is planar with 120° F-B-F bond angles. When the adduct with NH_3 is formed the boron acquires another shared electron pair. In F_3BNH_3 the boron has four shared electron pairs, and the bond angles are approximately tetrahedral.

8.33 In a dipole the centers of positive and negative charge do not coincide. When an electric field is applied the dipole rotates so that the positive and negative ends align toward the charge of opposite sign. However there is no overall net charge on the dipole, so there cannot be a net movement toward one electrode or the other.

8.34 (a) Ordinarily no, but occasionally the structure causes a slightly non-polar charge distribution. Ozone is an example; it is a bent molecule with a structure described by two resonance forms; as outlined in Section 7.6. The dipole moment of ozone is 0.53 debye; not very large, but still non-zero. However, all homonuclear diatomic molecules have zero dipole moment. (b) A molecule may have polar covalent bonds and still have an overall dipole moment of zero. This happens whenever the geometry of the molecule is such that the individual bond dipoles cancel, as in BF_3 (trigonal planar), CF_4 (tetrahedral) or SF_6 (octahedral).

8.35 $$\mu = Qr = (1.60 \times 10^{-19} \text{ coul})(1 \times 10^{-10} \text{ m})$$
$$= 1.60 \times 10^{-29} \text{ coul-m} \left(\frac{1 \text{ D}}{3.33 \times 10^{-30} \text{ coul-m}}\right) = 4.80 \text{ D}$$

8.36 The dichloroethylene molecules are planar; they are merely substituted ethylene. The structure in which the two C-Cl bonds are on the same side is polar. The two C-Cl bonds are oriented at 120° to one another, and thus the C-Cl bond dipoles do not cancel. However in the other form the C-Cl bonds are exactly oppositely directed, and the C-Cl bond dipoles exactly cancel.

8.37 (a) Br_2 - nonpolar; (b) BrCl - polar; (c) CO_2 - nonpolar; (d) H_2S - polar; (e) $HCCl_3$ - polar (The term polar indicates that the molecule has a non-zero dipole moment, nonpolar indicates that it has no dipole moment.)

8.38 (a) two sp hybrid orbitals - linear; (b) four sp^3 hybrid orbitals - tetrahedral; (c) six d^2sp^3 hybrid orbitals - octahedral

8.39 Five carbon atoms are sp^2 hybridized, with 120° bond angles about them. Five have sp^3 hybridization. None have sp hybridization. One nitrogen has sp^2 hybridization, the other has sp^3 hybridization.

8.40 (a) sp^3 - no orbitals available for π formation; (b) sp^2 - one p orbital available for π bond formation; (c) sp - two p orbitals available for π bond formation

8.41 (a) $AsCl_3$ - sp^3; $AsCl_4^-$ - dsp^3; (b) SF_6 - d^2sp^3; SF_5^+ - dsp^3

8.42 (a) A localized bond is one in which the shared electron pair is localized between two atoms. In a delocalized bond the electron pair is distributed between three or more atoms (as in the π bonds in benzene, C_6H_6.) (b) A hybrid orbital is one that is composed of some mixture of atomic orbitals with differing values of ℓ. For example, an sp hybrid is a mixture of s and p orbitals. An unhybridized orbital consists of just one original atomic orbital, for example a 2p orbital; (c) A σ orbital is one in which the overlap of the atomic orbitals is greatest directly along the internuclear axis. In a π orbital there is a nodal plane at the internuclear axis; the major overlaps lie above and below the nodal plane. (d) The σ^* is the anti-bonding component of a σ bond. In an antibonding orbital the electron density is not concentrated between the atoms as it is in a σ orbital. Rather it is concentrated behind the atoms thus destabilizing the bond if there should be electrons in the σ^* orbital.

8.43 The hexagon represents the core of six carbon atoms. It is understood that each corner of the hexagon represents an sp^2 hybridized carbon atom with a C-H bond. The circle represents the three pairs of delocalized π electrons.

8.44 A double or triple bond consists of a pair of σ bond electrons and one or two pairs of π bond electrons. The π bond electrons are non-directional. They are not effective in determining the geometry about the central atom because their energies do not change appreciably with changes in bond angles.

8.45 The potential energy is decreasing as we approach the minimum from large distance because the overlap of atomic orbitals is increasing. This means that the interaction of the bonding electrons with both nuclei is producing a net stabilization. However, when the internuclear distance grows still shorter, the nuclear-nuclear repulsion begins to increase rapidly. In part this happens because the bonding electrons no longer shield the nuclei from one another very well; the nuclei begin to move inside the average distance of the electrons from the nucleus. Repulsive forces between the nuclei thus destabilize the bond. (In many-electron atoms there are also strong electron-electron repulsions at short distances from electrons on adjacent atoms.)

8.46 (a) 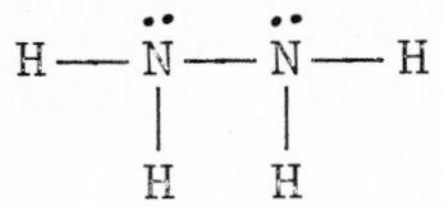

(b) The geometry about each nitrogen is an approximate trigonal pyramid. Thus, the molecule is shaped something like this:

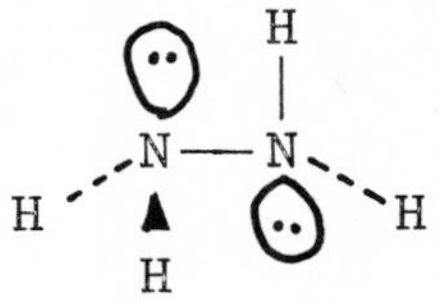

There is rotation about the N-N bond, so each end of the molecule can have varying orientation relative to the other end.

(c) The nitrogen may be considered to employ sp^3 hybrid orbitals, just as in ammonia. In fact, hydrazine can be viewed as two ammonia molecules joined by removing a hydrogen atom from each, and making the N-N bond in their place.

(d) All of the covalent bonds in hydrazine are σ in character.

8.47 (a) Si_2H_4 is unstable because Si does not form a strong π bond to another Si atom. (b) NF_5 is unstable because the nitrogen atom cannot accommodate an expanded valence shell of electrons. Further, there is not room for five fluorine atoms about the relatively small nitrogen. (c) GeF_6 does not form because there are insufficient valence-shell electrons on Ge to form single bonds to six fluorines. The electronic configuration of Ge beyond the Ar core is $[Ar]4s^23d^{10}4p^2$. The 4s and 4p electrons are readily available for bonding, but it would require considerable energy to promote two electrons out of the closed $3d^{10}$ sub-shell to form bonds to fluorine. Thus GeF_4 is stable, but GeF_6 would not be stable. (d) Mg_2O is not a stable compound; it would require that Mg have a valence of +1, which would leave a single electron in the 3s orbital. Such an oxidation state cannot be prepared because the Mg^{1+} ion would very readily react with more oxygen, forming the $Mg^{2+}O^{2-}$ lattice, or it would disproportionate:

$$2Mg^+ \rightarrow Mg^{2+} + Mg^\circ$$

8.48 Assume 100 g compound:

$$20.2 \text{ g Al} \left(\frac{1 \text{ mol Al}}{27.0 \text{ g Al}}\right) = 0.748 \text{ mol Al}$$

$$79.8 \text{ g Cl} \left(\frac{1 \text{ mol Cl}}{35.4 \text{ g Cl}}\right) = 2.25 \text{ mol Cl}$$

The ratio of Cl to Al is 3.0. Thus the empirical formula is $AlCl_3$. However, the molecular weight tells us that the molecular formula is twice this, Al_2Cl_6. Note that a molecule of formula $AlCl_3$ would involve only six electrons in the valence orbitals about aluminum. The Lewis structure is

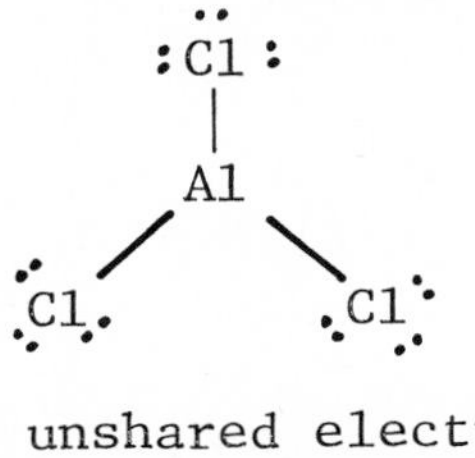

Note also that each chlorine has unshared electron pairs it might donate to a sufficiently strong electron pair acceptor. Thus, we can imagine two AlCl3 units coming together this way:

The structure observed for Al_2Cl_6 is just the one which is predicted by this model.

8.49 A triple bond consists of one σ bond and two π bonds.

8.50 These are illustrated in Figure 8.17 and the lowest portion of 8.23.

8.51 The CN^- ion is isoelectronic with N_2. Therefore, the orbital energy diagram looks just like that for N_2 in Table 8.7, as long as we assume the order of orbital energies of the homonuclear diatomic molecules. The carbon-nitrogen bond order is 3. The Lewis structure also predicts a bond order of 3:

$$[:C \equiv N:]^-$$

8.52 When an electron is added to O_2 it enters one of the π^*_{2p} orbitals. A second added electron enters the other π^*_{2p} orbital. Thus, each added electron reduces the O–O bond order. On the other hand, an electron removed from O_2 is taken from one of the π^*_{2p} orbitals. This has the effect of increasing the O–O bond order. Thus, we have:

molecule or ion:	O_2^+	O_2	O_2^-	O_2^{2-}
bond order:	2.5	2.0	1.5	1.0

increasing bond length →

8.53 The 2s electrons in Li_2 cannot attain the same degree of overlap before repulsive effects come into play as can be attained in forming H_2. The bond strength is roughly related to the overlap.

8.54

	NO^-	NO	NO^+
σ^*_{2p}	□	□	□
π^*_{2p}	↑ | ↑	↑ | □	□ | □
σ_{2p}	↑↓	↑↓	↑↓
π_{2p}	↑↓ | ↑↓	↑↓ | ↑↓	↑↓ | ↑↓
σ^*_{2s}	↑↓	↑↓	↑↓
σ_{2s}	↑↓	↑↓	↑↓

The predicted bond orders are NO^-, 2.0; NO, 2.5; NO^+, 3.0. The NO^+ ion should have the shortest N–O bond distance.

8.55 In a heteronuclear diatomic molecule the two atoms do not have atomic orbitals of the same energy. Thus the molecular orbitals formed from the two sets of atomic orbitals do not contain equal contributions from the orbitals of the two atoms. In the bonding molecular orbitals the contribution of the orbital of the more electronegative atom is greater.

Conversely, in the antibonding molecular orbital the contribution of the orbital of the less electronegative atom is greater.

8.56

$$H - \underset{\underset{\textstyle H}{|}}{\overset{\overset{\textstyle H}{|}}{C}} - \ddot{\underset{\cdot\cdot}{O}} - H \qquad \overset{H}{\underset{H}{>}} C = \ddot{\underset{\cdot\cdot}{O}} \qquad :C \equiv O:$$

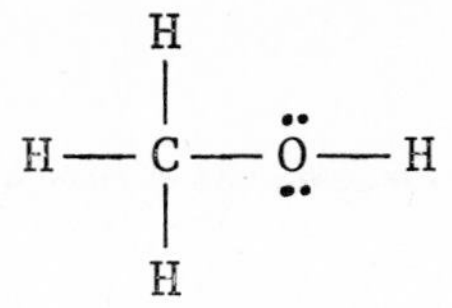

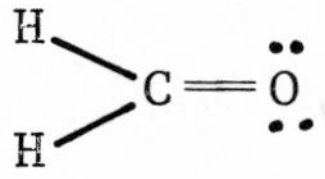

The bond order increases in this series, while at the same time the bond distance decreases. This is consistent with expectations; the atoms are drawn closer together by the overlap of orbitals forming an increasingly strong bond.

9

The properties of gases

9.1 Ethyl alcohol (boils at 78°C, freezes at -117°C). Carbon tetrachloride (boils at 77°C, freezes at -23°C); benzene (freezes at 6°C); sulfur (melts at 113°C); sodium metal (melts at 97°C); carbon dioxide (sublimes at room temperature); iodine (sublimes at room temperature).

9.2 The distances between gas molecules are very much greater than distances between molecules in liquids or solids. Thus it is relatively easy to compress the gas, reducing the distances between molecules still further.

9.3 Temperature (Kelvin); Pressure (atm; mm Hg; kilopascals); Volume (liter); quantity (number of moles)

9.4 In a column of Hg of 1 cm^2 cross-section, the mass of Hg is (1 cm^2) x (76.0 cm)(13.59 g Hg/1 cm^3) = 1.033 x 10^3 g Hg. The force which this exerts is given by:

$$\text{Force} = ma = (1.033\ \text{kg})(9.81\ \text{m/sec}^2) = 10.13\ \text{kg-m/sec}^2 = 10.13\ \text{N}$$

The pressure exerted is given by the force per unit area:

$$\text{Pressure} = \left(\frac{(10.13\ \text{N})}{1.00\ \text{cm}^2}\right)\left(\frac{10^2\ \text{cm}}{1\ \text{m}}\right)^2 = 1.013 \times 10^5\ \text{N/m}^2$$

$$\left(\frac{1.013 \times 10^5\ \text{N}}{\text{m}^2}\right)\left(\frac{1\ \text{Pa}}{1\ \text{N/m}^2}\right) = 1.013 \times 10^5\ \text{Pa} = 101.3\ \text{kPa}$$

9.5 The force which the column of liquid metal must exert at its base is the same as that exerted by a column of mercury 760 mm in height. The force exerted by the column is F = mg, where g is the acceleration due to earth's gravitational field. But the mass acting on a unit cross-sectional area is just the density, g/cm^3, times the height of the column, h. Thus, for two different substances we have $d_1h_1g = d_2h_2g$. Thus, $h_2/h_1 = d_1/d_2$.

Inserting the appropriate values, we find:

$$h_2 = 760\ \text{mm}\ (13.6\ \text{g/cm}^3)/(5.73\ \text{g/cm}^3) = 1.80 \times 10^3\ \text{mm} = 1.80\ \text{m}$$

9.6 The earth's atmosphere exerts a pressure that can push water up an evacuated pipe until the downward force of the water column equals atmospheric pressure. The density of water is 1.0 g/cm^3. Using the formula we derived in exercise 9.5 above, $h_2 = 76.0\ \text{cm}\ (13.6\ \text{g/cm}^3)/(1.0\ \text{g/cm}^3)$

$$= 1.03 \times 10^3\ \text{cm} \left(\frac{1\ \text{in.}}{2.54\ \text{cm}}\right)\left(\frac{1\ \text{ft}}{12\ \text{in.}}\right)$$

$$= 33.9\ \text{ft}$$

We must remember, however, that water has an appreciable vapor pressure, and the ordinary pitcher pump does not produce a really good vacuum in the pipe.

9.7 An equation of state is a relationship between all the variables that are needed to specify the state, or condition, of some system of interest. For an ideal gas these variables are pressure, temperature, volume and quantity of the gas.

9.8 Density = mass/volume. Doubling pressure at constant temperature means halving the volume, by Boyle's law. Thus, d = 3.67 g/1.26 L = 2.91 g/L

9.9 (a) False - pressure is directly proportional to the number of moles of gas. (b) correct; (c) correct; (d) false. At constant volume, the pressure is directly proportional to absolute temperature.

9.10 The reduction of pressure from 2250 lb/in^2 to 1 atm means an increase in volume. Thus, 25 L (2250/14.7) = 3826 L. Total volume at 1 atm required per year is 365 x 12.0 L = 4380. Thus, 4380/3826 = 1.14 tanks are needed.

9.11 We can use $V_1P_1 = V_2P_2$ since n and T are constant.
$V_1 = 0.25(1/8) = 0.031$ L

9.12 We can write $d_1 = m/V_1$, $d_2 = m/V_2$. Solve each equation for m, equate, to get $d_1V_1 = d_2V_2$; $d_2 = 1.45\ (6.50/3.20) = 2.94$ g/L.

9.13 Since n and T are constant, $P_1V_1 = P_2V_2$. $P_2 = P_1(V_1/V_2) = 0.372$ atm $(64.0\ \text{cm}^3/12.0\ \text{cm}^3) = 1.98$ atm. Note that reduction in volume causes increase in pressure.

9.14 The large increase in volume means a decrease in pressure.
$P = 127\ \text{atm}(0.172\ \text{cm}^3/125\ \text{cm}^3) = 0.175$ atm

9.15 Remember to convert to absolute temperature. The extrapolated value for temperature at zero volume is zero, within experimental error. This temperature is referred to as absolute zero. It represents the temperature at which the molecules of a gas would have zero kinetic energy. If you graphed temperature in °C vs. volume, your intercept should have occurred at near -273°C.

9.16 An increase in temperature at constant volume means an increase in pressure. 137 kPa (421K/300K) = 192 kPa. (Note the conversion to absolute temperature scale!)

9.17 Solve the ideal gas equation for volume, since that is the constant quantity in this problem. V/R = nT/P. Then,

$$\frac{n_2T_2}{P_2} = \frac{n_1T_1}{P_1} \qquad n_2 = \left(\frac{T_1}{T_2}\right)\left(\frac{P_2}{P_1}\right) n_1 = \left(\frac{298\ \text{K}}{373\ \text{K}}\right)\left(\frac{17.2\ \text{atm}}{1.50\ \text{atm}}\right)(0.182\ \text{mol})$$

$$= 1.67\ \text{mol}$$

9.18 (a) Use $P_1/T_1 = P_2/T_2$ $P_1 = 12.3 \text{ atm } (30.\text{ K}/547\text{ K}) = 6.77 \text{ atm}$

(b) $V_2 = 6.75 \text{ L} \left(\frac{89.0 \text{ atm}}{6.85 \text{ mm}}\right)\left(\frac{760 \text{ mm}}{1 \text{ atm}}\right) = 6.67 \times 10^4 \text{ L}$

(c) $P_2 = 1.65 \text{ atm } (0.0880 \text{ L}/6.85 \text{ L}) = 0.0212 \text{ atm}$

9.19 It is simplest to calculate the partial pressure of each gas as it expands into the total volume, then sum the partial pressures.

$$P_{N_2} = 635 \text{ mm Hg } (1.0 \text{ L}/2.5 \text{ L}) = 254 \text{ mm Hg}$$

$$P_{Ne} = 212 \text{ mm Hg } (1.0 \text{ L}/2.5 \text{ L}) = 85 \text{ mm Hg}$$

$$P_{H_2} = 418 \text{ mm Hg } (0.5 \text{ L}/2.5 \text{ L}) = \underline{84 \text{ mm Hg}}$$

$$\text{Total Pressure} = 423 \text{ mm Hg}$$

$$= 4.2 \times 10^2 \text{ mm Hg}$$

(We have just two significant figures.)

9.20 (a) $2.5 \text{ L } (635 \text{ mm Hg}/760 \text{ mm Hg})(273 \text{ K}/302 \text{ K}) = 1.9 \text{ L}$

(b) $d = \frac{(1 \text{ atm})(39.95 \text{ g/mol})}{(0.08206 \text{ L-atm/mol-K})(273.1 \text{ K})} = 1.783 \text{ g/L}$

(c) $n = \frac{(1.15 \text{ atm})(3.0 \times 10^7 \text{ m}^3)}{(0.0821 \text{ L-atm/mol-K})(285 \text{ K})}\left(\frac{10^3 \text{ L}}{1 \text{ m}^3}\right) = 1.5 \times 10^9 \text{ mol}$

9.21 $$P_{H_2} = \frac{(0.0821 \text{ L-atm/mol-K})(361 \text{ K})(1.5 \text{ g } H_2)}{3.20 \text{ m}^3}\left(\frac{1 \text{ mol } H_2}{2.0 \text{ g } H_2}\right)\left(\frac{1 \text{ m}^3}{10^3 \text{ L}}\right)$$

$$= 6.9 \times 10^{-3} \text{ atm}$$

$$P_t = \frac{(0.0821 \text{ L-atm/mol-K})(361 \text{ K})}{3.20 \text{ m}^3}\left(\frac{1 \text{ m}^3}{10^3 \text{ L}}\right)\left\{1.5 \text{ g } H_2\left(\frac{1 \text{ mol } H_2}{2.0 \text{ g } H_2}\right) + 85.2 \text{ g } O_2\left(\frac{1 \text{ mol } O_2}{32.0 \text{ } O_2}\right) + 17.0 \text{ g Ar}\left(\frac{1 \text{ mol Ar}}{40.0 \text{ g Ar}}\right)\right\} = 3.55 \times 10^{-2} \text{ atm}$$

9.22 $$MW = dRT/P = \left(\frac{1.104 \text{ g}}{\text{L}}\right)\left(\frac{0.0821 \text{ L-atm}}{\text{mol-K}}\right)(303.1 \text{ K})\left(\frac{1}{(720/760) \text{ atm}}\right)$$

$$= 29.0 \text{ g/mol}$$

(Three significant figures is all that is justified, since pressure is known to three significant figures.)

9.23 $P_2/P_1 = (0.46 \text{ L}/0.082 \text{ L})(693 \text{ K}/308 \text{ K}) = 12.6$

9.24 The mass of air displaced by the balloon is its volume times the density of air.

$$m = \frac{265 \text{ L}(28.9 \text{ g/mol})(1 \text{ atm})}{(0.0821 \text{ L-atm/mol-K})(298 \text{ K})} = 313 \text{ g}$$

The mass of helium in this volume is

$$m_{He} = \frac{265 \text{ L}(4.00 \text{ g/mol})(1 \text{ atm})}{(0.0821 \text{ L-atm/mol-K})(298 \text{ K})} = 43 \text{ g}$$

The mass of helium and balloon is thus 89 g. The lifting capacity is 313 - 89 = 224 g.

If the balloon can expand as it rises to lower pressure regions, it should in principle be capable of rising to a higher elevation. This is so because a larger balloon will displace a larger volume, and therefore a larger mass, of air.

9.25 Calculate the mass of water represented by the difference in water vapor pressures, 5 mm Hg.

$$\text{mass } H_2O = (MW)\ n = \frac{18.0\ g}{mol}\left(\frac{5}{760}\ atm\right)(1.3 \times 10^4\ ft^3)\left(\frac{12\ in.}{1\ ft}\right)^3$$

$$\times\left(\frac{2.54\ cm}{1\ in.}\right)^3\left(\frac{1\ L}{10^3\ cm^3}\right)\Big/\left(\frac{0.0821\ L\text{-}atm}{mol\text{-}K}\right)(295\ K) = 1.8 \times 10^3\ g\ H_2O$$

9.26 Calculate the number of moles of O_2:

$$n = \frac{(6.5 \times 10^{-6}\ mm\ Hg/760\ mm\ Hg)(0.283\ L)}{(0.0821\ L\text{-}atm/mol\text{-}K)(300\ K)} = 9.8 \times 10^{-11}\ mol$$

$$(9.8 \times 10^{-11}\ mol\ O_2)\left(\frac{2\ mol\ Mg}{1\ mol\ O_2}\right)\left(\frac{24.0\ g\ Mg}{1\ mol\ Mg}\right) = 4.8 \times 10^{-9}\ g\ Mg$$

9.27 We must first find which is the limiting reagent, by calculating the number of moles of NH_3 that would be formed by complete reaction of either the Mg_3N_2 or H_2O.

$$6.52\ g\ Mg_3N_2\left(\frac{1\ mol\ Mg_3N_2}{101\ g\ Mg_3N_2}\right)\left(\frac{2\ mol\ NH_3}{1\ mol\ Mg_3N_2}\right) = 0.129\ mol\ NH_3$$

$$0.185\ g\ H_2O\left(\frac{1\ mol\ H_2O}{18.0\ g\ H_2O}\right)\left(\frac{2\ mol\ NH_3}{3\ mol\ H_2O}\right) = 6.85 \times 10^{-3}\ mol\ NH_3$$

We see that H_2O is the limiting reagent; now we calculate the volume of gas formed:

$$V_{NH_3} = \frac{(6.85 \times 10^{-3}\ mol)(0.0821\ L\text{-}atm/mol\text{-}K)(303\ K)}{(715/760)\ atm} = 0.181\ L$$

9.28 $$3.25\ g\ KClO_3\left(\frac{1\ mol\ KClO_3}{123\ g\ KClO_3}\right)\left(\frac{3\ mol\ O_2}{2\ mol\ KClO_3}\right) = 0.0396\ mol\ O_2$$

We want to collect this gas over water at 22°C, so the partial pressure of O_2 is 740 mm - 20 mm = 720 mm Hg.

$$V_{O_2} = \frac{(0.0396\ mol)(0.0821\ L\text{-}atm/mol\text{-}K)(395\ K)}{(720/760)\ atm} = 1.01\ L$$

9.29 $$6.50\ g\ CuS\left(\frac{1\ mol\ CuS}{95.6\ g\ CuS}\right)\left(\frac{2\ mol\ O_2}{1\ mol\ CuS}\right) = 0.136\ mol\ O_2$$

$$V_{O_2} = \frac{(0.136\ mol)(0.0821\ L\text{-}atm/mol\text{-}K)(309\ K)}{(710/760)\ atm} = 3.69\ L\ O_2$$

9.30 (a) increasing temperature at constant volume; increase in pressure; (b) decrease in temperature; (c) increase in volume or decrease in n; (d) increase in temperature

9.31 Gases are highly compressible. More quantitatively, the pressure of a gas increases inversely with total container volume. The fact that we don't need to make a correction at ordinary pressures for the volumes of the gas molecules themselves suggest that these volumes are negligible relative to the container volume.

9.32 The ratio of average kinetic energies is exactly one as long as the temperatures are equal. Since $\varepsilon_{N_2} = \varepsilon_{H_2}$, $m_{N_2}u^2_{N_2} = m_{H_2}u^2_{N_2}$.

$$\frac{u_{N_2}}{u_{H_2}} = \left(\frac{2.02}{28.0}\right)^{1/2} = 0.269\ ; \quad \frac{u_{H_2}}{u_{N_2}} = 3.72$$

9.33 (a) The rms speed increases as $T^{1/2}$ (Equation 9.18). Thus it increases by 1.414. (b) doubles; (c) decreases - We know that the pressure is the same as before, but each collision results in more of a change in momentum since u is greater. Thus, there must be fewer collisions per unit time in each unit area, as a result of the increased volume. Since speed is greater by 1.414, the number of collisions per unit time is lower by 1/1.414.

9.34 $u_{HCl} = (1693 \text{ m/sec})(2.02/36.5)^{1/2} = 398 \text{ m/sec}$

9.35 (a) same; (b) more collision per unit time for N_2; (c) The N_2 distribution curve is spread out more toward higher speeds; (d) N_2 effuses $(146/28.0)^{1/2} = 2.28$ times faster.

9.36 The relative effusion rates are $r_{H_2}/r_{F_2} = (38/2.02)^{1/2} = 4.34$. Thus, one would need 4.34 times as many holes for F_2 gas.

9.37 (a) He; (b) N_2; (c) O_2 In each case we choose the molecule or atom with highest rms speed.

9.38 C_2H_6, the most complex molecule, and with the highest molecular mass, would be most likely to exhibit the largest intermolecular interactions, which are responsible for negative deviations. This substance should also require the largest correction for molecular volume.

9.39 In each case, the molecule with largest molecular mass is selected. (a) UF_6; (b) $SiCl_4$; (c) SO_2

9.40 At STP, the molar volume is $V = \dfrac{1 \text{ mol}(0.0821 \text{ L-atm/mol-K})(273 \text{ K})}{1 \text{ atm}} = 22.4 \text{ L}$

Dividing the value for b, 0.0322 L/mol, by 4, we obtain 0.0080 L. Thus, the volume of the Ar atoms is (0.0080/22.4)100 = 0.036% of the total volume. At 100 atm pressure the molar volume is 0.224 L, and the volume of the Ar atoms is 3.6% of the total volume.

9.41 The intermolecular forces that cause negative departures from the ideal gas equation are countered by the kinetic energies of gas molecules. As temperature increases these kinetic energies are relatively more important, since the intermolecular forces do not vary much with temperature.

As to the second point, we know that for a given pressure the volume of the gas container decreases as temperature decreases. Thus, for example, at 600 atm, the volume required for a mole of CH_4 at 200 K is lower than that required at say 500 K. This means that the volumes of the CH_4 molecules themselves are a larger fraction of the total volume at 200 K than at 500 K.

9.42 (a) force = 9.60 kg $(9.81 \text{ m/sec}^2) = 94.2 \text{ N}$

$$\text{pressure} = \frac{94.2 \text{ N}}{3.2 \times 10^3 \text{ cm}^2}\left(\frac{100 \text{ cm}}{1 \text{ m}}\right)^2 = 294 \text{ Pa} = 0.294 \text{ kPa}$$

(b) force = 2.5 g $(9.81 \text{ m/sec}^2)(1 \text{ kg}/1000 \text{ g}) = 2.5 \times 10^{-2} \text{ N}$

$$\text{pressure} = \frac{2.5 \times 10^{-2} \text{ N}}{2.7 \times 10^{-4} \text{ mm}^2}\left(\frac{10^3 \text{ mm}}{\text{m}}\right)^2 = 9.3 \times 10^7 \text{ Pa} = 9.3 \times 10^4 \text{ kPa}$$

9.43 (a) 760 mm Hg; (b) 1 mm Hg; (c) 59 mm Hg; (d) 646.9 mm Hg; (e) 120.9 kPa (convert h to kPa using 101.3 kPa = 76.0 cm Hg)

9.44 We use (0.752/13.6) = 0.0553 to convert oil column height to units of mercury column height. (a) 16.5 cm (0.0553) = 0.91 cm Hg = 9.1 mm Hg. Pressure in B is 760 - 9.1 = 751 mm Hg. (b) 0.65 m (0.0553) = 0.036 m Hg = 36 mm Hg. Vapor pressure = 135 - 36 = 99 mm Hg.

9.45 (a) 2.34 atm; (b) 8.22×10^{-10} atm; (c) 0.359 atm; (d) 112 kPa

9.46 (8.73 L)(735 mm Hg/1675 mm Hg) = 3.83 L

9.47 (a) $V = \dfrac{(3.45 \text{ mol})(0.0821 \text{ L-atm/mol-K})(320 \text{ K})}{835 \text{ kPa}}\left(\dfrac{101.3 \text{ kPa}}{1 \text{ atm}}\right) = 11.0 \text{ L}$

(b) $MW = dRT/P = \dfrac{1.32 \text{ g } (0.0821 \text{ L-atm/mol-K})(300 \text{ K})}{1.15 \text{ L } (92.4 \text{ kPa})}\left(\dfrac{101.3 \text{ kPa}}{1 \text{ atm}}\right)$

$= 31.0 \text{ g/mol}$

(c) $P = nRT/V = \dfrac{(6.4 \text{ mol})(0.0821 \text{ L-atm/mol-K})(310 \text{ K})}{25.0 \text{ L}} = 6.5 \text{ atm}$

9.48 $P = (600 \text{ mm Hg})\left(\dfrac{4.7 \times 10^{-2} \text{ cm}^3}{0.68 \text{ L}}\right)\left(\dfrac{1 \text{ L}}{10^3 \text{ cm}^3}\right) = 4.1 \times 10^{-2} \text{ mm Hg}$

9.49 $P = 125 \text{ atm}\left(\dfrac{1.86 \text{ L}}{2.8 \times 10^4 \text{ m}^3}\right)\left(\dfrac{298 \text{ K}}{1143 \text{ K}}\right)\left(\dfrac{1 \text{ m}^3}{10^3 \text{ L}}\right) = 2.2 \times 10^{-6} \text{ atm}$

9.50 temp = (63/42)(298 K) = 447 K ∿ 450 K (2 significant figures)

9.51 $d = \dfrac{(39.9 \text{ g/mol})(1.65 \text{ atm})}{(0.0821 \text{ L-atm/mol-K})(286 \text{ K})} = 2.80 \text{ g/L}$

9.52 Calculate density of the vapor: 0.912 g/0.342 L = 2.67 g/L

$MW = \dfrac{(2.67 \text{ g/L})(0.0821 \text{ L-atm/mol-K})(372 \text{ K})}{(742/760) \text{ atm}} = 83.5 \text{ g/mol}$

9.53 $1.25 \text{ g Mg}\left(\dfrac{1 \text{ mol Mg}}{24.3 \text{ g Mg}}\right)\left(\dfrac{1 \text{ mol } H_2}{1 \text{ mol Mg}}\right) = 0.0514 \text{ mol } H_2$

Since the vapor pressure of water at 28°C is 28 mm Hg, we will collect the H_2 gas at a partial pressure of 748 - 28 = 720 mm Hg.

$V = \dfrac{(0.0514 \text{ mol})(0.0821 \text{ L-atm/mol K})(301 \text{ K})}{(720/760) \text{ atm}} = 1.34 \text{ L}$

9.54 Calculate the number of moles of Ar in the vessel:
n = (339.712 - 337.428)/39.948 = 0.05717 mol
The total number of moles of the mixed gas is the same. Thus, the average atomic weight is (339.146 - 337.428)/0.05717 = 30.05. Let the mole fraction of Ar be X. Then, X(39.948) + (1 - X)(20.183) = 30.05. X = 0.499. Neon is thus 50.1 mole percent of the mixture.

9.55 (a) 62.2 g/mol; (b) 730 mm Hg; 1.15×10^3 cm^3; (c) 0.596 g/L (use the average molecular weight, 21.97 g/mol.)

9.56 $1.86 \text{ g } Pb(NO_3)_2\left(\dfrac{1 \text{ mol } Pb(NO_3)_2}{331 \text{ g } Pb(NO_3)_2}\right)\left(\dfrac{5 \text{ mol gas}}{2 \text{ mol } Pb(NO_3)_2}\right) = 0.0140 \text{ mol gas}$

$P = \dfrac{0.0140 \text{ mol}(0.0821 \text{ L-atm/mol-K})(303 \text{ K})}{1.62 \text{ L}} = 0.215 \text{ atm}$

9.57 If the molecules possessed appreciable attractive forces for one another at the distances normally encountered in gases, then energy would be needed to overcome these attractive forces when the distances grow larger, as a gas expands. This energy would come, initially at least, from the kinetic energies of the molecules themselves. They would thus suffer a decrease in temperature upon expanding. No large effect is noticed; the observed temperature changes are quite small.

9.58 (a) $r_{He}/r_{Ar} = (40.0/4.00)^{1/2} = 3.16$;

(b) $r_{CO}/r_{CO_2} = \left[MW_{CO_2}/MW_{CO}\right]^{1/2} = (44/28)^{1/2} = 1.25$

9.59 The time required is proportional to the reciprocal of the effusion rate. Thus, $r_X/r_{O_2} = 28/88 = 0.32 = [32/(MW)_X]^{1/2}$. $MW_X = 320$ g/mol (2 significant figures)

9.69 $r_1/r_2 = (29.01/28.01)^{1/2} = 1.018$ for CO

$r_1/r_2 = (45.01/44.01)^{1/2} = 1.011$ for CO_2

One could obtain a greater degree of separation using CO rather than CO_2.

9.61 The PV/RT relationship becomes more negative for ethylene because the intermolecular attractive forces are larger for this substance. At very high pressures the PV/RT relationship for ethylene becomes more positive because the molecules of this substance are larger than for N_2.

9.62 $V = nRT/P = \dfrac{1\text{ mol}(0.0821\text{ L-atm/mol-K})(273\text{ K})}{200\text{ atm}} = 0.112$ L

Using the van der Waals equation:

$$V - nb = \frac{nRT}{P + an^2/V^2} = \frac{1\text{ mol}(0.0821\text{ L-atm/mol-K})(273\text{ K})}{[200 + 1.34/(0.112)^2]\text{ atm}} \quad (A)$$

$V - 0.0322$ L $= 0.0730$ L

$V = 0.1052$ L. Replace V^2 in the denominator of Equation (A) with this new estimate, recalculate V.

$V - 0.0322$ L $= 0.0698$ L; $V = 0.1020$ L

Repeat the procedure with this new value, and so forth, until the change in V from one cycle to the next is sufficiently small. You should obtain a value for V of about 0.0985 L. Of this volume, 0.25 x 0.0322 = 0.0080 L are the volumes of the Ar atoms themselves. This is 8.1 percent of the total volume.

10

Chemistry of the atmosphere

10.1 Troposphere (0 - 11 km) temperature varies from 290 K to 220 K. Stratosphere (11 - 50 km) temperature varies from 220 K to 270 K. Mesosphere (50 - 85 km) temperature varies from 270 K to 180 K. Thermosphere (85 km - upwards) temperature increases from 180 K beyond 1200 K.

10.2 Tropopause (220 K); stratopause (270 K); mesopause (180 K).

10.3 The earth's atmosphere is attracted by gravitational forces toward earth. Each layer of gas at a particular elevation is compressed by the force acting downward, exerted by the atmosphere above that layer. The force is greatest (and the compression is greatest) at the surface, and decreases as one goes up in elevation. Thus, the atmosphere is most compressed (is densest) at the surface.

10.4 He partial pressure is 3.98×10^{-3} mm Hg. CH_4 partial pressure is 2×10^{-3} mm Hg.

10.5 $$\left(\frac{210 \times 10^3 \text{ J}}{\text{mol}}\right)\left(\frac{1 \text{ mol}}{6.022 \times 10^{23} \text{ moleclues}}\right) = 3.49 \times 10^{-19} \text{ J/molecule}$$

$\lambda = c/\nu$. We also have that $E = h\nu$, so $\nu = E/h$. Thus,

$$\lambda = \frac{hc}{E} = \frac{(6.625 \times 10^{-34} \text{ J-sec})(3.00 \times 10^{8} \text{ m/sec})}{3.49 \times 10^{-19} \text{ J}} = 5.69 \times 10^{-7} \text{m}$$

$= 569$ nm

10.6 $$\left(\frac{339 \times 10^3 \text{ J}}{\text{m}}\right)\left(\frac{1 \text{ mol}}{6.022 \times 10^{23} \text{ molecules}}\right) = 5.63 \times 10^{-19} \text{ J/molecule}$$

$$\lambda = \frac{hc}{E} = \frac{(6.625 \times 10^{-34}\ \text{J-sec})(3.00 \times 10^{8}\ \text{m/sec})}{5.63 \times 10^{-19}\ \text{J}} = 353\ \text{nm}$$

$$\left(\frac{293 \times 10^{3}\ \text{J}}{\text{mol}}\right)\left(\frac{1\ \text{mol}}{6.022 \times 10^{23}\ \text{molecules}}\right) = 4.87 \times 10^{-19}\ \text{J/molecule}$$

$$\lambda = \frac{(6.625 \times 10^{-34}\ \text{J-sec})(3.00 \times 10^{8}\ \text{m/sec})}{4.87 \times 10^{-19}\ \text{J}} = 408\ \text{nm}$$

Photons of wavelengths longer than 408 nm cannot cause rupture of the C-Cl bond in either CF_3Cl or CCl_4. Photons with wavelengths between 408 and 353 nm can cause C-Cl bond rupture in CCl_4, but not in CF_3Cl.

10.7 Temperature is about 200 K. Rearrange the gas equation to give n/V

$$\frac{n}{V} = \frac{P}{RT} = \frac{2.3 \times 10^{-3}\ \text{mm Hg}}{\left(\frac{0.0821\ \text{L-atm}}{\text{mol-K}}\right)(220\ \text{K})}\left(\frac{1\ \text{atm}}{760\ \text{mm Hg}}\right) = 1.7 \times 10^{-7}\ \text{mol/L}$$

$$\left(\frac{1.7 \times 10^{-7}\ \text{mol}}{1\ \text{L}}\right)\left(\frac{1\ \text{L}}{1000\ \text{cm}^3}\right)\left(\frac{6.022 \times 10^{23}\ \text{molecules}}{1\ \text{mol}}\right) = 1.0 \times 10^{14}\ \frac{\text{molecules}}{\text{cm}^3}$$

10.8 Avg. Molec. Wt. = 0.55 (40.0) + 0.45 (32.0) = 36.4. At 150 km, the fractions of particles are different because of the dissociation of O_2. We can "normalize" the fractions by saying that the 0.45 fraction of O_2 is converted into two 0.45 fractions of O atoms, then divide by the total fractions, 0.55 + 0.45 + 0.45 = 1.45:

$$\text{Av. Atomic Wt.} = \frac{0.55(40.0) + 2(0.45)(16)}{1.45} = 25.1$$

10.9 The energy of the photons must be large enough to cause O-O bond rupture, and the photon must be absorbed by the O_2 molecule.

10.10 $$\left(\frac{941 \times 10^{3}\ \text{J}}{\text{mol}}\right)\left(\frac{1\ \text{mol}}{6.022 \times 10^{23}\ \text{molecules}}\right) = 1.56 \times 10^{-18}\ \text{J/molecule}$$

Using the formula derived in the solution to 10.5,

$$\lambda = hc/E = \frac{(6.625 \times 10^{-34}\ \text{J-sec})(3.00 \times 10^{8}\ \text{m/sec})}{1.56 \times 10^{-18}\ \text{J}} = 127\ \text{nm}$$

10.11 During daytime solar irradiation is producing high energy particles such as oxygen atoms, and causing ionization, mainly of N_2, O_2 and O. As positive ions are formed, free electrons are also produced. At night these ionic species recombine by various pathways, so the concentration of ions decreases during the night. Similarly, high energy neutral species such as O undergo reactions (such as recombination) to form lower energy species.

10.12 (a) $O_2(g) + h\nu \rightarrow 2O(g)$

(b) $N_2^+(g) + e^- \rightarrow N(g) + N(g)$

(c) $O^+(g) + N_2(g) \rightarrow NO^+(g) + N(g)$

10.13 NO could be formed by combination of N and O. However, the process requires a third body to carry off extra energy:

$$N(g) + O(g) + M(g) \rightarrow NO(g) + M^*(g)$$

At the low pressures present in the upper atmosphere such a reaction is relatively slow. It could form also by electron capture by NO^+, but this process usually results in dissociation:

$NO^+(g) + e^-(g) \rightarrow N(g) + O(g)$. A very small fraction of the NO molecules formed this way may undergo collision or radiate a photon before they fall apart. In both these cases a higher total pressure would increase the rate of formation, by making a stabilizing collision more likely.

10.14 (a)

$O_2(g) \rightarrow O_2(g)^+ + e^-$	$\Delta E = 1205$ kJ
$NO^+(g) + e^- \rightarrow NO(g)$	$\Delta E = -890$ kJ
$NO^+(g) + O_2(g) \rightarrow NO(g) + O_2^+(g)$	$\Delta E = +315$ kJ

(b)

$N_2(g) \rightarrow N(g) + N(g)$	$\Delta E = 941$ kJ
$N(g) + O(g) \rightarrow NO(g)$	$\Delta E = -682$ kJ
$NO(g) \rightarrow NO^+(g) + e^-$	$\Delta E = 890$ kJ
$O^+(g) + e^- \rightarrow O(g)$	$\Delta E = -1313$ kJ
Overall: $N_2(g) + O^+(g) \rightarrow NO^+(g) + N(g)$	$\Delta E = -164$ kJ

(c)

$O_2^+(g) + e^- \rightarrow O_2(g)$	$\Delta E = -1205$ kJ
$O_2(g) \rightarrow O(g) + O(g)$	$\Delta E = 495$ kJ
$O_2^+(g) + e^- \rightarrow O(g) + O(g)$	$\Delta E = -710$ kJ

(d)

$O_2(g) \rightarrow O(g) + O(g)$	$\Delta E = 495$ kJ
$N(g) + O(g) \rightarrow NO(g)$	$\Delta E = -682$ kJ
$O_2(g) + N(g) \rightarrow NO(g) + O(g)$	$\Delta E = -187$ kJ

10.15 N_2 has a very high ionization energy, 1495 kJ/mol, as compared with 890 kJ/mol for NO. Any other molecules or ions that are ionized will not be able to undergo an electron transfer with N_2 because the process (for example, $O_2^+ + N_2 \rightarrow O_2 + N_2^+$) is endothermic. On the other hand, NO has such a low ionization energy that it readily transfers an electron on collision with nearly any other ion.

10.16 For oxygen to recombine with another O atom to form O_2, it must have present a third body to carry off the excess energy, as described in Section 10.4. This is much more likely to occur at 50 km than at 120 km because the atmosphere is much denser at the lower elevation; there are many more atoms or molecules per unit volume.

10.17 Ozone is formed by photodissociation of O_2, followed by a reaction of O and O_2 (with a stabilizing third body present to carry off the excess energy). The O_3 so formed absorbs solar radiation from the 200 to 310 nm wavelength regions. Such high energy photons are injurious to biological systems, both plant and animal. In humans they can produce skin cancers.

10.18 The asterisk denotes a high energy molecule, one that possesses excess energy from some source. On collision with M the excess energy is transferred to M, thus stabilizing the O_3 molecule. M can be any molecule or even an atom; the only limitation is that it does not readily react with the O_3.

10.19 $NO(g) + O_3(g) \rightarrow NO_2(g) + O_2(g)$

$\Delta H = 33.8 - 142.3 - 90.4 = -198.9$ kJ

$NO_2(g) + O(g) \rightarrow O_2(g) + NO(g)$

$\Delta H = 90.4 - 247.5 - 33.8 = -190.9$ kJ

10.20 $SO_2(g) + O_3(g) \rightarrow SO_3(g) + O_2(g)$

$\Delta H = -395.2 - 142.3 + 296.9 = -240.6$ kJ

$$SO_3(g) + O(g) \rightarrow SO_2(g) + O_2(g)$$

$$\Delta H = -296.9 - 247.5 + 395.2 = -149.2 \text{ kJ}$$

In comparing these values with those for NO, as in exercise 10.19, we see that both steps in the process are exothermic, and thus likely to proceed at a high rate. Thus, on this basis $SO_2(g)$ could act as a catalyst for O_3 decomposition just as NO(g) does.

10.21 Use of a compound containing a C-Br bond in place of a C-Cl bond would probably offer no solution at all. Photodecomposition of the CF_3Br compound would probably still take place in the stratosphere, and the Br atoms formed would be a catalyst for O_3 decomposition just as Cl atoms are. The fact that C-Br bonds are somewhat weaker would simply increase the rate of C-Br bond cleavage by solar irradiation.

10.22 The $CFCl_3$ molecule is approximately tetrahedral. The Cl-C-Cl bond angles should be slightly larger than tetrahedral, the Cl-C-F angles slightly smaller.

10.23 The geometry about carbon in CF_2Cl_2 is approximately tetrahedral. However, the Cl-C-Cl angle should be larger than the 109.5° tetrahedral angle, whereas the F-C-F angle should be slightly smaller.

10.24
$$CF_2Cl_2(g) + h\nu \rightarrow CF_2Cl(g) + Cl$$
$$CF_2Cl(g) + h\nu \rightarrow CF_2(g) + Cl$$
$$Cl(g) + O_3(g) \rightarrow ClO(g) + O_2(g)$$
$$ClO(g) + O(g) \rightarrow Cl(g) + O_2(g)$$

10.25 NO acts as a catalyst for decomposition of ozone in the stratosphere. The reactions are written in equation [10.17]. The increased destruction of ozone by NO would result in less absorption of short wavelength UV radiation now being largely screened out by the ozone. Radiation of this wavelength is known to be harmful to humans; it causes skin cancer. There is evidence that many plants don't tolerate it very well either, though more research is needed to test this idea.

10.26 Ozone partial pressure is $(0.26 \times 10^{-6})(740 \text{ mm Hg}) = 1.9 \times 10^{-4}$ mm Hg.

$$\frac{n}{V} = \frac{1.9 \times 10^{-4} \text{ mm Hg}}{\left(\frac{0.0821 \text{ L-atm}}{\text{mol-K}}\right)(299 \text{ K})} \left(\frac{1 \text{ atm}}{760 \text{ mm Hg}}\right)\left(\frac{10^3 \text{ L}}{1 \text{ m}^3}\right)\left(\frac{6.022 \times 10^{23} \text{ molecules}}{1 \text{ mol}}\right)$$

$$= 6.0 \times 10^{18} \text{ molecules/m}^3$$

10.27 Partial pressure is $(0.92 \times 10^{-6})(710 \text{ mm Hg}) = 6.5 \times 10^{-4}$ mm Hg.

$$\frac{n}{V} = \frac{(6.5 \times 10^{-4} \text{ mm Hg})}{\left(\frac{0.0821 \text{ L-atm}}{\text{mol-K}}\right)(303 \text{ K})} \left(\frac{1 \text{ atm}}{760 \text{ mm Hg}}\right)\left(\frac{10^3 \text{ L}}{1 \text{ m}^3}\right)\left(\frac{6.022 \times 10^{23} \text{ molecules}}{1 \text{ mol}}\right)$$

$$= 2.1 \times 10^{19} \text{ molecules/m}^3$$

10.28 CO in unpolluted air is typically 0.05 ppm; in urban air about 10 ppm. A major source is automobile exhaust. SO_2 is less than 0.01 ppm in unpolluted air, maybe on the order of 0.08 ppm in urban air. A major source is coal and oil-burning power plants, but there is also some SO_2 in auto exhausts. NO is about 0.01 ppm in unpolluted air, about 0.05 ppm in urban air. It comes mainly from auto exhausts.

10.29 $2ZnS(s) + 3O_2(g) \rightarrow 2ZnO(s) + 2SO_2(g)$

$$6.0 \times 10^5 \text{ tons Zn} \left(\frac{2000 \text{ lb}}{1 \text{ ton}}\right)\left(\frac{453 \text{ g}}{1 \text{ lb}}\right)\left(\frac{1 \text{ mol Zn}}{63 \text{ g Zn}}\right)\left(\frac{1 \text{ mol ZnS}}{1 \text{ mol Zn}}\right)\left(\frac{2 \text{ mol SO}_2}{2 \text{ mol ZnS}}\right)$$

$$\times \left(\frac{64 \text{ g SO}_2}{1 \text{ mol SO}_2}\right) = 5.5 \times 10^{11} \text{ g SO}_2$$

10.30 $Cu_2S(s) + O_2 \rightarrow 2Cu(s) + SO_2(g)$

$$1.6 \times 10^6 \text{ tons Cu} \left(\frac{2000 \text{ lb}}{1 \text{ ton}}\right)\left(\frac{453 \text{ g}}{1 \text{ lb}}\right)\left(\frac{1 \text{ mol Cu}}{63.5 \text{ g Cu}}\right)\left(\frac{1 \text{ mol SO}_2}{2 \text{ mol Cu}}\right)\left(\frac{64 \text{ g SO}_2}{1 \text{ mol SO}_2}\right)$$

$$= 7.3 \times 10^{11} \text{ g SO}_2$$

10.31 It may be oxidized following excitation by sunlight; upon adsorption onto dust and other particulates in the atmosphere; upon dissolving in rain droplets. The product of the oxidation, SO_3, is soluble in water to form H_2SO_4 solutions. This substance is the major component of "acid rain," and is responsible for much damage to human health and property.

10.32 (a) CO is a possible product of some vegetable matter decay. It may arise from the oceans. (b) SO_2 is released in volcanic gases, and also is produced by bacterial action on decomposing vegetable and animal matter. (c) Methane, CH_4, arises from decomposition of organic matter by certain microorganisms; it also escapes from underground natural gas deposits. (d) Nitric oxide, NO, results from oxidation decomposing organic matter, and is formed in lightning flashes.

10.33 The reactions involved are: $CaCO_3(s) \rightarrow CaO(s) + CO_2(g)$; $S(s) + O_2(g) \rightarrow SO_2(g)$; $CaO(s) + SO_2(s) \rightarrow CaSO_3(s)$

$$(1 \text{ ton oil})\left(\frac{0.017 \text{ ton S}}{1 \text{ ton oil}}\right)\left(\frac{2000 \text{ lb}}{1 \text{ ton}}\right)\left(\frac{453 \text{ g}}{1 \text{ lb}}\right)\left(\frac{1 \text{ mol S}}{32 \text{ g S}}\right)\left(\frac{1 \text{ mol SO}_2}{1 \text{ mol S}}\right)$$

$$\times \left(\frac{1 \text{ mol CaO}}{0.22 \text{ mol SO}_2}\right)\left(\frac{1 \text{ mol CaCO}_3}{1 \text{ mol CaO}}\right)\left(\frac{100 \text{ g CaCO}_3}{1 \text{ mol CaCO}_3}\right)$$

$$= 2.2 \times 10^5 \text{ g CaCO}_3$$

10.34 The volume of air space involved is:

$$(580 \text{ km}^2)(1.2 \text{ km})\left(\frac{10^3 \text{ m}}{1 \text{ km}}\right)^3\left(\frac{100 \text{ cm}}{1 \text{ m}}\right)^3\left(\frac{1 \text{ L}}{10^3 \text{ cm}^3}\right) = 7.0 \times 10^{14} \text{ L}$$

The total moles of gas in this volume under the stated conditions is

$$n = \frac{(740 \text{ mm Hg})(7.0 \times 10^{14} \text{ L})}{\left(\frac{0.0821 \text{ L-atm}}{\text{mol-K}}\right)(297 \text{ K})}\left(\frac{1 \text{ atm}}{760 \text{ mm Hg}}\right) = 2.8 \times 10^{13} \text{ mol}$$

The number of moles of SO_2 is $(0.087 \times 10^{-6})(2.8 \times 10^{13}) = 2.4 \times 10^6$. The mass of SO_2 is:

$$(2.4 \times 10^6 \text{ mol SO}_2)\left(\frac{64 \text{ g SO}_2}{1 \text{ mol SO}_2}\right) = 1.5 \times 10^8 \text{ g SO}_2$$

In slightly more familiar units, this is 170 tons of SO_2!

10.35 NO_2 builds up early in the day because auto traffic produces NO and this gets oxidized to NO_2. However, sunlight converts some of the NO_2 to NO + O; the oxygen atoms go on in part to form ozone. At the same time, the auto exhaust is also producing a concentration of unburned and partially burned hydrocarbons. These compounds begin to get oxidized by the atomic oxygen formed from photodissociation of NO_2. Thus, the cycle described by equations [10.29], [10.30] and [10.31] is broken, and the NO_2 level drops.

10.36 Major sources of CO are auto exhausts in urban areas, and burning tobacco in the inhalations of smokers. The solution to the latter problem is obvious, even if not easy for some people to implement. The reduction of CO in auto exhausts requires a catalyst in the exhaust system to convert the CO to CO_2, and improved engine designs to ensure more complete combustion.

10.37 Iron-containing hemoglobin in the blood is the binding site for O_2 in the lungs. This same site also binds CO, in fact much more tightly than it binds O_2. Thus, low levels of CO result in some substantial reduction of the blood's hemoglobin available for binding of O_2. This places an extra burden on the cardiovascular system.

10.38 The oxygen of the atmosphere is present to the extent of 209,000 parts per million. If the CO binds 210 times more effectively than O_2, then the effective concentration of CO is 210 x 120 ppm = 25,200 ppm. The fraction of carboxyhemoglobin in the blood leaving the lungs is thus $\frac{25{,}200}{25{,}200 + 209{,}000} = 0.11$. Thus, 11 percent of the blood is in the form of carboxyhemoglobin, 89 percent as the O_2-bound, or oxy-, hemoglobin.

10.39 The problem arises from the fact that continuous formation of CO_2 will cause the CO_2 content of the atmosphere to increase, with a resultant effect on earth's climate. The CO_2 of the atmosphere acts to reduce radiation of infrared rays from earth's surface, and thus produces a warmer climate. It is predicted that by about 2050 the CO_2 content of the atmosphere will have doubled, with a resultant increase of about 3°C in earth's average temperature.

10.40 Solar radiation causes dissociation to form atoms, or ionization. The products of these processes are high in energy. When they undergo subsequent reactions they cause release of a good deal of heat. For example, the dissociative recombination, $N_2^+ + e^- \rightarrow N + N$, produces two N atoms with a great deal of kinetic energy. Similarly, the process $N + O + M \rightarrow NO + M^*$ leaves M^* with a high energy content.

10.41 Solar dissociation of O_2 is a major factor. This results in formation of two oxygen atoms rather than O_2, so the average molecular weight is lowered. In addition, there is some separation in the earth's gravitational field. The lightest gases such as H_2 and He are relatively more abundant in the outer reaches of the atmosphere.

10.42 Since the moon has no atmosphere the water vapor from the seas would be subjected to the full effect of solar radiation. Thus, the water molecules would be subject to photodissociation:

$$H_2O + h\nu \rightarrow H + OH$$
$$OH + h\nu \rightarrow O + H$$

Over a long period of time, these atoms would recombine in various ways. The recombinations would be slow because collisions between the high energy atoms would be infrequent. However, some O_2 and H_2 could be expected to form. The O_2 would be subject to photoionization, so there should be some O_2^+ present. The likelihood of reaction of O_2 to O to form an O_3 molecule would doubtless be negligible. The H or H_2 species, being of low mass, would soon escape from the gravitational force of the moon. So also would O and O_2, but at a much slower rate. Thus, eventually, all the water placed on the moon would disappear.

10.43 (a) $N_2(g) + h\nu \rightarrow N_2^+(g) + e^-$

$N_2^+(g) + e^- \rightarrow N(g) + N(g)$ (dissociative recombination)

(b) $O^+(g) + N_2(g) \rightarrow NO^+(g) + N(g)$

(c) $NO^+(g) + e^- \rightarrow N(g) + O(g)$

(d) $N_2^+(g) + O(g) \rightarrow N(g) + NO^+(g)$

10.44 The temperature maximum arises from the "ozone machine." It begins with photodissociation of O_2 to form O atoms. Then $O(g) + O_2(g) + M(g) \rightarrow O_3(g) + M^*(g)$. Photodissociation of $O_3(g)$ and/or of $O_2(g)$ also leads to a steady concentration of O atoms which may also simply recombine to form O_2: $O(g) + O(g) + M(g) \rightarrow O_2(g) + M^*(g)$. Note that both O_2 and O_3 formation result in liberation of heat. In effect, the sun's energy is being converted to heat.

10.45 A catalyst is a substance that affects the speed of a chemical reaction without itself being consumed in the process. Cl acts as a catalyst in the two-step process as follows:

$Cl(g) + O_3(g) \rightarrow ClO(g) + O_2(g)$

$ClO(g) + O(g) \rightarrow Cl(g) + O_2(g)$

The sum of these two reactions is just $O_3(g) + O(g) \rightarrow 2O_2(g)$. Note that Cl is recycled continuously, without undergoing any permanent change.

10.46 Photoionization of O_2 requires 1205 kJ/mol. Photodissociation requires only 495 kJ/mol (Table 9.6). At lower elevations solar radiation of wavelengths corresponding to 1205 kJ/mol or shorter has already been absorbed. However, the longer wavelength radiation has passed through relatively well. In the much denser lower atmosphere it is capable of causing photodissociation of O_2. The points to note here are that photoionization of O_2 dominates over photodissociation in the upper atmosphere. As a result, there is relatively little O_2 around above 120 km. Photodissociation is less likely to occur, but in the stratosphere, where the density of O_2 is much higher, even this less likely process soaks up all the longer wavelength radiation of sufficiently high energy.

10.47 The ionization energies of the metal atoms are much lower than those of any of the atomic or molecular ions present at 120 km; that is, O_2^+, O^+, NO^+. Thus, any of these ions react with the metal atoms as illustrated for NO^+: $M + NO^+ \rightarrow M^+ + NO$.

10.48 (a) $O_2(g) + h\nu \rightarrow O(g) + O(g)$

(b) $H_2O(g) + h\nu \rightarrow H(g) + OH(g)$

(c) $NO_2(g) + h\nu \rightarrow NO(g) + O(g)$

(d) $O^+(g) + N_2(g) \rightarrow NO^+(g) + N(g)$

(e) $O(g) + O_2(g) + M \rightarrow O_3(g) + M^*(g)$

10.49 $$\left(\frac{150 \times 10^{-3}\ g}{m^2}\right)\left(\frac{1\ m}{100\ cm}\right)^2\left(\frac{2.54\ cm}{1\ in.}\right)^2\left(\frac{12\ in.}{1\ ft}\right)^2\left(\frac{5280\ ft}{1\ mi}\right)^2$$

$$= 3.9 \times 10^5\ g\ sulfate/mi^2$$

This corresponds to about 0.5 ton.

10.50 Both water vapor and CO_2 act as infrared absorbers. Absorption of long wavelength radiation emitted from earth's surface causes some of this radiation to be re-radiated back to earth, and thus helps to maintain a moderate, even temperature at the surface. Increased concentrations of infrared-absorbing materials would result in increased retention of heat at the earth's surface and, thus, produce a warmer climate. An increased concentration of CO_2 from burning of fossil fuels might accomplish such a result by mid-21st century.

<u>10.51</u> The origin of acid rain is mainly SO_2. This substance is oxidized by one means or another to SO_3, which in aqueous medium forms sulfuric acid, H_2SO_4. Nitric acid, HNO_3, may also contribute to acidity in the rain in some areas.

11

Liquids, solids, and intermolecular forces

11.1 Molecules diffuse by a process of random collisions with other molecules, moving first this way, then that. Bromine diffuses much more rapidly through the vessel containing air. This is so because the distance the Br_2 molecule travels between collisions is much longer in the gas state because of the lower density of molecules.

11.2 The solid is slightly denser because the molecules are packed more regularly in the solid state. Thus, there is less "free volume," or space between molecules than there is in the liquid, in which the molecules are more randomly oriented. H_2O is one of the rare exceptions to the rule.

11.3 There is more free volume in a liquid. Because the molecules are moving about with respect to one another, their orientations with respect to one another are not so regular as in a solid. Thus, on compression, there is more room for the molecules to move closer together than there is in the solid state.

11.4 In the condensed liquid phase molecules experience intermolecular attractive forces that keep them together. To vaporize a liquid, these intermolecular forces must be overcome, because they are comparatively much smaller in the gas state. It takes energy to do this.

11.5 To account for the transfer of radioactive iodine from one side to the other we must propose a transfer process involving the gas phase. Thus, on each side of the container solid iodine is in equilibrium with iodine vapor. Once in the gas phase the radioactive iodine can diffuse through the space to come into contact with solid iodine on the other side. The fact that radioactive iodine is found in the solid on the other side shows that the gas-solid equilibrium is indeed a two-way process. That is, iodine molecules can pass from the gaseous to solid phase.

11.6 Liquid ethyl chloride at room temperature is far above its boiling point. When the liquid boils, heat is required to overcome the intermolecular forces between molecules. This heat is extracted from the surface that contacts the liquid.

11.7

$$\begin{array}{lll} H_2O(s) \rightarrow H_2O(\ell) & \Delta H = H_f & \\ H_2O(\ell) \rightarrow H_2O(g) & \Delta H = H_v & \Delta H = 6.01 \text{ kJ/mol} + 44.86 \text{ kJ/mol} \\ \hline H_2O(s) \rightarrow H_2O(g) & \Delta H = \Delta H_f + \Delta H_v & \quad = 50.87 \text{ kJ/mol} \end{array}$$

11.8 The molar heat of sublimation contains the heat of melting and the heat of vaporization, so it is always larger than the heat of vaporization alone, measured at the same temperature.

11.9 The heat added to solid CuO is "stored" in the solid in the form of vibrational motions of the Cu^{2+} and O^{2-} ions in the lattice.

11.10 (a) no change; (b) increases with increasing temperature; (c) higher intermolecular forces mean lower vapor pressure; (d) no change

11.11 Melting involves an equilibrium between the solid and liquid phases. Since these two phases do not generally differ much in density, higher pressure does not favor one phase over the other, so the melting temperature is not much affected. However, pressure has a big effect on the density of a gas; a higher temperature is required to produce a vapor pressure equal to the higher gas pressure that results when pressure is applied to a gas-liquid equilibrium.

11.12 Diethyl ether, about 23°C; ethyl alcohol, 68°C; water, 90°C.

11.13 When a liquid is poured, the molecules must flow past one another. Thus viscosity, which measures the ease with which liquids flow, is related to the nature of the contacts between molecules in the liquid state. Molecules with strong intermolecular attractive forces may not easily move with respect to one another. Perhaps even more important is molecular shape and size. Large, long molecules that become entangled do not readily move with respect to one another, and viscosity is high. Motor oils furnish a good example.

11.14 Vaporization into a vacuum will be more rapid than into air. The molecules leaving the liquid surface to enter a vacuum will never return. On the other hand, when there is air present, collisions with air molecules cause many of the molecules that have just left the liquid surface to be deflected back into the liquid. It takes longer in air for the vapor molecules to diffuse far enough away so they will not return to the liquid.

11.15 Viscosity is related to the ease with which molecules move past one another. As chain length increases, intermolcular attractive forces increase (for example, the boiling points in this series increase), and the molecules are longer, thus more readily entangled.

11.16 High molecular mass means generally larger intermolecular attractions. Polar groups may lead to strong intermolecular attractions. A long, chain-like structure produces more entanglements of the molecules, which in turn leads to higher viscosity.

11.17 As the figure to the right shows, the linear relationship is obeyed.

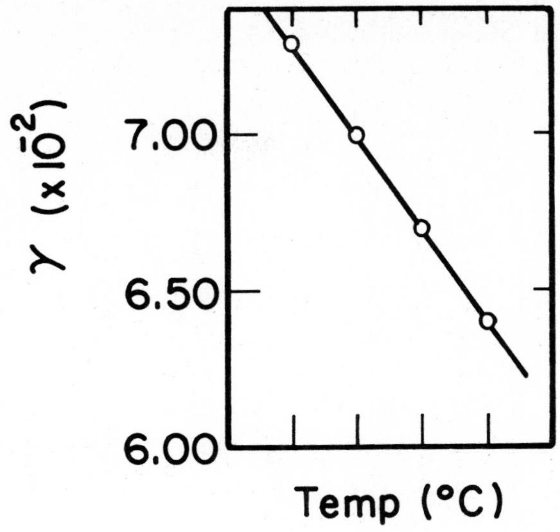

11.18 (a) decreases; (b) increase; (c) increase; (d) increase; (e) increases; (f) increases; (g) increases; (h) unaffected

11.19 The intermolecular forces are higher in CS_2 than in CO_2. Thus, the temperature required to attain the critical condition is higher, and the critical pressure is higher.

11.20

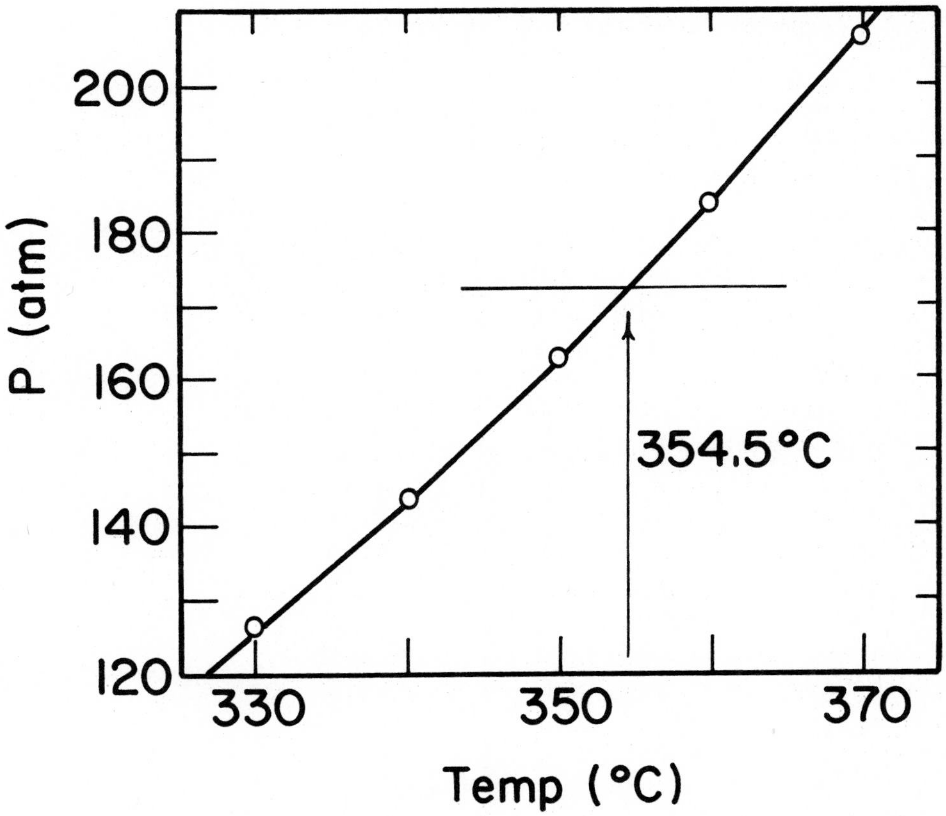

11.21 We need to graph log P vs. 1/T, where T is absolute temperature. If the Clausius-Clapeyron equation is obeyed, the relationship is linear. Here are the data in form for graphing.

Temp (K)	1/T	VP(mm Hg)	log(VP)
323	3.10×10^{-3}	1.267×10^{-2}	-1.897
333	3.00×10^{-3}	2.524×10^{-2}	-1.600
343	2.92×10^{-3}	4.825×10^{-2}	-1.316
353	2.83×10^{-3}	8.80×10^{-2}	-1.056
363	2.75×10^{-3}	1.582×10^{-1}	-0.801

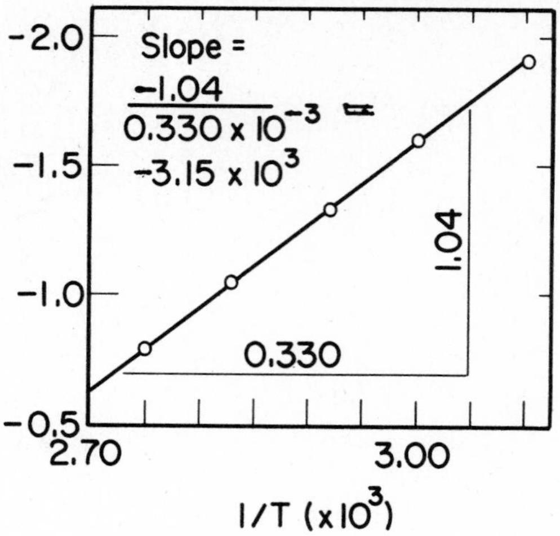

A graph of log(VP) vs. 1/T is linear, as you should determine for yourself (see figure on right). The slope, -3.15×10^3, equals $-\Delta H_V/2.3\ R$. Thus, $\Delta H_V = 60.3$ kJ/mol.

11.22 Graph log (VP) vs. 1/T, where T is absolute temperature.

Temp (K)	1/T	VP(mm Hg)	log VP
253.2	3.99×10^{-3}	0.640	-0.194
257.2	3.89×10^{-3}	1.132	0.0538
261.2	3.83×10^{-3}	1.632	0.213
265.2	3.77×10^{-3}	2.326	0.367
269.2	3.71×10^{-3}	3.280	0.516
273.2	3.66×10^{-3}	4.579	0.661

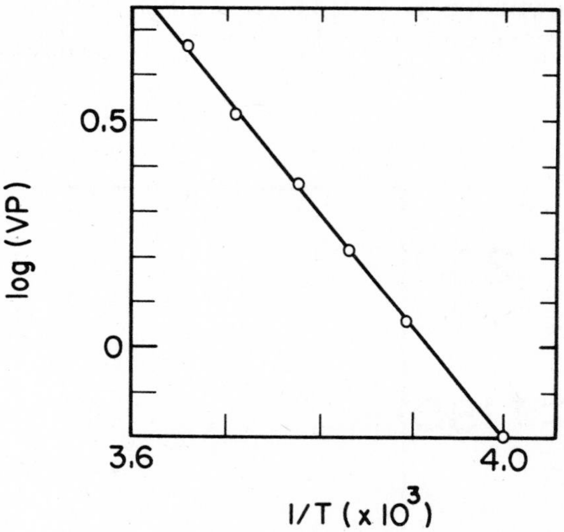

The graph is shown on the right. The slope can be estimated by reading log(VP) at two well-separated values of 1/T:

$$\text{Slope} = \frac{(0.680 - 0.090)}{3.65 \times 10^{-3} - 3.95 \times 10^{-3}}$$

$$= -2.57 \times 10^3$$

This slope equals $-\Delta H_s/2.3\ R$. Thus,

$$-\Delta H_s = 2.3(8.314)\ (-2.57 \times 10^{-3})$$

$$\Delta H_s = 49.1 \text{ kJ/mol}$$

11.23 It is an amorphous solid. A crystalline material would not be likely to have flexibility. Rather, it would cleave when pressure is applied.

11.24 Well-defined X-ray patterns might be seen for (a) sugar; (b) KBr; (d) pure iron; (e) ice.

11.25 $n\lambda = 2d \sin \theta$ $\quad \sin \theta = \frac{n\lambda}{2d} = \frac{n(1.65)}{2(2.68)}$

n	sin θ	θ
1	0.3078	17.9°
2	0.6156	38.0°
3	0.9234	67.4°

11.26 In this case n = 1; $d = \lambda/2 \sin \theta = 1.68/2(\sin 8.7°) = 5.55$ Å.

11.27 A body-centered unit cell contains one CsBr unit. Thus, it contains

$$\left(\frac{212.8 \text{ g CsBr}}{1 \text{ mol}}\right)\left(\frac{1 \text{ mol}}{6.022 \times 10^{23} \text{ CsBr units}}\right)\left(\frac{1 \text{ CsBr unit}}{\text{unit cell}}\right)$$

$= 3.53 \times 10^{-22}$ g CsBr/unit cell

Then, $\frac{3.53 \times 10^{-22} \text{ g CsBr}}{\text{unit cell}}\left(\frac{1 \text{ cm}^3}{4.428 \text{ g CsBr}}\right) = \frac{7.98 \times 10^{-23} \text{ cm}^3}{\text{unit cell}}$

Since the unit cell is cubic, the length of a side is just the cube root of this number, or 4.30×10^{-8} cm.

11.28 There are four formula units per face-centered unit cell. The volume of the krypton unit cell is $(5.59 \times 10^{-8} \text{ cm})^3 = 1.75 \times 10^{-22} \text{ cm}^3$. The mass of krypton in this volume is $(4 \times 83.80 \text{ g})/6.022 \times 10^{23} = 5.57 \times 10^{-22}$ g. Density is thus $5.57 \times 10^{-22} \text{ g}/1.75 \times 10^{-22} \text{ cm}^3 = 3.18 \text{ g/cm}$.

11.29 Each unit cell contains 4 KF units. The mass of a unit cell is thus $4(39.1 + 19.0) = 232.4 \text{ g}/(6.022 \times 10^{23}) = 3.86 \times 10^{-22}$ g. The volume is thus

$$\left(\frac{1 \text{ cm}^3}{2.468 \text{ g KF}}\right)\left(\frac{3.86 \times 10^{-22} \text{ g KF}}{\text{unit cell}}\right) = 1.56 \times 10^{-22} \text{ cm}^3/\text{unit cell}$$

The unit cell is thus $(1.56 \times 10^{-22})^{1/3}$ on each side, or 5.39×10^{-8} cm. The distance between K^+ and F^- is one-half the unit cell length (see Figure 11.13(a)): $1/2(5.39 \times 10^{-8}) = 2.70 \times 10^{-8}$ cm.

11.30 The area bounded by the lines represents the unit cell of this two-dimensional array. Note that the total number of circles in the lines is exactly one. This unit is the smallest unit which can, by repetition in either dimension, reproduce the entire two-dimensional structure.

11.31 The volume of the simple cubic cell is $(2r)^3 = 8r^3$, where r is the radius of the units that occupy each corner. There is a total of 1 sphere in the unit cell. The volume of this sphere is $4\pi r^3/3$. The ratio of this volume to the unit cell volume is $(4\pi/3)/8 = 0.523$.

11.32

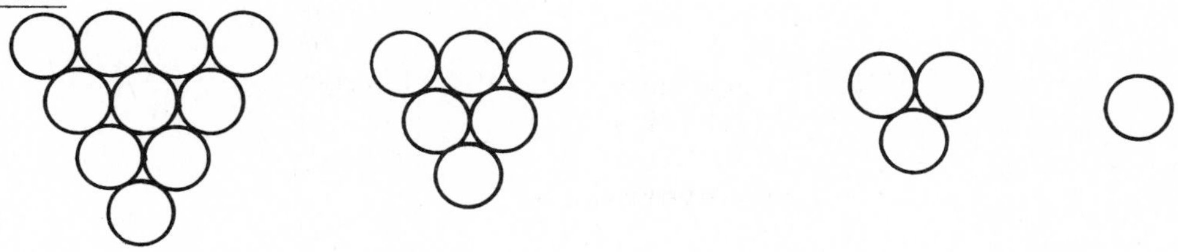

layer 1 2 3 4

By drawing each layer lightly over the others you can see that the stacking is of the form ABCA, rather than ABAB. Thus the cannonballs are cubic close-packed. A total of 20 cannonballs is required.

11.33 There are four Ar atoms in the face-centered unit cell. We know that the Ar atoms are in contact. Thus the radius of the Ar atom is 1/4 of the unit cell diagonal. The diagonal is just $\sqrt{2}$ times the unit cell length. Thus,

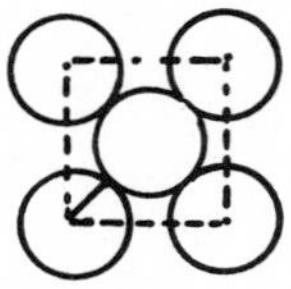

$$r = \frac{1.414 \times 5.25\ \text{Å}}{4} = 1.86\ \text{Å}$$

11.34 Intermolecular forces are much weaker than intramolecular bonding forces; they are less directional; they often operate over shorter distances; they are more diverse in character, ranging from ion-ion interactions to London dispersion forces.

11.35 $Br_2(\ell) \rightarrow Br_2(g)$ $\quad \Delta H = 30.71$ kJ

$Br_2(g) \rightarrow 2Br(g)$ $\quad \Delta H = 192.9$ kJ

The first process represents an overcoming of the intermolecular attractive energies. The second requires rupture of the Br-Br bond, an intramolecular interaction. As is usual, the intermolecular interactions are weaker than the intramolecular ones.

11.36 The ionic substance $FeCl_2$ is most likely to exist as a solid; the non-polar, low molecular weight F_2 is most likely to exist as a gas. Remember that HF is strongly hydrogen-bonded, SO_2 is polar and PCl_3 is much more massive. These three substances would liquefy under pressure at room temperature more readily than F_2.

11.37 High boiling point: RbCl, as the only ionic compound in the list. Lowest boiling point: Ar, because it has the lowest mass and least complexity among all the non-polar species listed.

11.38 (a) London dispersion forces (the I_2 molecule is non-polar); (b) ion-ion attractions; (c) dipole-dipole; (d) hydrogen bonding.

11.39 The X-H bond in the X-H...Y hydrogen bond must be polar covalent so that the hydrogen has some positive character associated with it. The Y atom must have a lone electron pair on it for interaction with the hydrogen. Strong hydrogen bonds exist only between two 2nd row elements N, O or F.

11.40 $C_6H_5OH > CH_3NH_2 > H_2S$

11.41 We expect that ion-ion attractive forces will be strongest when the distances between charge centers is least. The ionic radii of the anions increase in the order $F^- < Cl^- < Br^- < I^-$. Thus the distances between the cation and anion increase in this same order, and the ion-ion forces decrease in this order: LiF > LiCl > LiBr > LiI. The order of boiling points is as expected.

11.42 The graph at right shows that the effective ionic radius for the non-spherical OH^- ion is (2.14 - 0.68) = 1.46 Å. Thus, OH^- behaves as though it were a bit larger than F^-.

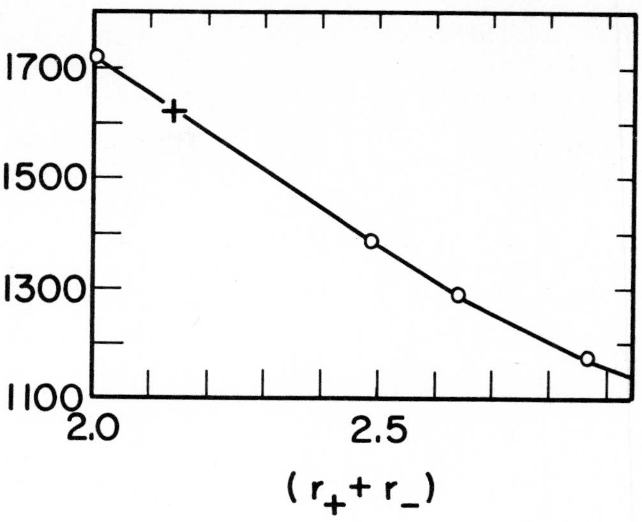

11.43 The most important inter-particle force in these liquids is the ion-ion force. These should decrease as the cation grows larger, because the coulombic attraction decreases as the distance between centers increases. We observe the expected order except that LiCl boils at a lower temperature than expected. The reason for this is probably that the Li^+ ion is really quite small, and is not large enough to keep chloride ions from experiencing anion-anion contacts. These are repulsive in character, and serve to weaken the overall ion-ion attractions.

11.44 (a) The most important intermolecular attractive forces in CH_3OH - H_2O are, in order of decreasing importance: hydrogen-bonding, dipole-dipole interactions, and dispersion-force interactions. (b) In Xe(ℓ) London dispersion forces are the only attractive forces. (c) London dispersion forces (the electrons of C_6H_6 are quite polarizable); (d) In Cl-F, both dipole-dipole and London dispersion forces are important. (e) In $Ca(NO_3)_2$ ion-ion forces are dominant, though a kind of ion-induced dipole could also play a role.

11.45 propane 1,3-diol > propanol > propane The hydrogen-bonding interactions are greatest in the diol, but hydrogen bonding is operative also in propanol. Because the intermolecular forces tend to tie molecules together, they cause a more restricted flow of molecules past one another. In propane, intermolecular forces are comparatively much weaker, and viscosity is much lower as well.

11.46 In LiF(ℓ) the dominant intermolecular force is ion-ion. This is true also for BeF_2, but there is probably some fractional covalent character in the Be-F bond. This and the fact that Be^{2+} is very small, thus allowing anion-anion contacts, causes the boiling point of BeF_2 to be higher. BF_3 and CH_4 are non-polar molecules for which London dispersion forces provide the major source of intermolecular attraction. NF_3 is slightly polar (it has a dipole moment of only 0.2 D); thus, there are weak dipole-dipole forces operating in addition to London dispersion forces. OF_2 is similarly of low polarity, so it also has a low boiling point. F_2 is a completely non-polar molecule for which the London dispersion forces are the only source of intermolecular attraction. It has the lowest mass of the covalent substances listed, and along with that the lowest boiling point.

11.47 In dibromomethane, CH_2Br_2, the dispersion force contribution will be larger than for CH_2Cl_2, because bromine is more polarizable than the lighter element chlorine. At the same time, the dipole-dipole contribution for CH_2Cl_2 is greater than for CH_2Br_2 because CH_2Cl_2 has a larger dipole moment. Just the opposite comparisons apply to CH_2F_2, which is less polarizable and of higher dipole moment than CH_2Cl_2.

11.48 The critical temperature increases with the magnitude of the intermolecular forces. The critical temperature is abnormally high for HF because of hydrogen-bonding interactions. It increases in the series HCl < HBr < HI because of increasing dispersion forces. These are sufficiently large to overcome the trend of decreasing dipole-dipole forces, which decrease in the order HCl > HBr > HI.

11.49 The network solid diamond is the hardest of the four substances, followed, in all probability by Ni, $BaCl_2$ and SCl_2, in that order. Among the four, only Ni should exhibit any conductivity. The melting points vary in the order diamond > nickel > $BaCl_2$ > SCl_2.

11.50 (a) $CuBr_2$; because the ion-ion forces are much stronger than the London dispersion forces that would dominate in solid Br_2. (b) SiO_2; because it is a three-dimensional network solid of great stability. (c) Cr; because the metal bonding produces a stable, efficiently packed solid with metal-metal bonds between the fundamental particles making up the lattice. In S (in the form S_8), only London dispersion forces operate between the units of the solid lattice. (d) CaF_2; because a solid with ions of charge 2 experiences stronger ion-ion forces than one in which only charges of 1 are involved. Also because the ions in CaF_2 are smaller than in CsBr; thus, the ion-ion forces are larger.

11.51 C_2H_3Cl; It should be largely amorphous in character because the long chains do not pack in a very regular way. However, regions of crystalline character (regular alignments) do exist.

11.52 The difference between an ionic substance and one that involves polar covalent bonds is not easy to determine. In the case of ZnS the character of the three-dimensional solid is not so evident from its "bulk" properties. The two bonding types differ in the degree to which the bonding electron pairs are transferred from the less electronegative to the more electronegative atoms. Both types of lattice may show good cleavage properties along certain directions. Other properties such as color and so forth can give a clue to the more sophisticated investigator, but there are no obvious distinctions. The electronegativity difference is perhaps the best indicator. In the case of ZnS, the electronegativity difference between Zn^{2+} and S is 1.0 (Figure 7.6). Since the electronegativity of Zn would be less than that of Zn^{2+}, the Zn-S bond should be quite polar, and can be most simply visualized as ionic, with Zn^{2+} and S^{2-} ions.

11.53 Hexachlorobenzene has the same shape as benzene, except for the fact that the chlorines are much larger than the hydrogens they replace. C_6Cl_6 is much more massive, and has many more electrons per molecule. Thus, the London dispersion forces are much greater in this compound.

11.54 Our guesses: pentachlorobenzene, m.p. = 100°C; b.p. = 280°C; tetrachlorobenzene, m.p. = 120°C; b.p. = 248°C.

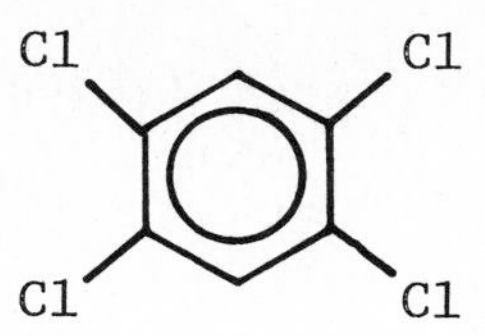

If one graphs boiling point vs. the number of chlorines, one obtains a smooth curve, nearly a straight line. It is easy to interpolate the boiling points from this curve. The melting points are much more sensitive to the geometry. One can guess that the symmetrical tetrachlorobenzene will melt at a higher temperature than the pentachlorobenzene, and that it will melt at a temperature between that for trichloro and hexachlorobenzene. The intermediate temperature between the melting points of these two substances is about 120°C. The observed melting and boiling points of the tetrachlorobenzene with structure shown are 138°C and 243°C, respectively. The melting point of pentachlorobenzene we can simply guess will be lower than for hexachlorobenzene, but it should still be fairly high to reflect the large molecular mass; say 100°C. The observed values of melting and boiling point are 85°C and 276°C, respectively.

11.55 The temperature in the pan containing boiling water is 100°C; in the pan containing water and ice it is 0°C. Temperature is constant in each case because the heat being added is being used to effect a phase change at constant temperature. In the one pan, liquid water is being converted to steam at 100°C. In the other, solid water is being converted to liquid water at 0°C.

11.56 The heat added at 80°C is going to convert liquid benzene to benzene vapor. The amount of heat required per mole is the heat of vaporization of liquid benzene at 80°C.

11.57 The heating curve looks something like the figure at right. The key to determining the lengths of the horizontal sections and the slopes of the lines is the heat capacity. Thus, for example, it requires

$$\left(\frac{0.17\ \text{J}}{\text{mol-°C}}\right)(224°\text{C})$$

$$= 38.1\ \text{kJ/mol}$$

to increase the temperature from −127°C to 97°C. Thus, the time required at constant heat input is just a bit less than the time required to convert one mole of propanol to vapor, 41.7 kJ.

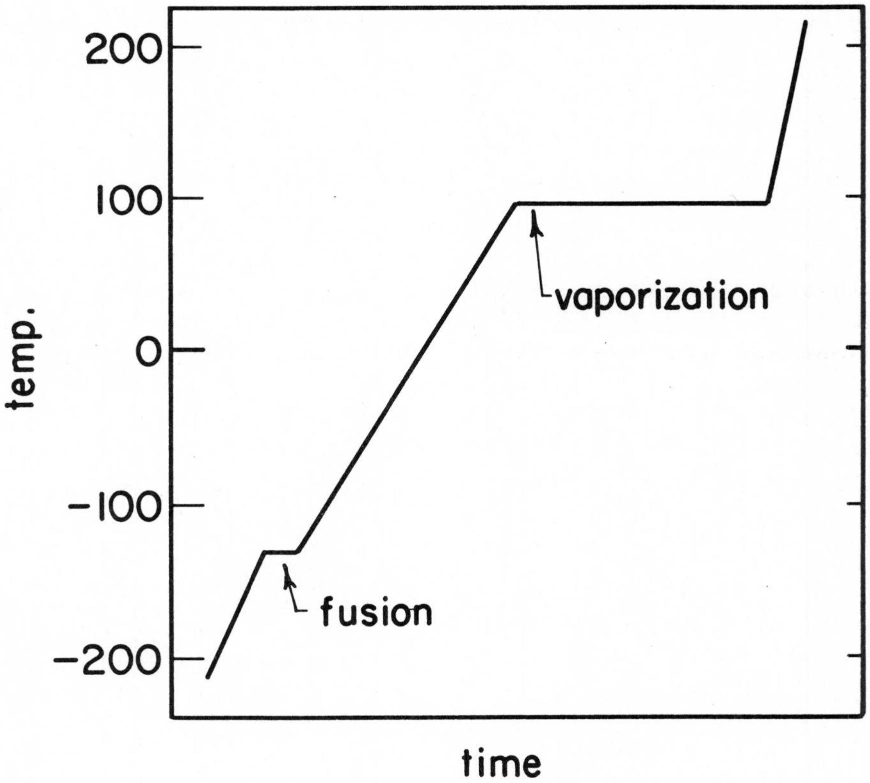

11.58 $\Delta H = (26\ g)\left[\left(\frac{2.092\ J}{g\text{-}°C}\right)(20°C) + \left(\frac{4.184\ J}{g\text{-}°C}\right)(100°C) + \left(\frac{1.841\ J}{g\text{-}°C}\right)(20°C)\right.$

$$\left. + \left(\frac{6.02 \times 10^3 J}{mole}\right)\left(\frac{1\ mol}{18\ g}\right) + \left(\frac{40.67 \times 10^3\ J}{mole}\right)\left(\frac{1\ mol}{18\ g}\right)\right]$$

$= 8.04 \times 10^4\ J = 80.4\ kJ$

11.59

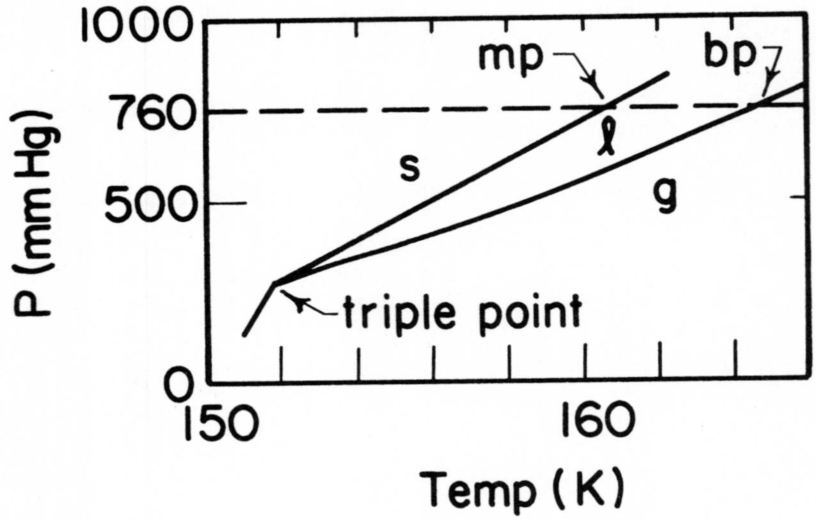

11.60 Benzene is the only non-polar molecule in this group. The polarities of the others decrease in the order $C_6H_5F > C_6H_5Cl > C_6H_5Br > C_6H_5I$. This order comes from the fact that the C-X bond polarity decreases in this order. Note, however, that ΔH_s does not follow this order, but simply increases with increasing molecular mass. This suggests that most of the variation in ΔH_s arises from the increasing magnitude of London dispersion forces with increasing molecular mass, due to the greater polarizabilities of the heavier halogens.

11.61 Each corner atom in the face-centered cubic structure contributes 1/8 atom. There are eight of these for a total of 1. Each face-centered atom contributes 1/2. There are six of these for a total of 3. Thus, there are four atoms per face-centered unit cell. In the body-centered there are again 8 corner atoms for a total of 1, then one in the center for a total of 2. If the center and corner atoms are different, say X and Y, there is one XY unit per body-centered unit cell.

11.62 $\lambda = 2(2.10\ Å) \sin(7.60°)/n = 0.555\ Å,\ 0.278\ Å$, etc.

11.63 In raising the temperature of liquid water from 0°C to 100°C, additional kinetic energy is imparted to the molecules. The added kinetic energy serves to partially disrupt the hydrogen-bonding interactions that are a major source of the attractive forces between H_2O molecules in the liquid. Thus, less energy is needed to separate a water molecule from its neighbors in vaporization at 100°C as compared with 0°C.

11.64 A graph of log P vs. 1/T (in K, of course) is shown at right. We can estimate the slope by reading off log P at two convenient values of 1/T.

$$\text{slope} = \frac{-0.740 - (-0.105)}{(4.75 - 4.25) \times 10^{-3}}$$
$$= -1.27 \times 10^3/K$$

$$\Delta H = (1.27 \times 10^3/K)(2.3)(8.314\ J/mol\text{-}k)$$
$$= 24.3\ kJ/mol$$

Changing units of pressure would have no effect on the slope, and thus on calculated ΔH_v.

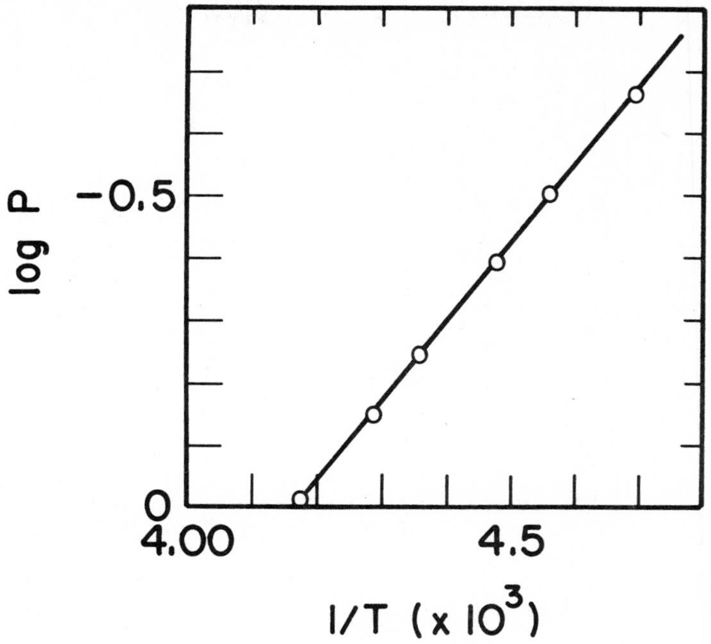

11.65 The higher critical temperature and pressure for $CClH_3$ suggest that the intermolecular attractive forces are higher in this compound than in $CClF_3$. This seems surprising at first glance, since neither molecule is really very polar, and one would suppose that London dispersion forces would constitute a major portion of the intermolecular attractive force. Since fluorine has many more electrons than H, it would be reasonable to expect that $CClF_3$ would have larger intermolecular attractive forces than $CClH_3$. However, fluorine covalently bound to carbon seems not to be a very polarizable center. Note for example that CF_4 boils at -124°C, not much higher than the boiling point of CH_4, -161°C. In the present case it seems likely that the greater polarity of $CClH_3$ over $CClF_3$ produces enough dipole-dipole interaction energy to more than compensate for any difference in attractive energies due to London dispersion energies.

11.66 $\gamma = a - bT$. At 28°C, $\gamma = 24.05 \times 10^{-3}\ J/m^2 - (0.0832 \times 10^{-3}\ J/m^2 - °C)\ 28°C = 21.72 \times 10^{-3}\ J/m^2$. At 60°C, $\gamma = 24.05 \times 10^{-3}\ J/m^2 - (0.0832 \times 10^{-3}\ J/m^2 - °C)\ 60°C = 19.06 \times 10^{-3}\ J/m^2$. The surface tension is lower at higher temperature because the increased kinetic energies of the molecules tend to counter the intermolecular attractive energies that draw the molecules together, producing the surface tension.

11.67 The major source of intermolecular attractive energy in glycerol is hydrogen bonding. As temperature is raised, the increasing kinetic energies of the molecules result in more disruption of hydrogen bonds. There are three OH groups on each glycerol molecule capable of hydrogen bonding. The substance is thus quite viscous, because these hydrogen bonds keep molecules from moving easily with respect to their neighbors. Increasing temperature diminishes the effect of the hydrogen bonding, and thus viscosity decreases.

11.68 The smallest part of the array that reproduces the whole must contain a minimum of one white circle. Further, it must have a ratio of one white circle to three black circles to correspond to the overall ratio of white to black. The unit cell shown can be duplicated in four directions in the plane to reproduce the two-dimensional pattern.

11.69 In CaF_2, 4 CaF_2 units; in CsCl, 1 CsCl unit; in ZnS, 4 ZnS units; in TiO_2, 2 TiO_2 units

11.70 $n\lambda = 2d \sin\theta$; $\sin\theta = n\lambda/2d$

$$\sin\theta = n(1.78)/2(4.06) = 0.219,\ \theta = 12.7°C,;\ 0.438,\ \theta = 26.0°;$$

$$0.658,\ \theta = 41.1°$$

11.71 A most important attractive force would be that due to hydrogen bonding. In addition, there should be strong dipole-dipole interactions that bind a polar molecule such as CH_3OH to the polar surface.

11.72 H_2O_2 is not a great deal more massive than H_2O. Thus, the melting and boiling points are exceptionally high, just as they are for H_2O. The data strongly suggest that hydrogen bonding is very important for determining the properties of H_2O_2. The O-H...O hydrogen bonds extend in a more or less random manner throughout the liquid.

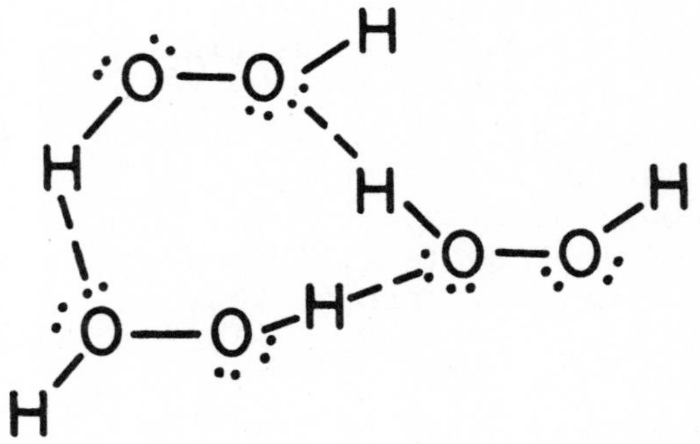

11.73 The critical temperature for any substance is the highest temperature at which it is possible to liquefy that substance. Formation of the liquid from the gaseous state occurs when the intermolecular attractive energies are sufficiently large to overcome the kinetic energies of the molecules. (Application of pressure brings the molecules into more frequent contact, thus promoting formation of the condensed state.) If the intermolecular forces are not large the kinetic energies of the molecules will become sufficiently large at a low temperature to prevent condensation at <u>any</u> pressure. If intermolecular forces are large, the kinetic energies required to prevent condensation are much greater; thus, the critical temperature is high.

11.74 The variation in boiling points reflects the influence of increasing molecular mass, which parallels larger London dispersion forces. The exception is HF, a strongly hydrogen-bonded substance. The heats of vaporization of HCl, HBr and HI reflect increasing London dispersion and dipole-dipole attraction energies. However, ΔH_v for HF is anomalously low. This substance is also hydrogen-bonded in the gas phase at temperatures near the normal boiling point. Thus, it is not necessary to break all hydrogen-bonding interactions to proceed from the liquid to gaseous state. We can imagine "molecules" of $(HF)_2$, $(HF)_3$ and so forth passing from the liquid to gaseous state. In a mole of HF there are thus fewer than a mole of vaporizing "molecules", and ΔH_v is lower than expected.

11.75 All three molecules are planar-trigonal in shape and non-polar. The increasing critical temperature simply reflects increasing intermolecular attractive forces in the series. These are due to London dispersion interactions.

11.76 Cross-linking has the effect of increasing the molecular weight enormously, since long chains are bound together by the cross-links. The chains cannot move past one another as readily, so the material becomes much less flexible, more brittle and higher melting.

11.77 A, solid; B, liquid and gas in equilibrium; C, solid, liquid and gas in equilibrium (the triple point); D, gas.

12

Solutions

12.1 (a) $256\ \text{mL}\left(\frac{1\ \text{L}}{1000\ \text{mL}}\right)\left(\frac{0.358\ \text{mol Ca(NO}_3)_2}{1\ \text{L}}\right) = 0.0916\ \text{mol Ca(NO}_3)_2$

(b) $4.00 \times 10^4\ \text{L}\left(\frac{0.0567\ \text{mol HBr}}{1\ \text{L Soln}}\right) = 2.27 \times 10^3\ \text{mol HBr}$

(c) $\left(\frac{0.565 \times 10^{-2}\ \text{g NaCl}}{1\ \text{g Soln}}\right)(450\ \text{g Soln})\left(\frac{1\ \text{mol NaCl}}{58.5\ \text{g NaCl}}\right)$

$= 4.35 \times 10^{-2}\ \text{mol NaCl}$

12.2 (a) $\left(\frac{16.0\ \text{g CaCl}_2}{4.00 \times 10^2\ \text{mL}}\right)\left(\frac{1\ \text{mol CaCl}_2}{111\ \text{g CaCl}_2}\right)\left(\frac{1000\ \text{mL}}{1\ \text{L}}\right) = 0.360\ M\ \text{CaCl}_2$

(b) $\left(\frac{6.0\ \text{mol H}_3\text{PO}_4}{1\ \text{L}}\right)\left(\frac{0.016\ \text{L}}{0.462\ \text{L}}\right) = 0.21\ M\ \text{H}_3\text{PO}_4$

(c) We must convert mole fraction, X_{HCl}, to weight fraction, then recalculate the number of moles of HCl per liter.

$$\text{wt. fraction HCl} = \frac{0.160(35.5)}{0.160(35.5) + 0.840(18.0)} = 0.273$$

$$\left(\frac{1180\ \text{g Soln}}{1\ \text{L}}\right)\left(\frac{0.273\ \text{g HCl}}{1\ \text{g Soln}}\right)\left(\frac{1\ \text{mol HCl}}{35.5\ \text{g HCl}}\right) = 9.07\ M\ \text{HCl}$$

12.3 (i) $\text{wt. fraction} = \frac{3.6\ \text{g}}{(3.6 + 340)\text{g}} = 0.010$

$$X_{NaCl} = \frac{(3.6\ \text{g}/58.5)}{(3.6/58.5) + (340/18.0)} = 3.2 \times 10^{-3}$$

$$X_{solvent} = 1 - 0.32 \times 10^{-3} = 0.968$$

$$\underline{m} = \left(\frac{3.6\ \text{g NaCl}}{340\ \text{g H}_2\text{O}}\right)\left(\frac{1000\ \text{g H}_2\text{O}}{1\ \text{kg H}_2\text{O}}\right)\left(\frac{1\ \text{mol NaCl}}{58.5\ \text{g NaCl}}\right) = 0.18\ \underline{m}\ \text{NaCl}$$

(ii) wt. fraction $= \dfrac{4.45\ g}{4.45\ g + 564\ g} = 7.81 \times 10^{-3}$

$X_{KBr} = \dfrac{(4.45/119.0)}{(4.45/119.0) + (564\ g/46.1)} = 3.05 \times 10^{-3}$

$X_{solvent} = 1 - 3.05 \times 10^{-3} = 0.997$

$\underline{m} = \left(\dfrac{4.45\ g\ KBr}{564\ g\ C_2H_6O}\right)\left(\dfrac{1000\ g\ C_2H_6O}{1\ kg\ C_2H_6O}\right)\left(\dfrac{1\ mol\ KBr}{119.0\ g\ KBr}\right) = \dfrac{0.065\ mol\ KBr}{1\ kg\ C_2H_6O}$

$= 0.0663\ \underline{m}\ KBr$

(iii) wt. fraction $= \dfrac{35.2\ g}{35.2 + 65.0\ g} = 0.351$

$X_{solute} = \dfrac{(35.2\ g/62.1)}{(35.2/62.1) + (65.0/18.0)} = 0.136$

$X_{solvent} = 1 - 0.136 = 0.864$

$\underline{m} = \left(\dfrac{35.2\ g\ C_2H_6O_2}{65.0\ g\ H_2O}\right)\left(\dfrac{1000\ g\ H_2O}{1\ kg\ H_2O}\right)\left(\dfrac{1\ mol\ C_2H_6O_2}{62.1\ g\ C_2H_6O_2}\right)$

$= 8.72\ mol\ C_2H_6O_2/1\ kg\ H_2O = 8.72\ \underline{m}\ C_2H_6O_2$

(iv) wt. fraction NaCl $= \dfrac{2.4}{2.4 + 3.6 + 82} = 0.027$

wt. fraction KBr $= \dfrac{3.6}{2.4 + 3.6 + 82} = 0.041$

$X_{NaCl} = \dfrac{(2.4/58.5)}{(2.4/58.5) + (3.6/119) + (82/18)} = 8.9 \times 10^{-3}$

$X_{KBr} = \dfrac{3.6/119}{(2.4/58.5) + (3.6/119) + (82/18)} = 6.5 \times 10^{-3}$

$X_{solvent} = \dfrac{(82/18)}{(2.4/58.5) + (3.6/119) + (82/18)} = 0.98$

$\underline{m}_{NaCl} = \left(\dfrac{2.4\ g\ NaCl}{82\ g\ H_2O}\right)\left(\dfrac{1000\ g\ H_2O}{1\ kg\ H_2O}\right)\left(\dfrac{1\ mol\ NaCl}{58.5\ g\ NaCl}\right) = \dfrac{0.50\ mol\ NaCl}{1\ kg\ H_2O}$

$= 0.50\ \underline{m}\ NaCl$

Similarly, $\underline{m}_{KBr} = 0.37\ \underline{m}$ KBr; total molality = 0.87 $\underline{m}$.

<u>12.4</u> (i) wt. fraction $= \dfrac{12.0}{12.0 + 75.0} = 0.138$

$X_{C_3H_6O} = \dfrac{(12.0/58.1)}{(12.0/58.1) + (75.0/18.0)} = 0.0472$

$X_{H_2O} = 1 - 0.0472 = 0.953$

$\underline{m} = \left(\dfrac{12.0\ g\ C_3H_6O}{75.0\ g\ H_2O}\right)\left(\dfrac{1000\ g\ H_2O}{1\ kg\ H_2O}\right)\left(\dfrac{1\ mol\ C_3H_6O}{58.1\ g\ C_3H_6O}\right) = 2.76\ \underline{m}$

(ii) wt. fraction $= \dfrac{3.22\ g}{3.22\ g + 1.46 \times 10^3\ g} = 2.20 \times 10^{-3}$

$X_{La(NO_3)_3} = \dfrac{(3.22/324)}{(3.22/324) + (1.46 \times 10^3/18)} = 1.22 \times 10^{-4}$

$X_{H_2O} = 1 - 1.22 \times 10^{-4} = 0.99988$

$\underline{m} = \left(\dfrac{3.22\ g\ La(NO_3)_3}{1.46 \times 10^3 g\ H_2O}\right)\left(\dfrac{1000\ g\ H_2O}{1\ kg\ H_2O}\right)\left(\dfrac{1\ mol\ La(NO_3)_3}{324\ g\ La(NO_3)_3}\right) = 6.81 \times 10^{-3}\ \underline{m}$

(iii) wt. fraction $= \dfrac{1.26\ g}{1.26\ g + 560\ g} = 2.24 \times 10^{-3}$

$$X_{KBr} = \frac{(1.26/119)}{(1.26/119) + (560/17.0)} = 3.21 \times 10^{-4}$$

$$X_{NH_3} = 1 - 3.2 \times 10^{-4} = 0.99968$$

$$\underline{m} = \left(\frac{1.26\ g\ KBr}{560\ g\ NH_3}\right)\left(\frac{1000\ g\ NH_3}{1\ kg\ NH_3}\right)\left(\frac{1\ mol\ KBr}{119\ g\ KBr}\right) = 1.89 \times 10^{-2}\ \underline{m}$$

(iv) wt. fraction acetone $= \dfrac{16.8}{16.8 + 1.65 + 142} = 0.105$

wt. fraction $Al(NO_3)_3 = 1.65/160 = 1.03 \times 10^{-2}$

$$X_{C_3H_6O} = \frac{(16.8/58.1)}{(16.8/58.1) + (1.65/213) + (142/18.0)} = 0.0353$$

$$X_{Al(NO_3)_3} = (1.65/213)/8.19 = 9.46 \times 10^{-4}$$

$$X_{H_2O} = 1 - (0.0353 + 9.46 \times 10^{-4}) = 0.964$$

$$\underline{m}_{C_3H_6O} = \left(\frac{16.8\ g\ C_3H_6O}{142\ g\ H_2O}\right)\left(\frac{1000\ g\ H_2O}{1\ kg\ H_2O}\right)\left(\frac{1\ mol\ C_3H_6O}{58.1\ g\ C_3H_6O}\right) = 2.04\ \underline{m}$$

$$\underline{m}_{Al(NO_3)_3} = \left(\frac{1.65\ g\ Al(NO_3)_3}{142\ g\ H_2O}\right)\left(\frac{1000\ g\ H_2O}{1\ kg\ H_2O}\right)\left(\frac{1\ mol\ Al(NO_3)_3}{213\ g\ Al(NO_3)_3}\right)$$

$$= 0.0546\ \underline{m}$$

12.5 (a) We need $\left(\dfrac{2.00 \times 10^{-2}\ mol\ KBr}{1\ L}\right)(2.40\ L)\left(\dfrac{119\ g\ KBr}{1\ mol\ KBr}\right) = 5.71\ g\ KBr$
Weigh out this much KBr, dissolve in water, dilute with stirring to a volume of 2.40 L.

(b) We need to determine the weight fraction of KBr.

$$\left(\frac{0.420\ mol\ KBr}{1000\ g\ H_2O}\right)\left(\frac{119\ g\ KBr}{1\ mol\ KBr}\right) = \frac{50.0\ g\ KBr}{1000\ g\ H_2O}$$

Thus, wt. fraction $= \dfrac{50\ g\ KBr}{1000 + 50} = 0.0476$

In 150 g of the 0.420 $\underline{m}$ solution, there are thus

$$(150\ g\ Soln)\left(\frac{0.0476\ g\ KBr}{1\ g\ Soln}\right) = 7.14\ g\ KBr$$

Thus, we weigh out 7.14 g KBr, dissolve it in 150 - 7.14 = 142.86 g H_2O to make exactly 150 g of 0.420 $\underline{m}$ solution.

(c) Since we don't know the density of the KBr solution exactly, we can't determine the quantities needed to give us exactly 2.5 L. However, we can easily make up a slight excess of solution, then take exactly 2.5 L of this. We can guess that the density of the solution will not be much larger than that of pure water; let's say 1.1 kg/L. Then to prepare about 2.6 L of solution, we will need a total of

$$2.6\ L\left(\frac{1.1\ kg}{1\ L}\right)\left(\frac{0.14\ kg\ KBr}{1\ kg\ Soln}\right) = 0.400\ kg\ KBr$$

Weigh out 0.400 kg KBr. To make the solution 0.14 weight fraction in KBr, we need an amount of water given by

$\dfrac{400\ g\ KBr}{400\ g\ KBr + wt.\ H_2O} = 0.14.$ Solve for wt. H_2O: wt. H_2O = 2.46 kg

We dissolve 0.400 kg KBr in 2.46 kg H_2O, then measure out the desired 2.5 L. There is a small excess of solution.

(d) Calculate the number of moles of AgBr to be precipitated.

$$26.0 \text{ g AgBr}\left(\frac{1 \text{ mol AgBr}}{188 \text{ g AgBr}}\right)\left(\frac{1 \text{ mol KBr}}{1 \text{ mol AgBr}}\right) = 0.138 \text{ mol KBr}$$

$$0.138 \text{ mol KBr}\left(\frac{1 \text{ L Soln}}{0.200 \text{ mol KBr}}\right) = 0.690 \text{ L Soln}$$

Thus, we weigh out 0.138 mol KBr, which is 16.4 g, dissolve it in a few hundred mL water, dilute to 0.690 L.

12.6 (a) $$\left(\frac{0.050 \text{ mol La(NO}_3)_3}{1 \text{ kg H}_2\text{O}}\right)(0.100 \text{ kg H}_2\text{O})\left(\frac{324 \text{ g La(NO}_3)_3}{1 \text{ mol La(NO}_3)_3}\right) = 1.62 \text{ g La(NO}_3)_3$$

(b) $$0.0105 = \frac{(\text{mol La(NO}_3)_3)}{\text{mol La(NO}_3)_3 + \text{mol(H}_2\text{O)}}.$$ Let mol H_2O = 100/18.0 = 5.56.

$$0.0105 = \frac{\text{mol La(NO}_3)_3}{\text{mol La(NO}_3)_3 + 5.56}\ ;\ \text{mol La(NO}_3)_3 = 0.0590$$

$$\text{wt. La(NO}_3)_3 = 0.0590 \text{ mol La(NO}_3)_3\left(\frac{324 \text{ g La(NO}_3)_3}{1 \text{ mol La(NO}_3)_3}\right)$$
$$= 19.1 \text{ g La(NO}_3)_3$$

(c) We have that $$\frac{\text{wt. La(NO}_3)_3}{\text{wt. La(NO}_3)_3 + 100 \text{ g H}_2\text{O}} = 0.0126$$

$$\text{wt. La(NO}_3)_3 = 1.28 \text{ g}$$

12.7 (a) $$\text{wt. percentage acetone} = \left(\frac{66.0}{46.0 + 66.0}\right) 100 = 58.9\%$$

$$\text{wt. percentage H}_2\text{O} = 100 - 58.9 = 41.1\%$$

$$\text{mole percentage C}_3\text{H}_6\text{O} = \left(\frac{(66.0/58.1)}{(66.0/58.1) + (46.0/18.0)}\right) 100 = 30.8\%$$

$$\text{mole percentage H}_2\text{O} = 69.2\%$$

(b) $$\left(\frac{0.926 \text{ g Soln}}{1 \text{ mL}}\right)\left(\frac{0.589 \text{ g C}_3\text{H}_6\text{O}}{1 \text{ g Soln}}\right)\left(\frac{1000 \text{ mL}}{1 \text{ L}}\right)\left(\frac{1 \text{ mol C}_3\text{H}_6\text{O}}{58.1 \text{ g C}_3\text{H}_6\text{O}}\right) = 9.39 \text{ M}$$

12.8 (a) $$(1.68 \text{ g C}_7\text{H}_6\text{O}_2)\left(\frac{1 \text{ mol C}_7\text{H}_6\text{O}_2}{122 \text{ g C}_7\text{H}_6\text{O}_2}\right) = 0.0138 \text{ mol C}_7\text{H}_6\text{O}_2$$

$$(206 \text{ mL CCl}_4)\left(\frac{1.59 \text{ g CCl}_4}{1 \text{ mL CCl}_4}\right)\left(\frac{1 \text{ mol CCl}_4}{154 \text{ g CCl}_4}\right) = 2.13 \text{ mol CCl}_4$$

$$(206 \text{ mL C}_2\text{H}_5\text{OH})\left(\frac{0.782 \text{ g C}_2\text{H}_5\text{OH}}{1 \text{ mL C}_2\text{H}_5\text{OH}}\right)\left(\frac{1 \text{ mol C}_2\text{H}_5\text{OH}}{46.1 \text{ g C}_2\text{H}_5\text{OH}}\right) = 3.49 \text{ mol C}_2\text{H}_5\text{OH}$$

$$\text{mole fraction C}_7\text{H}_6\text{O}_2 \text{ in CCl}_4 = \frac{1.38 \times 10^{-2}}{1.38 \times 10^{-2} + 2.13} = 6.45 \times 10^{-3}$$

$$\text{molality in CCl}_4 = \left(\frac{0.0138 \text{ mol C}_7\text{H}_6\text{O}_2}{206 \text{ mL CCl}_4}\right)\left(\frac{1 \text{ mL CCl}_4}{1.59 \text{ g CCl}_4}\right)\left(\frac{1000 \text{ g CCl}_4}{1 \text{ kg CCl}_4}\right)$$
$$= 4.21 \times 10^{-2}\ \underline{\text{m}}$$

$$\text{mole fraction C}_7\text{H}_6\text{O}_2 \text{ in C}_2\text{H}_5\text{OH} = \frac{1.38 \times 10^{-2}}{1.38 \times 10^{-2} + 3.49} = 3.94 \times 10^{-3}$$

$$\text{molality in } C_2H_5OH = \left(\frac{0.0138 \text{ mol } C_7H_6O_2}{206 \text{ mL } C_2H_5OH}\right)\left(\frac{1 \text{ mL } C_2H_5OH}{0.782 \text{ g } C_2H_5OH}\right)$$

$$\times \left(\frac{1000 \text{ g } C_2H_5OH}{1000 \text{ g } C_2H_5OH}\right) = 8.57 \times 10^{-2} \ \underline{m}$$

(b) mass of CCl_4 solution = 1.68 g + 209(1.59) = 330 g

volume of CCl_4 solution = $(330 \text{ g})\left(\frac{1 \text{ mL}}{1.59 \text{ g}}\right)$ = 208 mL

$$M = \frac{0.0138 \text{ mol } C_7H_6O_2}{0.208 \text{ L}} = 0.0663$$

In C_2H_5OH solution, mass of Soln = 1.68 + 161 g = 163 g.

volume = $(163 \text{ g})\left(\frac{1 \text{ mL}}{0.782 \text{ g}}\right)$ = 208 mL; $M = \frac{0.0138 \text{ mol}}{0.208 \text{ L}} = 0.0663$ M

Within the round-off error the molarities are the same, as they should be if we have the same total volume of solution in each case. However, molalities are not the same; they vary with the densities of the solvent when a given quantity of solute is dissolved in each case in the same volume of solvent.

12.9 (a) The normality is twice the molarity, since two equivalents (that is, two times Avogadro's number) of electrons are gained by each mol Sn^{+4}. Thus a 0.36 M solution is 0.72 N; a 0.42 N solution is 0.21 M. (b) Three protons are transferred by H_3PO_4 in the reaction with NaOH. Thus a 0.105 N solution is 0.0350 M.

12.10 (a) N = 3(0.1 M) = 0.3 M; (b) N = 2(0.1 M) = 0.2 N; (c) N = M = 0.1 N

12.11 $\left(\frac{0.031 \text{ g MMT}}{1 \text{ gal gas}}\right)\left(\frac{1 \text{ gal}}{4 \text{ qt}}\right)\left(\frac{1.06 \text{ qt}}{1 \text{ L}}\right)\left(\frac{1 \text{ mol MMT}}{218 \text{ g MMT}}\right) = 3.8 \times 10^{-5}$ M

$$\left(\frac{3.8 \times 10^{-5} \text{mol MMT}}{1 \text{ L Soln}}\right)\left(\frac{1 \text{ L Soln}}{0.78 \text{ kg gas}}\right) = 4.9 \times 10^{-5} \ \underline{m}$$

12.12 (a) ion-dipole; (b) London dispersion force; (c) F—H•••O hydrogen bonding; (d) dipole-dipole (both molecules are quite polar.)

12.13 The dissolving of CsCl in water involves the undoing of the CsCl lattice, in which strong ion-ion attractive energies contribute. Thus, the dissolving process contains the lattice energy of the ionic lattice as an endothermic contribution.

12.14 The energy that is required to separate the K^+ and Br^- ions of KBr in forming a solution must be made up, or nearly all made up, by the ion-solvent interactions. Because CCl_4 is a non-polar molecule, with stable C-Cl bonds, there is no possibility for large, exothermic enthalpy terms corresponding to solvation of either K^+ or Br^- in CCl_4.

12.15 (a) Ethyl alcohol is a polar molecule, with an ability to solvate metal ions much like that of H_2O. It has an oxygen at the negative end of the dipole, as does water; it is capable of intermolecular hydrogen bonding as is water, and so forth. Ethyl alcohol is not as good a solvent for ionic substances as is water, but it does resemble it. (b) The oxygen atoms in dioxane are capable of acting as electron-pair donors in hydrogen bonding to water. No such attractive interaction with H_2O is possible for cyclo-hexane. (c) Ethyl alcohol, though much like water, does consist in large part of a nonpolar portion that is compatible with solute molecules of relatively low polarity. Chloroform, though it has a dipole moment, is not highly polar, and is thus able to mix with substances that are also not very polar.

12.16 Evaporation of $I_2(s)$ and mixing of $Br_2(\ell)$ and $CCl_4(\ell)$ are processes which are either endothermic, (a), or proceed with $\Delta H \sim 0$, (b). They proceed because the system moves spontaneously toward states of larger uncertainty or disorder. Processes (c) and (d) both lead to more ordering, not less. These processes proceed because they result in a lower energy for the system.

12.17 The formation of ice from liquid water at a temperature below 0°C; condensation of a vapor into a liquid; alignment of iron filings in a magnetic field.

12.18 Ethyl alcohol is miscible because it is capable of hydrogen-bonding interactions just as is water. Ethyl cyanide has a large dipole moment (note that it should be like CH_3CN, Table 11.6), and thus can experience dipole-dipole, as well as hydrogen-bonding interactions with water. Ethyl chloride is much less polar than ethyl cyanide, and there are no hydrogen-bonding interactions of importance. In CH_3CH_2SH a weak S—H•••O hydrogen bonding could contribute to the intermolecular attraction with water.

12.19 (a) Methyl alcohol is completely miscible with water and gasoline, and thus could be expected to help dissolve water that forms in the gas line. (b) A water-repellent substance will be non-polar and incapable of hydrogen bonding; as little like water as possible. The solvent that will dissolve such a substance must itself be unlike water; non-polar and non-hydrogen bonding. Methylene chloride, CH_2Cl_2 satisfies this criterion. (c) Waxes are high molecular weight organic substances. A solvent that will dissolve them must be capable of moderately strong intermolecular attraction forces, of the London dispersion type, such as the aromatic hydrocarbons.

12.20 (a) miscible-hydrogen bonding interactions; (b) nonmiscible - the outer periphery of $C(CH_2OH)_4$ is dominated by the OH groups that will engage in hydrogen bonding. There is no intermolecular attractive force possible between $C(CH_2OH)_4$ and C_7H_{16} that can compare with the intermolecular attractive energies between $C(CH_2OH)_4$ molecules. (c) nonmiscible - Molten NaI is dominated by ion-ion attractions, whereas in liquid mercury metallic bonding prevails. There are no possibilities for attractive interactions of comparable magnitude between $Hg(\ell)$ and $NaI(\ell)$. (d) miscible. Both molecules are slightly polar, so the intermolecular attractions within each liquid are expected to be a mixture of dipole-dipole and London dispersion force.

12.21 Henry's law states that the solubility of a gas in a liquid is directly proportional to the pressure of the gas over the solution. The Henry's law constant for oxygen is larger than that for nitrogen, since for the same gas pressure O_2 is more soluble than N_2.

12.22 The value of solubility for O_2 at 1 atm pressure can be taken as the Henry's law constant 1.38×10^{-3} M/atm.

$$C_{O_2} = (13.8 \times 10^{-3}\ \text{M/atm})(0.209\ \text{atm}) = 2.88 \times 10^{-4}\ \text{M}$$

12.23 $C_{CO_2} = (0.0414\ \text{M/atm})(3.30 \times 10^{-4}\ \text{atm}) = 1.37 \times 10^{-5}\ \text{M}$

12.24 $C_{He} = (3.7 \times 10^{-4}\ \text{M/atm})(0.5\ \text{atm}) = 1.8 \times 10^{-4}\ \text{M}$

$$C_{N_2} = (6.0 \times 10^{-4}\ \text{M/atm})(0.5\ \text{atm}) = 3.0 \times 10^{-4}\ \text{M}$$

12.25 The Henry's law constant is just the solubility when the external gas pressure is 1 atm. For CO the constants at 20°C and 40°C are 1.05×10^{-3} M/atm and 7.8×10^{-4} M/atm, respectively. For CH_4 at 20°C and

40°C, the constants are 1.48 x 10^{-3} M/atm and 1.06 x 10^{-3} M/atm, respectively.

12.26 The solubility of a gas is determined in part by the attractive forces between the gas and solvent molecules. At higher temperatures the increased kinetic energies of the molecules counter the effects of these attractive interactions.

12.27 A process that is exothermic proceeds in the forward direction when the system is subjected to a lower temperature; under such conditions the equilibrium shifts to provide more heat. Thus, we predict that $CuCl_2$ will be more soluble at lower temperature. However, we must be careful about such predictions because the data available are not always adequate. See G. Bodner, J. Chem. Educ., 57, p. 117, 1980.

12.28 (a) nonelectrolyte; (b) nonelectrolyte; (c) weak electrolyte; (d) strong electrolyte; (e) strong electrolyte; (f) weak electrolyte (This compound, methylamine, is a close relative to ammonia, NH_3.); (g) weak electrolyte; (h) nonelectrolyte; (i) strong electrolyte. The acids are HF, H_2SO_4, HClO. The bases are CH_3NH_2 and LiOH. $Ba(NO_3)_2$ is the only salt.

12.29 The $HgCl_2$ solution is a relatively poor conductor of electricity. The origin of this effect is the incomplete dissociation of dissolved $HgCl_2$ into ions. That is, the equilibrium $HgCl_2(aq) \rightleftarrows HgCl^+(aq) + Cl^-(aq)$ does not go completely to the right.

12.30 (a) molecular form; (b) As Ca^{2+} and Cl^- ions; (c) As Ca^{2+} and Cl^- ions; (The difference between (b) and (c) is that in molten NaCl the Ca^{2+} ions are surrounded by Cl^- ions; in H_2O, they are surrounded by H_2O molecules, as in Figure 12.2. (d) As separated Na^+ and OH^- ions, each solvated (hydrated) by H_2O molecules. (e) As the hydrated H^+ ion, which we write as $H^+(aq)$ and the separated ClO_4^- ion.

12.31 (a) 0.2 M (0.1 M Na^+, 0.1 M OH^-); (b) 1.05 M (0.35 M Ca^{2+} + 2(0.35 M) Br^-); (c) 0.14 M; (d) 0.40 M (0.050 M K^+ + 0.050 M ClO_3^- + 0.20 M Na^+ + 0.10 M SO_4^{2-}).

12.32 1. A precipitate is formed, e.g., $Ag^+(aq) + Br^-(aq) \rightarrow AgBr(s)$. 2. A gas is formed, e.g., $CO_3^{2-}(aq) + 2H^+(aq) \rightarrow H_2CO_3(aq) \rightarrow H_2O(\ell) + CO_2(g)$. 3. A nonelectrolyte is formed, e.g., $H^+(aq) + OH^-(aq) \rightarrow H_2O(\ell)$, or $H^+(aq) + NH_2^-(aq) \rightarrow NH_3(aq)$. 4. A weak electrolyte is formed, e.g., $H^+(aq) + ClO^-(aq) \rightarrow HClO(aq)$.

12.33 (a) $K^+(aq) + OH^-(aq) + H^+(aq) + ClO_4^-(aq) \rightarrow H_2O(\ell) + K^+(aq) + ClO_4^-(aq)$

net: $OH^-(aq) + H^+(aq) \rightarrow H_2O(\ell)$

(b) $Mg^{2+}(aq) + 2Cl^-(aq) + 2K^+(aq) + 2OH^-(aq) \rightarrow Mg(OH)_2(s) + 2K^+(aq) + 2Cl^-(aq)$

net: $Mg^{2+}(aq) + 2OH^-(aq) \rightarrow Mg(OH)_2(s)$

(c) $BaCO_3(s) + 2H^+(aq) + 2ClO_4^-(aq) \rightarrow Ba^{2+}(aq) + 2ClO_4^-(aq) + H_2O(\ell) + CO_2(g)$

net: $BaCO_3(s) + 2H^+(aq) \rightarrow H_2O(\ell) + CO_2(g) + Ba^{2+}(aq)$

12.34 (a) $Ba^{2+}(aq) + 2Cl^-(aq) + 2H^+(aq) + SO_4^{2-}(aq) \rightarrow BaSO_4(s) + 2H^+(aq) + 2Cl^-(aq)$

net: $Ba^{2+}(aq) + SO_4^{2-}(aq) \rightarrow BaSO_4(s)$

(b) $2Ag^+(aq) + 2NO_3^-(aq) + 2Na^+(aq) + CrO_4^{2-}(aq) \rightarrow Ag_2CrO_4(s) + 2Na^+(aq) + 2NO_3^-(aq)$

net: $2Ag^+(aq) + CrO_4^{2-}(aq) \rightarrow Ag_2CrO_4(s)$

(c) $Ag_2S(s) + 2H^+(aq) + 2Cl^-(aq) \rightarrow 2AgCl(s) + H_2S(g)$

net: $Ag_2S(s) + 2H^+(aq) + 2Cl^-(aq) \rightarrow H_2S(g) + 2AgCl(s)$

(d) $2K^+(aq) + SO_3^{2-}(aq) + 2H^+(aq) + SO_4^{2-}(aq) \rightarrow 2K^+(aq) + SO_4^{2-}(aq) + H_2O(\ell) + SO_2(g)$

net: $SO_3^{2-}(aq) + 2H^+(aq) \rightarrow H_2O(\ell) + SO_2(g)$

(e) $K^+(aq) + OH^-(aq) + H^+(aq) + NO_3^-(aq) \rightarrow K^+(aq) + NO_3^-(aq) + H_2O(\ell)$

net: $OH^-(aq) + H^+(aq) \rightarrow H_2O\ (\ell)$

12.35 (a) $BaSO_3(s) + 2HBr(aq) \rightarrow BaBr_2(aq) + H_2O(\ell) + SO_2(g)$

$BaSO_3(s) + 2H^+(aq) + 2Br^-(aq) \rightarrow Ba^{2+}(aq) + 2Br^-(aq) + H_2O(\ell) + SO_2(g)$

$BaSO_3(s) + 2H^+(aq) \rightarrow Ba^{2+}(aq) + H_2O(\ell) + SO_2(g)$

(b) $Cd(NH_2)_2(s) + 2HCl(aq) \rightarrow CdCl_2(aq) + 2NH_3(aq)$

$Cd(NH_2)_2(s) + 2H^+(aq) + 2Cl^-(aq) \rightarrow Cd^{2+}(aq) + 2Cl^-(aq) + 2NH_3(aq)$

$Cd(NH_2)_2(s) + 2H^+(aq) \rightarrow Cd^{2+}(aq) + 2NH_3(aq)$

(c) $Mn(OH)_2(s) + 2HClO_4(aq) \rightarrow Mn(ClO_4)_2(aq) + 2H_2O(\ell)$

$Mn(OH)_2(s) + 2H^+(aq) + 2ClO_4^-(aq) \rightarrow Mn^{2+}(aq) + 2ClO_4^- + 2H_2O(\ell)$

$Mn(OH)_2(s) + 2H^+(aq) \rightarrow Mn^{2+}(aq) + 2H_2O\ (\ell)$

12.36 The two precipitates formed are due to AgCl(s) and $SrSO_4(s)$. The fact that no precipitate forms on addition of chromate ion indicates that the other two possibilities, Ni^{2+} and Mn^{2+}, are absent.

12.37 Addition of acid (aqueous HBr) should cause dissolution of SrCO with evolution of CO_2, $MnSO_3$ with evolution of SO_2, and CaF_2 with formation of aqueous HF, a moderately weak electrolyte.

12.38 Properties (c) and (e) are colligative properties; they change in accordance with the concentration of solute particles, not according to their particular identities. Note especially that boiling point, (b), is not a colligative property, but boiling-point elevation is one.

12.39 The vapor pressure of a volatile liquid is lowered in proportion to the mole fraction of nonvolatile solute. The vapor-pressure lowering is not determined by the particular chemical nature of the nonvolatile solute, but only by the total number of solute particles. This is what is meant by the term colligative property.

12.40 We first calculate the number of moles of sucrose in 120 g:

$$(120\ g\ C_{12}H_{22}O_{11})\left(\frac{1\ mol\ C_{12}H_{22}O_{11}}{342\ g\ C_{12}H_{22}O_{11}}\right) = 0.351\ mol\ C_{12}H_{22}O_{11}$$

We require 0.351 mol of solute particles in each of the other solutes.

(a) $$(0.351\ mol\ solute\ particles)\left(\frac{1\ mol\ KBr}{2\ mol\ solute\ particles}\right)\left(\frac{119\ g\ KBr}{1\ mol\ KBr}\right) = 20.9\ g\ KBr$$

Recall that KBr ionizes to form $K^+(aq) + Br^-(aq)$.

(b) $(0.351 \text{ mol solute particles})\left(\frac{1 \text{ mol } SrCl_2}{3 \text{ mol solute particles}}\right)\left(\frac{159 \text{ g } SrCl_2}{1 \text{ mol } SrCl_2}\right)$

$= 18.6 \text{ g } SrCl_2$

(c) $(0.351 \text{ mol solute particles})\left(\frac{1 \text{ mol HI}}{2 \text{ mol solute particles}}\right)\left(\frac{128 \text{ g HI}}{1 \text{ mol HI}}\right)$

$= 22.4 \text{ g HI}$

12.41 We first calculate the mole fraction of water in the solution:

$$X_{H_2O} = \frac{(148 \text{ g } H_2O)\left(\frac{1 \text{ mol } H_2O}{18.0 \text{ g } H_2O}\right)}{(148 \text{ g } H_2O)\left(\frac{1 \text{ mol } H_2O}{18.0 \text{ g } H_2O}\right) + (465 \text{ g } C_3H_9O_3)\left(\frac{1 \text{ mol } C_3H_9O_3}{92.1 \text{ g } C_3H_9O_3}\right)} = 0.620$$

From Appendix C, $P^\circ_{H_2O}(60^\circ) = 149.4$ mm Hg. Thus, $P_{H_2O} = (0.620)(149.4)$

$= 92.6$ mm Hg

12.42 At 24°C, $P^\circ_{H_2O} = 22.4$ mm Hg

$P_{H_2O} = X_{H_2O}\, P^\circ_{H_2O}$; thus, $X_{H_2O} = P_{H_2O}/P^\circ_{H_2O} = 14.5/22.4 = 0.647$

Thus, we want 1 kg of a solution of glycerin in water in which X_{H_2O} is 0.647. We need to calculate the mass fraction of H_2O that corresponds to this mole ratio. We know that in a single mol of mixture we have

$$(0.647 \text{ mol } H_2O)\left(\frac{18.0 \text{ g } H_2O}{1 \text{ mol } H_2O}\right) = 11.6 \text{ g } H_2O$$

$$(0.353 \text{ mol } C_3H_9O_3)\left(\frac{92.1 \text{ g } C_3H_9O_3}{1 \text{ mol } C_3H_9O_3}\right) = 32.5 \text{ g } C_3H_9O_3$$

The mass fraction of H_2O is thus $11.6/(11.6 + 32.5) = 0.263$. To prepare 1 kg of the desired solution we thus want 263 g H_2O and 737 g $C_3H_9O_3$.

12.43 The mole fraction of methanol desired is $X_{CH_3OH} = P_{CH_3OH}/P^\circ_{CH_3OH}$ $= 15/77 = 0.085$. In 48 kg of the paraffin, there are

$$4.8 \times 10^4 \left(\frac{1 \text{ mol } C_{24}H_{50}}{338 \text{ g } C_{24}H_{50}}\right) = 142 \text{ mol } C_{24}H_{50}$$

Let the number of moles of CH_3OH required = y. Then,

$$0.085 = \frac{y}{y + 142} \qquad y = 13$$

$$13 \text{ mol } CH_3OH \left(\frac{32 \text{ g } CH_3OH}{1 \text{ mol } CH_3OH}\right) = 420 \text{ g } CH_3OH$$

12.44 In order of increasing boiling point the solutions are 0.030 m glycerin < 0.030 m benzoic acid < 0.020 m KBr < 0.018 m $MgCl_2$.

12.45 (a) $(1.2 \text{ m Soln})\left(\frac{1.22^\circ C}{m}\right) + 78.4 = 79.9^\circ C$ b.p.

$(1.2 \text{ m Soln})\left(\frac{-1.99^\circ C}{m}\right) + (-114.6^\circ) = -117.0^\circ C$ f.p.

(b) $$\frac{3.5 \text{ g } CCl_4}{128 \text{ g } C_6H_6}\left(\frac{1 \text{ mol } CCl_4}{154 \text{ g } CCl_4}\right)\left(\frac{1000 \text{ g } C_6H_6}{1 \text{ kg } C_6H_6}\right) = 0.177 \text{ \underline{m}}$$

$$(0.177 \text{ \underline{m} Soln})\left(\frac{2.53°C}{\text{m}}\right) + 80.1°C = 80.5°C \text{ b.p.}$$

$$(0.177 \text{ \underline{m} Soln})\left(\frac{-5.12°C}{\text{m}}\right) + 5.5°C = 4.6°C \text{ f.p.}$$

(c) $$\left(\frac{1.4 \text{ g } KNO_3}{33.6 \text{ g } H_2O}\right)\left(\frac{1 \text{ mol } KNO_3}{101 \text{ g } KNO_3}\right)\left(\frac{2 \text{ mol particles}}{1 \text{ mol } KNO_3}\right)\left(\frac{1000 \text{ g } H_2O}{1 \text{ kg } H_2O}\right) = 0.82 \text{ \underline{m}}$$

$$(0.82 \text{ \underline{m} Soln})\left(\frac{0.52°C}{1 \text{ \underline{m} Soln}}\right) + 100.0°C = 100.43°C \text{ b.p.}$$

$$(0.82 \text{ \underline{m} Soln})\left(\frac{-1.86°C}{1 \text{ \underline{m} Soln}}\right) + 0.0°C = -1.5°C \text{ f.p.}$$

(d) $$\left(\frac{1.86 \text{ g } Li_2CrO_4}{60.0 \text{ g } H_2O}\right)\left(\frac{1 \text{ mol } Li_2CrO_4}{130 \text{ g } Li_2CrO_4}\right)\left(\frac{3 \text{ mol particles}}{1 \text{ mol } Li_2CrO_4}\right)\left(\frac{1000 \text{ g } H_2O}{1 \text{ kg } H_2O}\right)$$
$$= 0.715 \text{ \underline{m}}$$

(Keep in mind that the molality we calculate is the total molality of the dissolved particles.)

$$(0.715 \text{ \underline{m} Soln})\left(\frac{0.52°C}{1 \text{ \underline{m} Soln}}\right) + 100.0°C = 100.37°C \text{ b.p.}$$

$$(0.715 \text{ \underline{m} Soln})\left(\frac{-1.86°C}{1 \text{ \underline{m} Soln}}\right) + 0.0°C = -1.33°C \text{ f.p.}$$

12.46 We have 1.50 mol $C_3H_6O_3$ per 1000 g H_2O. In 1000 g H_2O there are 1000/18.0 = 55.5 mol H_2O. Thus, the mole fraction of H_2O in the solution is

$$X_{H_2O} = \frac{55.5}{55.5 + 1.50} = 0.974$$

Then, $P_{H_2O} = 0.974(149.4) = 146$ mm Hg.

12.47 $$\pi = MRT \;;\; M = \pi/RT = \frac{2.86 \text{ atm}}{\left(\frac{0.0821 \text{ L-atm}}{\text{mol-K}}\right)(298 \text{ K})} = 0.117 \text{ M}$$

We can assume as an approximation that molarity and molality are the same. We have

$$\frac{8.0 \text{ g sugar}}{200 \text{ g } H_2O}\left(\frac{1000 \text{ g } H_2O}{1 \text{ kg } H_2O}\right) = 40 \text{ g sugar/kg } H_2O$$

Thus, 40 g sugar = 0.117 mol sugar; MW = 340 g/mol

12.48 $$M = \frac{57.1 \text{ mm Hg}}{\left(\frac{0.0821 \text{ L-atm}}{\text{mol-K}}\right)(298 \text{ K})}\left(\frac{1 \text{ atm}}{760 \text{ mm Hg}}\right) = 3.07 \times 10^{-3} \text{ M}$$

$$\left(\frac{0.036 \text{ g Solute}}{100 \text{ g } H_2O}\right)\left(\frac{1000 \text{ g } H_2O}{1 \text{ kg } H_2O}\right) = 0.36 \text{ g solute/kg } H_2O$$

Assuming molarity and molality are the same in this dilute solution, we can then say 0.36 g solute = 3.07×10^{-3} mol ; MW = 117 g/mol. Because the salt is completely ionized, the formula weight of the lithium salt is twice this calculated value, or 234 g/mol. The organic portion, $C_nH_{2n-1}O_2^-$, has a formula weight of 234 - 7 = 227 g. Subtracting 32 for the oxygens, and adding 1 to make the formula C_nH_{2n}, we have C_nH_{2n}, MW = 196. Since each CH_2 unit has a mass of 14, n = 14. Thus the formula for our salt is $LiC_{14}H_{27}O_2$.

12.49 (a) hydrophobic; (b) hydrophilic; (c) hydrophobic; (d) hydrophilic

12.50 The polar end of the stearate ion would become involved in hydrogen-bonding to the OH groups of the glucose; the non-polar hydrocarbon ends would project into the benzene solvent. Something like the sketch on the right, in which just one of the hydrogen bond interactions is shown.

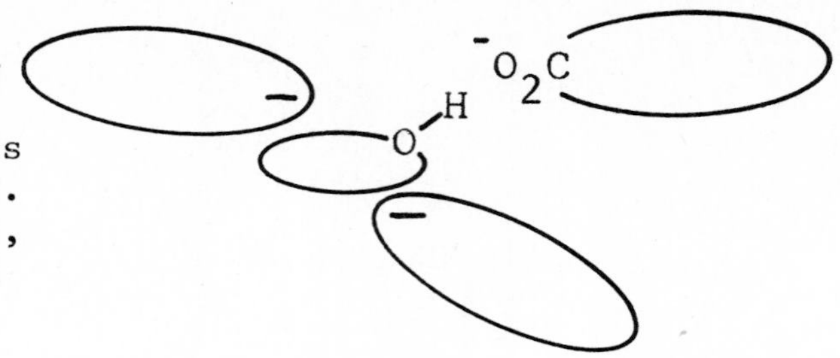

12.51 The polar ends of the soap molecules would be directed inward, with the long, non-polar hydrocarbon chains projecting into the oil.

12.52 To remain dispersed the gold particles must be extremely small, so that their kinetic energies of motion can be affected by collisions with solvent molecules. Under these conditions, the gravitational force will not be large relative to kinetic energies, and the particles will remain in suspension. Secondly, the particles must acquire a common electrical charge, by adsorption of ions, so that they are repelled by one another. This keeps them from aggregating to form larger particles that could settle out.

12.53 If the dirt particles have hydrophobic character, the detergent molecules adsorb onto the dirt so that their non-polar ends are bound to the particle, while the polar end projects into the water. The particle can thus be suspended in water by the action of the detergent in forming a kind of interface to the water, as illustrated in Figure 12.20. The detergent solubilizes oils in a similar way; the soil is covered with detergent molecules that project a non-polar end into the oil, and a polar end into the water phase.

12.54 $$\left(\frac{18 \text{ mol } H_2SO_4}{1 \text{ L Soln}}\right)\left(\frac{98 \text{ g } H_2SO_4}{1 \text{ mol } H_2SO_4}\right) = 1.76 \text{ kg } H_2SO_4/\text{L}$$

There are 1.84 kg of H_2SO_4 plus H_2O per liter; thus, the mass of water is 1.84 kg - 1.76 kg = 0.08 kg. wt. percentage H_2SO_4 = (1.76/1.84)100 = 96%. The number of moles of H_2O per liter of solution is (80 g H_2O)(1 mol H_2O/18 g H_2O) = 4.4 mol H_2O. Mole fraction H_2SO_4 = 18/(18 + 4.4) = 0.80.

12.55 (a) $$\left(\frac{12.0 \text{ g } H_2SO_4}{0.600 \text{ L Soln}}\right)\left(\frac{1 \text{ mol } H_2SO_4}{98.1 \text{ g } H_2SO_4}\right) = 0.204 \text{ M}$$

(b) $$2.0 \text{ N } H_2SO_4 \left(\frac{1 \text{ mol } H_2SO_4}{2 \text{ equivalents } H_2SO_4}\right) = 1.0 \text{ M } H_2SO_4$$

(c) $$3.0 \text{ M } H_2SO_4 \left(\frac{8.0 \text{ mL}}{35.0 \text{ mL}}\right) = 0.69 \text{ M}$$

12.56 $$(1.10 \times 10^9 \text{ gal gas})\left(\frac{0.060 \text{ gal toluene}}{1 \text{ gal gas}}\right)\left(\frac{4 \text{ qt}}{1 \text{ gal}}\right)\left(\frac{1 \text{ L}}{1.06 \text{ qt}}\right)$$

$$\times \left(\frac{0.87 \text{ kg toluene}}{1 \text{ L}}\right) = 2.2 \times 10^8 \text{ kg toluene}$$

12.57 (a) Caffeine contains several nitrogens, each of which has an unshared electron pair (these are conventionally not shown in the structure). Each of these lone pairs can engage in hydrogen bonding to the water: H—O—H•••N. The carbonyl (C=O) group can also engage in hydrogen bonding to water. The caffiene molecule as a whole is polar, so there is probably some attractive interaction with water resulting from dipole-dipole interactions. (b) From the fact that the solubility of caffeine increases with increasing temperature, we can infer that ΔH of solution is positive; that is, the dissolving of caffeine is an endothermic process.

12.58 The most important aspect of this experiment is to ensure that equilibrium is attained. Since most solution processes are endothermic (that is, most salts are more soluble at higher temperature), it would be useful to stir an excess of KNO_3 with ethyl alcohol at a temperature above 25°C for some time, then reduce the temperature to 25°C. Continue stirring of the solution in contact with solid KNO_3 for some time, say 24 hours. Then allow the solution to stand until all solids have settled. Withdraw a known volume of the solution and measure the amount of KNO_3 in it. Resume stirring at 25°C for another 24 hours and repeat the measurement. If the quantity of KNO_3 in the known volume withdrawn remains the same, equilibrium has probably been attained. If it differs, stirring should be continued for additional periods until the concentration of KNO_3 becomes constant with time.

The experiment as described could be carried out at another temperature, say 35°C. If the solubility is higher at the higher temperature, the solution process is endothermic; if it is lower at the higher temperature, the solution process is exothermic.

12.59 The Ca^{2+} ion, because it carries a 2+ charge, and has the smallest ionic radius (Table 7.2), is the most strongly hydrated of these ions.

12.60 (a) A decrease in randomness or disorder results from the formation of an ordered ionic lattice. (b) The disorder increases as the CCl_4 molecules expand into the larger volume of the gas state. (c) Melting causes the sulfur molecules to move with respect to one another in a random fashion. It thus results in an increased disorder or randomness.

12.61 The fact that solubility decreases with increasing temperature indicates that the solution process is exothermic. The reverse process, the loss of Rn from the solution to the gas phase, is endothermic. According to LeChatelier's principle, this endothermic process is favored by raising the temperature.

12.62 We make use of Dalton's Law of Partial Pressures, which states that the partial pressure of a gas is proportional to its mole fraction:

$$80 \text{ atm gas} \left(\frac{3.5 \times 10^{-6} \text{ atm}}{1 \text{ atm gas}}\right)\left(\frac{7.27 \times 10^{-3} \text{ M}}{1 \text{ atm Rn}}\right) = 2.0 \times 10^{-6} \text{ M}$$

12.63 The outer periphery of the BHT molecule is mostly hydrocarbon-like groups, such as $-CH_3$. The lone OH group is rather buried inside, and probably does little to enhance solubility in water. Thus, BHT accumulates in fats, and has a fairly long residence time in the body.

12.64 The equilibrium between a solid ionic substance and its solution does not involve much of a change in volume. For example, the total volume of 1 L of pure water plus 0.2 mol KBr is not very much different from the solution formed by mixing these two components. According to LeChatelier's principle, upon application of increased pressure the equilibirum will shift in a direction that results in a smaller volume, since this would counter the increased pressure. However, if the volume change is small, the shift in equilibrium will also be small.

12.65 (a) Molarity measures moles of solute per liter of solution; molality measures moles of solute per 1 kg of solvent. (b) A solution represents a distribution of one substance within another that is homogeneous on the molecular level. An emulsion is a distribution of tiny droplets of one liquid within another. The droplets are much larger than molecular size. (c) A hydrophilic substance is one that gives rise to relatively strong adhesive energy with water. A hydrophobic substance exhibits little or no attractive interaction with water. (d) A saturated solution contains the maximum possible equilibrium concentration of a solute. A solution that is supersaturated contains more than this maximum equilibrium concentration; it is not at equilibrium. (e) An electrolyte is a substance which, when dissolved in a polar solvent (we nearly always mean water), produces ions in the solution. A nonelectrolyte is a solute that produces no ions, e.g., sugar or ethyl alcohol. (f) A strong acid is essentially completely ionized in solution. A weak acid is only partially ionized in solution. (g) Osmosis refers to the passage of solvent molecules through a semi-permeable membrane which does not allow passage of solutes such as electrolytes. Dialysis involves use of semi-permeable membranes that allow passage of ionic solutes consisting of small ions, but not of large molecules or colloids.

12.66 $Cu(CN)_2(s) + 2H^+(aq) \rightarrow Cu^{2+}(aq) + 2HCN(aq)$

12.67 (a) $Mn^{2+}(aq) + 2OH^-(aq) \rightarrow Mn(OH)_2(s)$

(b) $CaCO_3(s) + 2H^+(aq) \rightarrow Ca^{2+}(aq) + H_2O(\ell) + CO_2(g)$

(c) $Ba^{2+} + 2ClO_3^-(aq) + 2H^+(aq) + SO_4^{2-}(aq) \rightarrow$
$BaSO_4(s) + 2H^+(aq) + 2ClO_3^-(aq)$. (This, of course, is the full ionic equation; the net ionic equation does not show the chloric acid at all!)

$Ba^{2+}(aq) + SO_4^{2-}(aq) \rightarrow BaSO_4(s)$

12.68 First calculate molarity.

$$\left(\frac{0.64\text{ g } C_9H_8O_4}{0.200\text{ L Soln}}\right)\left(\frac{1\text{ mol } C_9H_8O_4}{180\text{ g } C_9H_8O_4}\right) = 0.0178\text{ M}$$

$$\pi = \left(\frac{0.0178\text{ mol}}{\text{L}}\right)\left(\frac{0.0821\text{ L-atm}}{\text{mol-K}}\right)(298\text{ K}) = 0.43\text{ atm}$$

12.69 These salts dissolve in water at the surface of the ice to form concentrated solutions of electrolytes, with freezing points lowered substantially from that for pure water.

12.70
$$\left(\frac{163\text{ g } C_3H_9O_3}{3.46\text{ kg } H_2O}\right)\left(\frac{1\text{ mol } C_3H_9O_3}{92.1\text{ g } C_3H_9O_3}\right) = 0.511\text{ mol } C_3H_9O_3/\text{kg } H_2O$$

$$\text{freezing point} = 0.0°\text{C} + 0.511\ \underline{m}\left(\frac{-1.86°\text{C}}{1\ \underline{m}}\right) = -0.95°\text{C}$$

$$\text{boiling point} = 100.0°\text{C} + 0.511\ \underline{m}\left(\frac{0.52°\text{C}}{1\ \underline{m}}\right) = 100.26°\text{C}$$

12.71 Arctic ocean, molality $= -1.98°\text{C}\left(\frac{1\ \underline{m}}{-1.86°\text{C}}\right) = 1.06\ \underline{m}$

Atlantic ocean, molality $= -2.08°\text{C}\left(\frac{1\ \underline{m}}{-1.86°\text{C}}\right) = 1.11\ \underline{m}$

12.72 $\left(\frac{0.125 \text{ g S}}{3.62 \text{ g camphor}}\right)\left(\frac{1000 \text{ g camphor}}{1 \text{ kg camphor}}\right) = 34.5 \text{ g S/1 kg camphor}$

$5.08°C(1 \text{ m Soln}/37.7°C) = 0.135 \text{ mol S/1 kg camphor}$

$34.5 \text{ g S} = 0.135 \text{ mol S}$ MW = 256 g/mol

This molecular weight corresponds to molecular formula S_8.

12.73 One would expect the alcohol to accumulate at the surface of water, with the hydrocarbon tails pointed outward, and the polar, OH-containing end down into water and hydrogen bonding with it. Alternatively, the cetyl alcohol molecules can form spherical clusters in the water in which the non-polar ends are together on the inside and the polar ends of the molecules pointed outward, and forming hydrogen bonds with water. This occurs at higher concentrations.

12.74 $\left(\frac{0.64 \text{ g A}}{36.0 \text{ g } CCl_4}\right)\left(\frac{1000 \text{ g } CCl_4}{1 \text{ kg } CCl_4}\right) = 17.8 \text{ g A/kg } CCl_4$

$0.49°C \left(\frac{1 \text{ m Soln}}{5.02°C}\right) = 0.098 \text{ m}$

$0.098 \text{ mol A} = 17.8 \text{ g A}$; $MW = 1.8 \times 10^2$

Empirical formula:

$(0.590 \text{ g C})\left(\frac{1 \text{ mol C}}{12.0 \text{ g C}}\right) = 0.0491 \text{ mol C}$

$(0.262 \text{ g O})\left(\frac{1 \text{ mol O}}{16.0 \text{ g O}}\right) = 0.0164 \text{ mol O}$

$(0.071 \text{ g H})\left(\frac{1 \text{ mol H}}{1.0 \text{ g H}}\right) = 0.071 \text{ mol H}$

$0.076 \text{ g N} \left(\frac{1 \text{ mol N}}{14.0 \text{ g N}}\right) = 0.0054 \text{ mol N}$

The elements are present in mole ratios 1N, 9C, 3O, 13H. This gives us $C_9H_{13}O_3N$, for which the formula wt. is 183. Thus, this appears to also be the molecular weight.

13

Rates of chemical reactions

13.1

Time (min)	Time Interval (min)	Concentration (M)	Conc. Change	Rate (M/min)
0		1.85		
	79		0.18	2.3×10^{-3}
79		1.67		
	79		0.15	1.9×10^{-3}
158		1.52		
	158		0.22	1.4×10^{-3}
316		1.30		
	316		0.30	0.95×10^{-3}
632		1.00		

13.2 From the slopes of the three tangents (figure at right) we obtain the following rates:

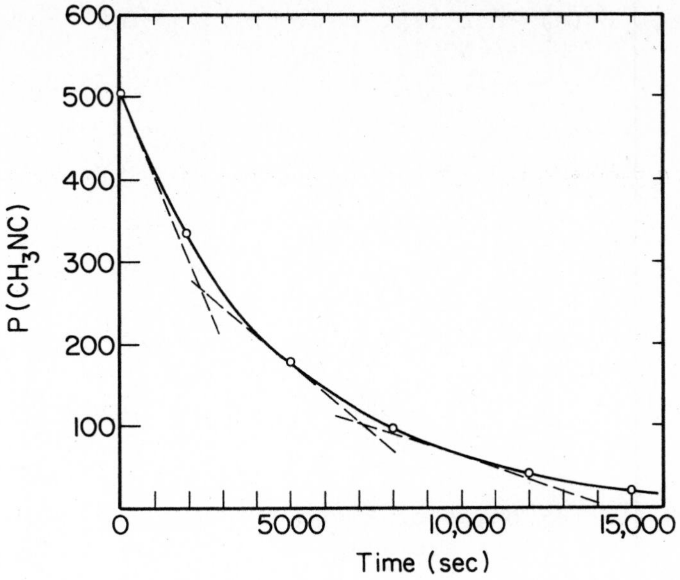

t (sec)	Rate (mm/sec)	Rate Const. (sec^{-1})
0	0.101	2.0×10^{-4}
5,000	0.0348	1.9×10^{-4}
10,000	0.0138	2.1×10^{-4}

For a first-order process, we have that $k = -\Delta[P]/\Delta t/[CH_3NC]$, so to obtain k we divide the rate (that is, the slope) by the concentration of CH_3NC at that time. The concentration is just the pressure. We have the pressures at times 0 and 5,000 sec from the table. We estimate $P(CH_3NC)$ at 10,000 sec from our graph as being 66 mm. The first-order rate constants are reasonably constant. The variations in value come from errors in graphing and estimating the slopes of the tangents. Two significant figures is as many as we can claim by these procedures.

13.3 From the slopes of the lines in the figure at right, the rates are 1.9×10^{-3} M/min at t = 0, 9.4×10^{-4} M/min at t = 500.

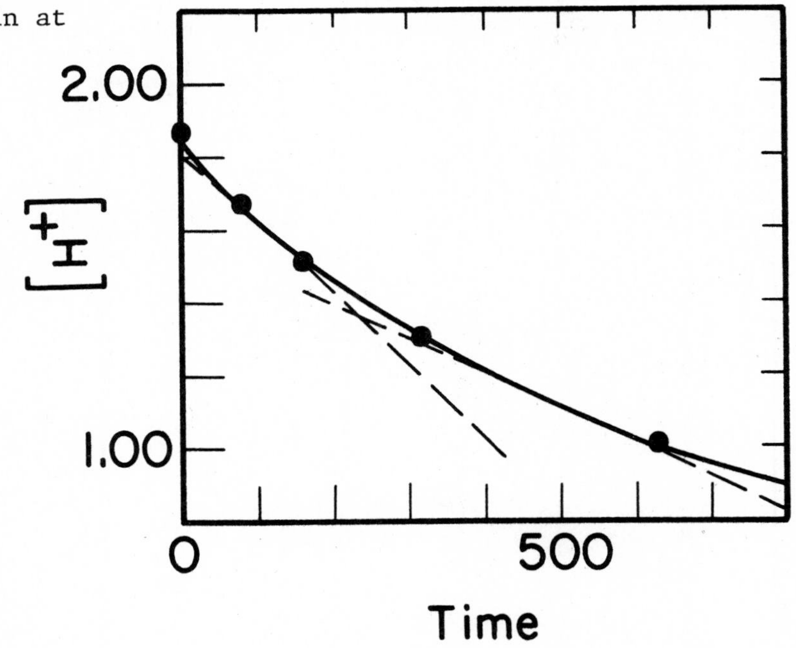

13.4 (a) $-\Delta[H_2]/\Delta t = (3/2)\Delta[NH_3]/\Delta t$; $-\Delta[N_2]/\Delta t = (1/2)\Delta[NH_3]/\Delta t$
(b) $-\Delta[NOCl]/\Delta t = \Delta[NO]/\Delta t = 2\Delta[Cl_2]/\Delta t$
(c) $-\Delta[HI]/\Delta t = -\Delta[CH_3I]/\Delta t = \Delta[CH_4]/\Delta t = \Delta[I_2]/\Delta t$

13.5 The rate of appearance of O_2 in the reaction is 1.5 times the rate of disappearance of O_3, because this is the ratio of their coefficients in the balanced equation. Thus,

$$\Delta[O_2]/\Delta t = (1.5)\ 9.0 \times 10^{-3}\ \text{atm/sec} = 13.5 \times 10^{-3}\ \text{atm/sec}.$$

13.6 (a) NO disappears at twice the rate of appearance of N_2.
(b) The rate law is $-(1/2)\Delta[NO]/\Delta t = k[H_2][NO]^2$
(c) k has units $\frac{M/sec}{(M)(M)^2} = 1/M^2\text{-sec}$, where M, of course, is moles/liter.

13.7 (a) first order in A, first order in B, second order overall.
(b) second order in A, zero order in B, second order overall.
(c) first order in A, second order in B, third order overall.

13.8 (a) Units of k are 1/M-sec. (b) Units of k are 1/M-sec.
(c) Units of k are $1/M^2$-sec.

13.9

Rate Law	Effect of Doubling [A]	Effect of Doubling [B]
$k[A][B]$	double rate	double rate
$k[A]^2$	four-fold increase	no effect
$k[A][B]^2$	double rate	four-fold increase

13.10 (a) no effect; (b) doubles rate; (c) four-fold rate increase; (d) eight-fold increase; (e) 1.414 increase

13.11 (a) rate = $k[B]^2$; (b) reaction is zero order with respect to A, second order with respect to B;

$$\text{(c)}\ k = \frac{7.0 \times 10^{-5}\ M/sec}{[0.10M]^2} = 7.0 \times 10^{-3}/M\text{-sec}$$

13.12 (a) The rate increases four-fold on doubling [NO], it doubles on doubling $[Br_2]$. Thus, rate = $k[NO]^2[Br_2]$.

$$\text{(b)}\ k = \frac{12\ M/sec}{[0.10M]^2[0.10M]} = 1.2 \times 10^4/M^2\text{-sec}$$

(c) From the balanced equation, $\Delta[NOBr]/\Delta t = -2\Delta[Br_2]/\Delta t$. That is, NOBr concentration increases at twice the rate the Br_2 concentration decreases.

13.13 For a first-order reaction, we have from equation [13.17] that $t_{1/2} = 0.693/k$. Thus, $k = 0.693/t_{1/2} = 0.693/(2.5 \times 10^3\ \text{sec}) = 2.8 \times 10^{-4}/\text{sec}$.

13.14 $t_{1/2} = 0.693/k = 0.693/(8.6 \times 10^{-3}/\text{sec}) = 81$ sec

13.15 Graph log $[SO_2Cl_2]$ vs. time. (Pressure is a satisfactory concentration unit for a gas, since the concentration in moles/liter is proportional to P.) The graph is linear with slope -9.53×10^{-6}/sec, as shown on the figure. The rate constant

$k = -2.30(-9.53 \times 10^{-6}/\text{sec})$
$= 2.19 \times 10^{-5}/\text{sec}.$

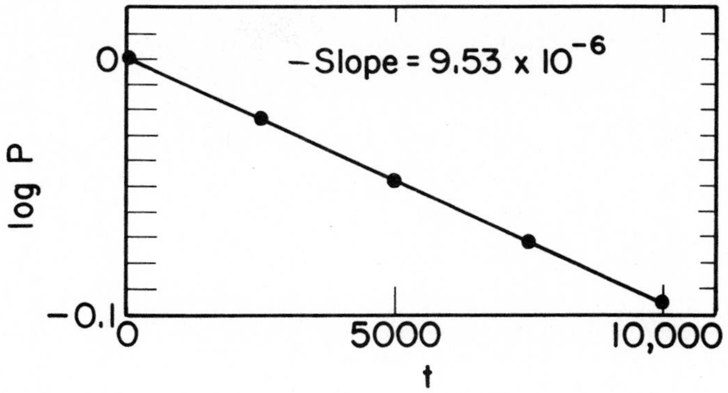

13.16 A graph of log $[C_{12}H_{22}O_{11}]$ vs. time is linear (at right). Thus, the reaction is first order. For the slope, we obtain -1.60×10^{-3}/min.

$k = -2.30(-1.60 \times 10^{-3}/\text{min})$
$= 3.68 \times 10^{-3}/\text{min}$
$= 6.13 \times 10^{-5}/\text{sec}.$

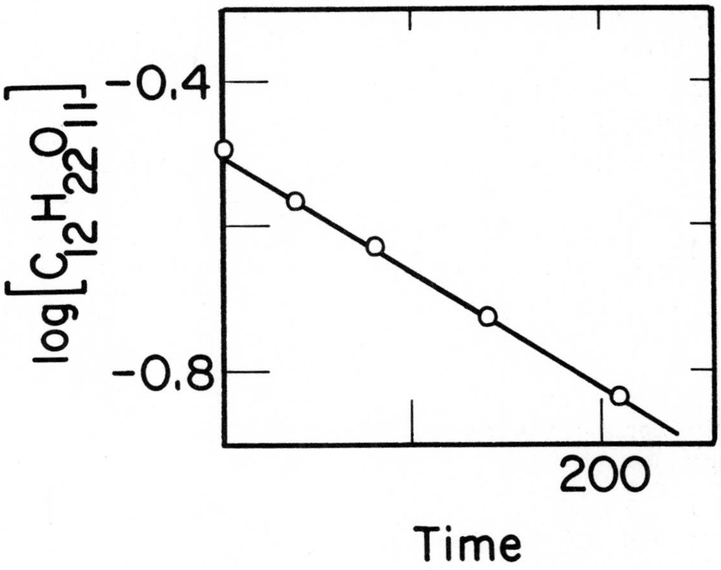

13.17 For second-order reaction, $t_{1/2} = 1/k[A]_o$. Thus, $k = 1/t_{1/2}[A]_o$ $= 1/350 \text{ min}[1.35 \text{ M}] = 2.11 \times 10^{-3}/\text{M-min.} = 3.53 \times 10^{-5}/\text{M-sec}$

13.18 We use equation [13.15]:

(a) $\log[N_2O_5]_t = -(k/2.3)t + \log[N_2O_5]_0$

$= -(4.87 \times 10^{-3}/2.3 \text{ sec})\ 300 \text{ sec} + \log[0.500 \text{ mol}/0.500\text{L}]$

$= -0.635 + 0$

$[N_2O_5] = 0.232 \text{ mol } N_2O_5/\text{L};\ \left(\dfrac{0.232 \text{ mol } N_2O_5}{1 \text{ L}}\right)(0.500 \text{ L})$

$= 0.115 \text{ mol } N_2O_5$

(b) $[N_2O_5] = 0.100 \text{ mol}/0.500 \text{ L} = 0.200 \text{ M}$

$\log[0.200] = \left(\dfrac{-4.87 \times 10^{-3}}{2.3 \text{ sec}}\right)t + \log[0.500 \text{ mol}/0.500 \text{ L}]$

$t = -4.72 \times 10^2[-0.699] = 330 \text{ sec}$

13.19 (a) $\log[C] = \left(\dfrac{-1.29}{2.3 \text{ yr}}\right)\dfrac{1 \text{ yr}}{12} + \log(5.0 \times 10^{-3})$

$= -0.0467 - 2.30 = -2.35$

$[C] = 4.49 \times 10^{-3}$ M after 1 month

$\log[C] = \left(\dfrac{-1.29}{2.3 \text{ yr}}\right) 1 \text{ yr} + \log(5.0 \times 10^{-3})$

$[C] = 1.38 \times 10^{-3}$ M after 1 yr.

(b) Using [13.14],

$t = (\log[Co/C])2.30/k$

$\dfrac{2.3(\log[5.0 \times 10^{-3}/1.00 \times 10^{-3}])}{1.29/\text{yr}}$

$= 1.24 \text{ yr} = 15 \text{ months}$

13.20

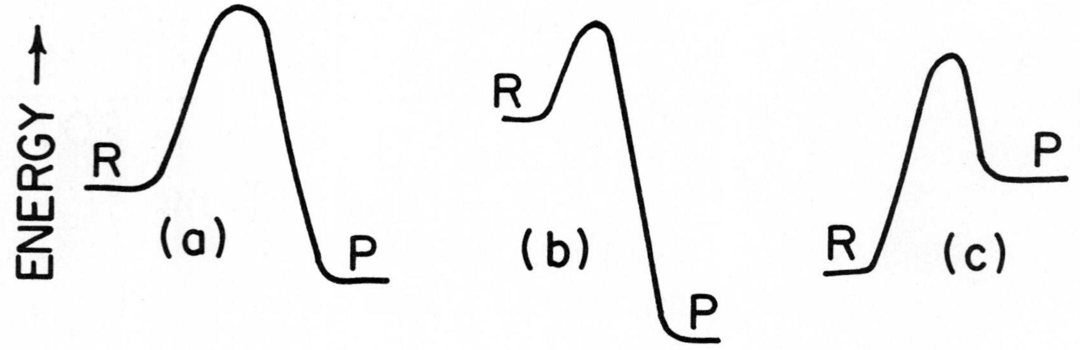

In these diagrams, the vertical axis represents energy, the symbols R and P represent reactants and products, respectively, and the horizontal axis represents the progress of the reaction.

13.21 Reaction (b) would be fastest, (c) slowest; the reverse reaction would be fastest in (c), slowest in (b).

13.22

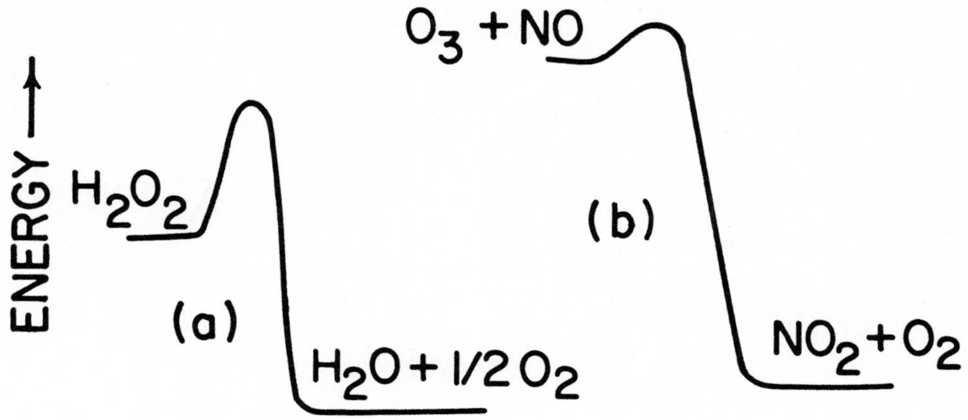

13.23

k	log k	T(K)	$1/T (\times 10^3)$
0.0521	-1.283	288	3.47
0.101	-0.996	298	3.36
0.184	-0.735	308	3.25
0.332	-0.479	318	3.14

The slope, -2.47×10^3, equals $-E_a/2.3\ R$. Thus,

$$E_a = (2.47 \times 10^3)2.30(8.314\ \text{J/mol})$$
$$= 47\ \text{kJ/mol.}$$

(We round to 2 significant figures because $\Delta(1/T)$ is known to only 2 significant figures.)

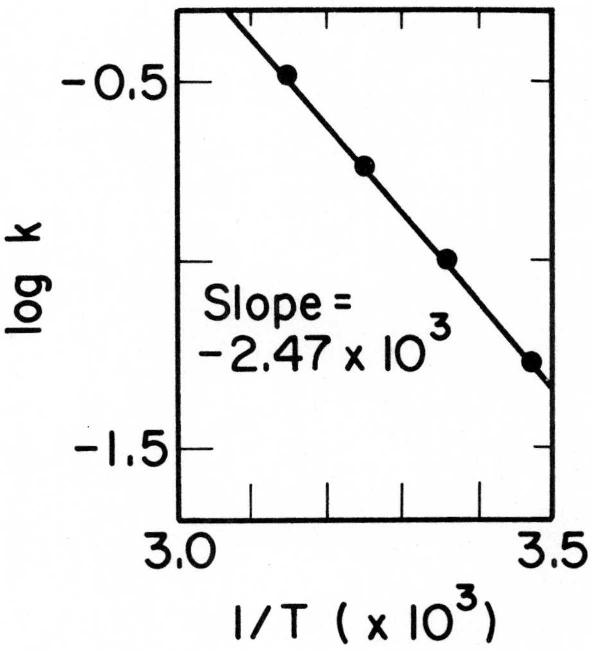

13.24

k	log k	T(K)	1/T(x 10^3)
0.028	-1.55	600	1.67
0.22	-0.658	650	1.54
1.3	0.114	700	1.43
6.0	0.778	750	1.33
23	1.362	800	1.25

The slope, -6.83×10^3, equals $-E_a/2.3\ R$. Thus, $E_a = (6.83 \times 10^3)(2.30)(8.314\ J/mol) = 130\ kJ/mol$. To calculate A, let us use the rate data at 700 K. From equation [13.20], $0.114 = \log A - 6.83 \times 10^3/700$; $\log A = 0.114 + 9.75$. $A = 7.3 \times 10^9$.

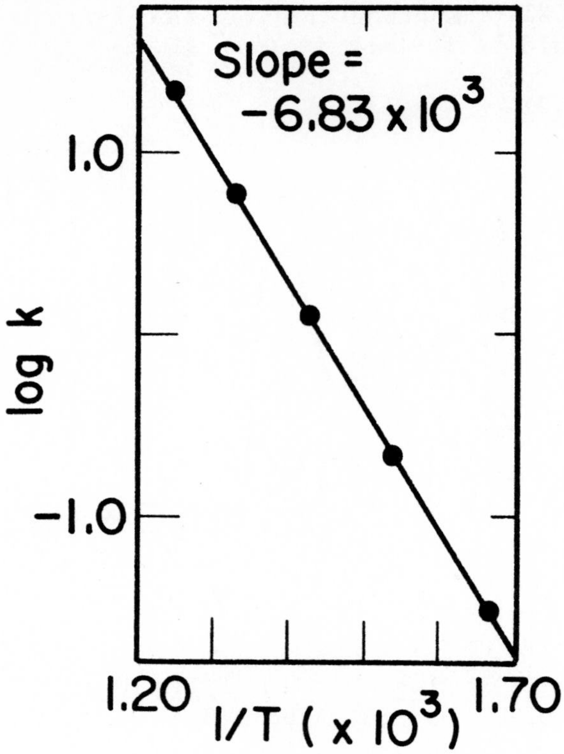

13.25 (a) Using eq. [13.21], $\log(k_{308}/k_{298}) = \dfrac{70.0 \times 10^3\ J/mol}{2.30(8.314\ J/mol)}\left(\dfrac{1}{298} - \dfrac{1}{308}\right)$

$= 0.398$. $k_{308}/k_{298} = 2.51$

(b) $\log(k_{383}/k_{373}) = \dfrac{70.0 \times 10^3\ J/mol}{2.30(8.314\ J/mol)}\left(\dfrac{1}{373} - \dfrac{1}{383}\right) = 0.256.$

$k_{383}/k_{373} = 1.80$

13.26 (a) $\log(k_{873}/k_{973}) = \dfrac{182 \times 10^3\ J/mol}{2.30(8.314\ J/mol)}\left(\dfrac{1}{973} - \dfrac{1}{873}\right) = -1.12$

$k_{873}/k_{973} = 0.0759$. $k_{873} = (7.59 \times 10^{-2})(1.57 \times 10^{-3}\ M/sec)$

$= 1.19 \times 10^{-4}\ M/sec$

(b) $\log(k_{1073}/k_{973}) = \dfrac{182 \times 10^3\ J/mol}{2.30(8.314\ J/mol)}\left(\dfrac{1}{973} - \dfrac{1}{1073}\right) = 0.911$

$k_{1073}/k_{973} = 8.16$. $k_{1073} = (8.16)(1.57 \times 10^{-3}\ M/sec)$

$= 1.28 \times 10^{-2}\ M/sec$

13.27 (a) $\log(k_{373}/k_{298}) = \dfrac{67.0 \times 10^3\ J/mol}{2.30(8.314\ J/mol)}\left(\dfrac{1}{298} - \dfrac{1}{373}\right) = 2.364$

$k_{373} = (1.0 \times 10^{-3}/sec)(231) = 0.231/sec$

(b) $\log(k_{373}/k_{298}) = \dfrac{134 \times 10^3\ J/mol}{2.30(8.314\ J/mol)}\left(\dfrac{1}{298} - \dfrac{1}{373}\right) = 4.728$

$k_{373} = (1.0 \times 10^{-3}/sec)(5.35 \times 10^4) = 53.5/sec$

Note that doubling the activation energy much more than doubles the rate of change of rate constant with temperature.

13.28 $\log(1.7 \times 10^{-2}/9.5 \times 10^{-2}) = \frac{E_a}{2.30(8.314\ J/mol)}\left(\frac{1}{450} - \frac{1}{430}\right)$

$(-5.40 \times 10^{-6}\ mol/J)\ E_a = 0.747$

$E_a = 138\ kJ/mol$

13.29 (a) $NOBr_2(g)$ is the intermediate; it is neither one of the original reactants nor a final product. (b) First step: rate = $\Delta[NOBr_2]/\Delta t = k[NO][Br_2]$. Second step: rate = $-\Delta[NOBr_2]/\Delta t = k[NOBr_2][NO]$. (c) If the first step is slow, the rate law for the overall reaction is just the rate law for the first step written in (b). (d) If the rate law is $k[NO]^2[Br_2]$, it is evident that the second step in the two step reaction is slow. This then means that the first step is probably reversible, so that $NOBr_2$ is able to decompose to reform NO and Br_2 or go on to react with more NO to form product. During the reaction, not much $NOBr_2$ is present at any one time.

13.30 If the second step is rate determining, we can write that $\Delta[NO_2]/\Delta t = k[NO_3][NO]$. However, the concentration of NO_3 is determined by the fast pre-equilibrium in the first step. We can see that the NO_3 concentration will be proportional to [NO] and $[O_2]$: $[NO_3] = K[NO][O_2]$. Substituting this for $[NO_3]$ in the rate expression above, and lumping all the constants together, we obtain $\Delta[NO_2]/\Delta t = k'[NO]^2[O_2]$.

13.31

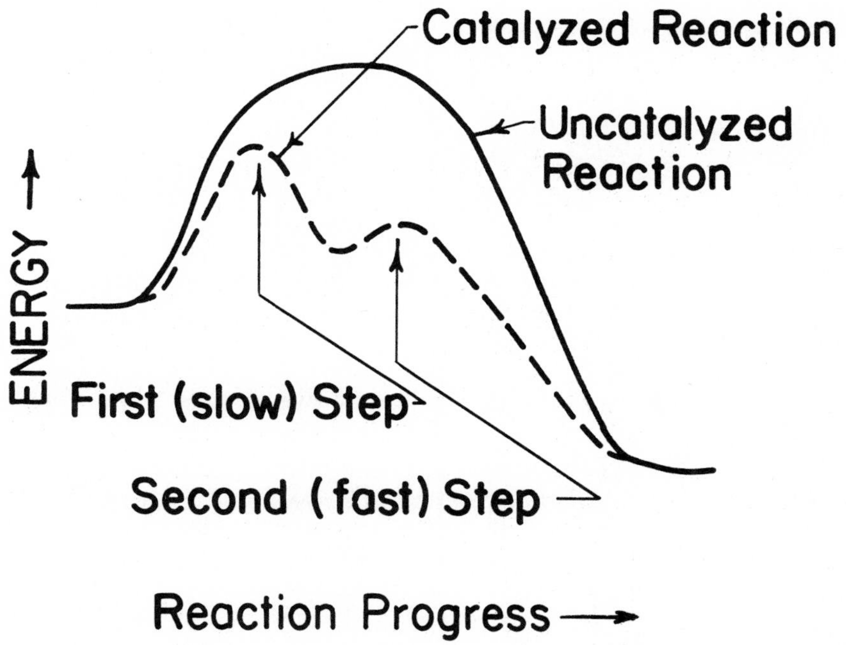

13.32 Most heterogeneous catalysts operate by providing a surface on which reaction can occur. The amount of surface per unit mass of the catalyst will vary with the method of preparation. In general, we want materials with very large surface area/mass ratios. Also, the chemical activity of the surface may depend on how the catalyst has been handled. Many active catalysts can be "poisoned" by exposure to a substance that blocks the active sites on the surface.

12.33 As illustrated in Figure 13.13, the two C-H bonds that exist on each carbon of the ethylene molecule before adsorption are retained in the process in which a D atom is added to each C (assuming we use D_2 rather than H_2). To put two deuteriums on a single carbon, it is necessary that one of the already existing C-H bonds in ethylene be broken while the molecule is adsorbed, so the H atom moves off as an adsorbed atom, and is replaced by a D. This requires a larger activation energy than simply adsorbing C_2H_4 and adding one D atom to each carbon.

13.34 (a) Concentrations of reactants; temperature; addition of a catalyst. (b) Increases in concentrations of reactants result in more frequent collisions per unit time, thus in a greater probability of reaction. An increase in temperature results in an increase in average kinetic energy of the reactants, so that a larger fraction may possess sufficient energy to form product. A catalyst lowers the activation energy for reaction, so that a larger fraction of the encounters between reactants can lead to products.

13.35 The energies of reactants must be sufficiently large (that is, their relative kinetic energies on collision if it is a bimolecular reaction, or the energy derived from collisions that is retained within a reactant if it is a unimolecular reaction). Secondly, geometrical requirements must be met. A molecule undergoing a unimolecular reaction may need to be distorted in a certain way. In a bimolecular reaction, the colliding molecules may need to have a certain orientation with respect to one another for reaction to occur.

13.36 average reaction rate = $\Delta[C]/\Delta t = \dfrac{[0.200 - 0.187]M}{100 \text{ sec}} = 1.3 \times 10^{-4}$ M/sec

13.37 (a) $-\Delta[N_2O_5]/\Delta t = (1/2)\Delta[NO_2]/\Delta t = 2\Delta[O_2]/\Delta t$

(b) k is not altered by whether we express the rate in terms of the loss of N_2O_5 or appearance of NO_2, when we use the expressions in (a).

13.38 (a) Doubling [A] changes [A], of course, and rate, but not k or [B]. (b) Addition of a catalyst changes rate and k. (c) $[A]^2$ increases by four-fold, [B] decreases by four-fold, rate and k are unchanged. (d) Increase in temperature increases k and rate.

13.39 Rate = $k[NO][O_3] = (1.5 \times 10^7/M\text{-sec})(2 \times 10^{-8})^2 = 6.0 \times 10^{-9}$ M/sec

13.40 Reactions often proceed through several steps. The rate at which product is formed will depend on which step in the overall reaction is slowest. There is no unique relation between the "mechanism", or detailed path taken by the reactants, and the overall balanced equation for the reaction.

13.41 Since glucose is not an acid, the reaction could perhaps be "followed" by titrating with base (Section 3.9) to measure the lactic acid formed. One would do this by having a large volume of solution undergoing the reaction at 25°C, then removing small samples of accurately known volume (called aliquots) at regular time intervals and promptly titrating with a standardized base solution.

13.42 Rate = $k[(CH_3)_3CBr]$. $k = (1.0 \times 10^{-3}$ M/sec$)/(0.10$ M$) = 1.0 \times 10^{-2}$/sec

13.43 (a) $k = (8.56 \times 10^{-5}$ M/sec$)/(0.200$ M$) = 4.28 \times 10^{-4}$/sec

(b) $\log[\text{Urea}] = -\left(\frac{4.28 \times 10^{-4}/\text{sec}}{2.3}\right) 5.00 \times 10^{3} \text{ sec} + \log(0.500)$

$= -0.931 - 0.301 = -1.23$

$[\text{Urea}] = 0.059\ M$

13.44 $k = \left(\frac{0.693}{75 \text{ min}}\right)\left(\frac{1 \text{ min}}{60 \text{ sec}}\right) = 1.54 \times 10^{-4}/\text{sec}$

13.45 The assumption must be made that the reaction is proceeding essentially to completion. Assuming this, note that the first half life is 2.0 hr. The second half life; that is, the time to go from 0.50 M to 0.25 M, takes 4.0 hours. Thus the reaction cannot be first order. For a second-order reaction $t_{1/2} = 1/k[A]_0$. Thus, the second half life, which starts with A one-half as large as for the first half-life, should be twice as long as the first half-life. The reaction is therefore second order; rate = $k[HI]^2$.

13.46 Removal rate would increase in proportion to the concentration of dissolved oxygen.

$$-\Delta[H_2S]/\Delta t = \left(\frac{4.5 \times 10^{-5}}{M\text{-sec}}\right)(5 \times 10^{-6}\ M)(2.6 \times 10^{-4}\ M)$$
$$= 5.8 \times 10^{-14}\ M/\text{sec}$$

13.47 $\Delta[HCl]/\Delta t = 2\left(\frac{3.5 \times 10^{-2}}{M\text{-sec}}\right)(5 \times 10^{-5}\ M)(5 \times 10^{-2}\ M) = 2 \times 10^{-7}\ M/\text{sec}$

13.48

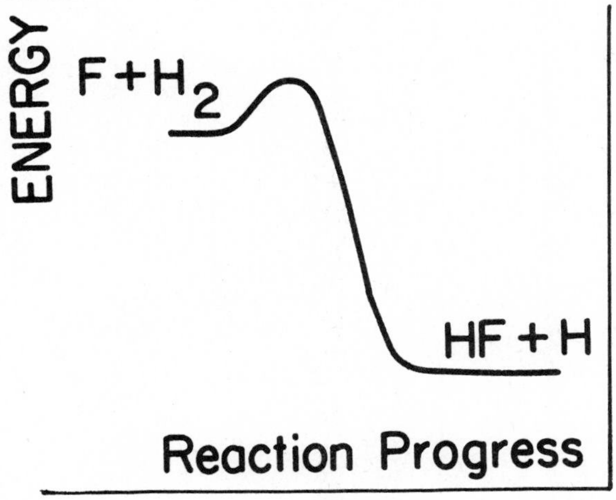

13.49

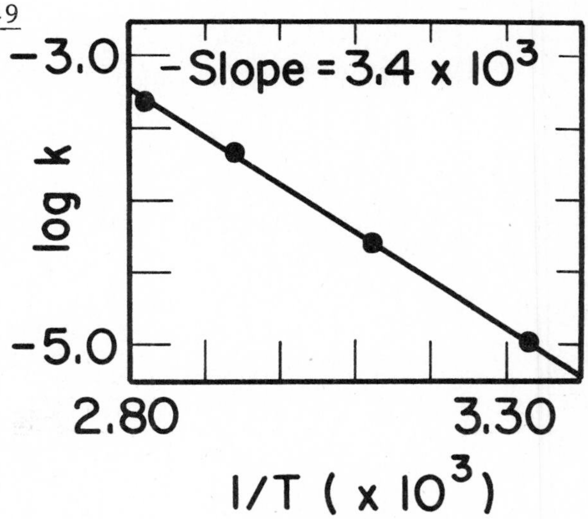

The activation, energy, E_a, equals (-slope)(2.30)(8.314 J/mol). Thus, $E_a = 3.4 \times 10^3(2.30)(8.314) = 6.5 \times 10^4$ J/mol = 65 kJ/mol.

13.50 We must assume that the rate of the control reaction is inversely proportional to the period between flashes. Thus, $k_{294.1} \propto 1/16.3$ sec, $k_{300.9} \propto 1/13.0$ sec. Using equation [13.21],

$$\log(k_{300.9}/k_{294.1}) = \log(16.3/13.0) = 0.0982$$
$$= \frac{E_a}{(2.30)(8.314 \text{ J/mol})}\left(\frac{1}{294.1} - \frac{1}{300.9}\right)$$
$$E_a = 24.4 \text{ kJ/mol}$$

13.51 We employ equation [13.21]; convert temperatures to K, then:

$$\log(k_{773}/k_{873}) = \frac{E_a}{2.3\ R}\left(\frac{1}{873} - \frac{1}{773}\right)$$

$$\log k_{773} - \log(0.75) = \frac{-114 \times 10^3\ \text{J/mol}}{(2.30)(8.314\ \text{J/mol})}(1.48 \times 10^{-4})$$

$$\log k_{773} = -0.883 - 0.125$$

$$k_{773} = 9.82 \times 10^{-2}/\text{M-sec}$$

13.52 (a) The first elementary reaction is unimolecular; the second is bimolecular. (b) Cl(g) is the intermediate; it is neither one of the original reactants nor one of the products. (a) If the first step is slow, the overall rate law is Rate = $k[NO_2Cl]$.

13.53 The third, or final step must be rate-determining. One can argue in the following way: If the first step were the slow, rate-determining step, the rate law would be simply Rate = $k[Cl_2]$. If the second step were rate-determining, the rate law would be Rate = $k[CO][Cl_2]^{1/2}$ (refer to the discussion surrounding equations [13.38], [13.39] and [13.40]). The chlorine dependence is higher than 1/2 because in the third step we again have a Cl_2 molecule. If this step is rate determining, the dependence on Cl_2 is raised to 3/2 order. Briefly,

$$\text{rate} = k[Cl_2][COCl]$$

$$\text{but } [COCl] = K'[Cl][CO]$$

$$\text{and } [Cl] = K^{1/2}[Cl_2]^{1/2}$$

Substituting these last two equations into the first and lumping together all the constants we obtain the observed rate law. Thus, the third step is rate determining.

13.54 Only mechanism (b) is consistent, because only in this mechanism is there a dependence in the slow step and all the ones preceeding it on both [NO] and $[O_3]$.

14

Chemical equilibrium

14.1 (a) $K = \frac{[SO_3]^2}{[SO_2]^2[O_2]}$ (b) $K = \frac{[H_2O]}{[H_2]}$ (c) $K = \frac{[CO][H_2O]}{[CO_2][H_2]}$

(d) $K = \frac{[N_2]^2[H_2O]^6}{[NH_3]^4[O_2]^3}$ (e) $K = [N_2][H_2O]^2$

14.2 (a) $K = \frac{[SO_2]^2[H_2O]^2}{[O_2]^3[H_2S]^2}$; a homogeneous equilibrium

(b) $K = \frac{[O_2][NO]^2}{[NO_2]^2}$; homogeneous (c) $K = [CO_2]$; heterogeneous

(d) $K = \frac{[SO_2]}{[O_2]}$; heterogeneous (e) $K = \frac{[HCl]^2}{[H_2][Cl_2]}$; homogeneous

14.3 $K_c = \frac{[H_2][I_2]}{[HI]^2} = \frac{(4.79 \times 10^{-4})(4.79 \times 10^{-4})}{(3.53 \times 10^{-3})^2} = 1.82 \times 10^{-2}$

14.4 $K_P = \frac{P^2_{Cl}}{P_{Cl_2}} = \frac{(2.97 \times 10^{-2})^2}{1.00} = 8.82 \times 10^{-4}$

14.5 $K_c = \frac{[SO_3]}{[SO_2][O_2]^{1/2}} = 20.4$

(a) K_c for the reaction written is

$$K_c' = \frac{[O_2]^{1/2}[SO_2]}{[SO_3]}$$

This is just the reciprocal of K_c written above. Thus, $K_c' = 1/K_c = 4.90 \times 10^{-2}$. The equilibrium constant for a reaction that is the reverse of another is just the reciprocal.

(b) $$K_c'' = \frac{[SO_3]^2}{[SO_2]^2[O_2]}$$

This is just the square of the expression for K_c above. Thus, $K_c'' = (20.4)^2 = 416$

(c) $$K_p = \frac{P^2_{SO_3}}{P_{O_2} \times P^2_{SO_2}}$$

This is the same equilibrium expression as in (b). We know that $K_p = K_c(RT)^{\Delta n}$ where Δn is the moles of gas of product less moles of gas of reactants in the balanced equation. In our case, Δn for the equation in part (c) is -1. Thus, $K_p = 416(0.0821 \times 973)^{-1} = 5.21$.

14.6 $K_p = K_c(RT)^{\Delta n}$. Here $\Delta n = 0$, so K_p equals $K_c = 0.11$.

14.7 $K_p = K_c(RT)^{\Delta n}$. Here $\Delta n = -1$. Thus, $K_p = 278(0.0821 \times 1000)^{-1} = 3.39$

14.8 $K_p = K_c(RT)^{\Delta n}$. Here $\Delta n = +1$. Thus, $167.5 = K_c(0.0821 \times 1273)$

$K_c = 1.603$

14.9 $$K_c = \frac{[H_2][CO]}{[H_2O]} = \frac{(4.0 \times 10^{-2})(4.0 \times 10^{-2})}{(1.0 \times 10^{-2})} = 0.16$$

$$K_p = K_c(RT)^{\Delta n} = 0.16(0.0821 \times 1073) = 14$$

14.10 (a) From the balanced equation we know that one mole of H_2 reacts for each mole of NO that does so. The number of moles of NO reacting is 0.100 - 0.070 = 0.030. Thus, the quantity of H_2 remaining is 0.050 - 0.030 = 0.020 moles. By reaction of 0.030 moles of NO we produce 0.030 moles of H_2O, and half as many, 0.015 moles, of N_2. We had initially 0.100 mol H_2O.

(b) $$K_c = \frac{(0.015)(0.130)^2}{(0.070)^2(0.020)^2} = 129$$

14.11 From the balanced equation we can see that each mole of SO_3 formed means consumption of 1 mole of SO_2 and 0.5 moles of O_2. Thus, at equilibrium the amount of SO_2 remaining is 1.000 - 0.919 - 0.081 mol. The amount of O_2 remaining is 1.000 - (0.919/2) = 0.540 mol.

$$K_c = \frac{(0.919)^2}{(0.081)^2(0.540)} = 238$$

14.12 We need to calculate the number of moles of each component present.

$$3.22 \text{ g NOBr} \left(\frac{1 \text{ mol NOBr}}{110 \text{ g NOBr}}\right) = 0.0293 \text{ mol NOBr}$$

$$3.08 \text{ g NO} \left(\frac{1 \text{ mol NO}}{30.0 \text{ g NO}}\right) = 0.103 \text{ mol NO}$$

$$4.19 \text{ g } Br_2 \left(\frac{1 \text{ mol } Br_2}{160 \text{ g } Br_2}\right) = 0.0262 \text{ mol } Br_2$$

(a) In calculating K_c we divide each number of moles by 5 to convert to moles/L, the insert into the expression for K_c to obtain:

$$K_c = \frac{[Br_2]^2[NO]^2}{[NOBr]^2} = \frac{(5.24 \times 10^{-3})(2.06 \times 10^{-2})^2}{(5.86 \times 10^{-3})^2} = 6.42 \times 10^{-2}$$

(b) $K_p = (6.42 \times 10^{-2})(0.0821 \times 373) = 1.98$

(c) The total moles of gas present is 0.0293 + 0.103 + 0.0262 = 0.158

$$P = (0.158 \text{ mol})\left(\frac{0.0821 \text{ L-atm}}{\text{mol-K}}\right)\left(\frac{373 \text{ K}}{5.00 \text{ L}}\right) = 0.968 \text{ atm}$$

14.13 The moles of H_2 reacted is 0.34 - 0.14 = 0.20. The number of moles of Br_2 reacted is the same. Thus, there are 0.22 - 0.20 = 0.02 moles Br_2 remaining. From the balanced equation we can see that there must have been 0.40 moles HBr formed. Because the container is 1 L in volume, molarity equals the number of moles present in each case:

$$K_c = \frac{(0.40)^2}{(0.02)(0.14)} = 57$$

14.14 Calculate the reaction quotient in each case, compare with

$$K_p = \frac{P^2_{NH_3}}{P_{N_2} \times P^3_{H_2}} = 4.51 \times 10^{-5}$$

(a) $Q = \dfrac{(100)^2}{(30)(500)^3} = 2.7 \times 10^{-6}$

Since $Q < K_p$, reaction will shift to right to attain equilibrium.

(b) $Q = \dfrac{(30)^2}{(0)(600)^3} = \infty$

Since $Q > K_p$, reaction must shift to left to attain equilibrium. There must be <u>some</u> N_2 present to attain equilibrium. This can only come in this case from some decomposition of NH_3 to form N_2 and H_2.

(c) $Q = \dfrac{(26)^2}{(42)^3(202)} = 4.52 \times 10^{-5}$ $Q = K_p$. Reaction is at equilibrium.

(d) $Q = \dfrac{(100)^2}{(5)(60)^3} = 9 \times 10^{-3}$ $Q > K_p$.

Reaction will proceed to the left to attain equilibrium.

14.15 The reaction quotient calculated in each case is to be compared with the value of K:

$$K_c = \frac{[Cl_2][CO]}{[COCl_2]} = 2.19 \times 10^{-10}$$

$$\text{(a) } Q = \frac{(1.0 \times 10^{-3})(1.0 \times 10^{-3})}{(2.19 \times 10^{-1})} = 4.6 \times 10^{-6};\ Q > K$$

Therefore, reaction will shift to the left, forming more $COCl_2$, to attain equilibrium.

$$\text{(b) } Q = \frac{(3.31 \times 10^{-6})(3.31 \times 10^{-6})}{(5.00 \times 10^{-2})} = 2.19 \times 10^{-10};\ Q = K$$

Therefore, the system is at equilibrium.

$$\text{(c) } Q = \frac{(4.50 \times 10^{-7})(5.73 \times 10^{-6})}{(8.57 \times 10^{-2})} = 3.00 \times 10^{-11};\ Q < K$$

Therefore, reaction will shift to the right, forming more CO and Cl_2, to attain equilibrium.

14.16 $K_c = \frac{[NO]^2}{[N_2][O_2]} = 4.1 \times 10^{-4} = \frac{(0.050)^2}{(0.10)[O_2]}$. Solve for $[O_2] = 61$ M

14.17 $K_p = 3.5 \times 10^4 = \frac{P^2_{HBr}}{(0.10)(0.40)}$. $P_{HBr} = 37$ atm

14.18 $K_c = \frac{[Br]^2}{[Br_2]} = 1.04 \times 10^{-3} = \frac{[Br]^2}{(4.53 \times 10^{-2}/0.200)}$. $[Br] = 1.53 \times 10^{-2}$ M

Since the volume is 0.200 L,

$$\text{mol Br} = \left(\frac{1.53 \times 10^{-2}\ \text{mol}}{1\ \text{L}}\right)(0.200\ \text{L}) = 3.07 \times 10^{-3}\ \text{mol}$$

14.19 $K_p = \frac{P^2_{CO}}{P_{CO_2}} = 167.5$. When $P_{CO} = 0.500$ atm,

$$P_{CO_2} = (0.500)^2/167.5 = 1.49 \times 10^{-3}\ \text{atm}$$

14.20 We know that the pressures of I_2 and Br_2 will be the same. Call that quantity P. Then

$$K = \frac{[IBr]^2}{[I_2][Br_2]} = 280 = \frac{(0.20)^2}{P^2}. \quad P = 1.2 \times 10^{-2}\ \text{atm}$$

14.21 $K_c = \frac{[NO]^2}{[N_2][O_2]} = 2.5 \times 10^{-3}$

Because we have the same number of moles of gaseous reactants and products, K is really dimensionless. We can thus simply work directly with the number of moles of substance. Let us call the number of moles of NO formed x. Then the number of moles of N_2 remaining at equilibrium is $(0.20 - x/2)$; the number of moles of O_2 is similarly $(0.20 - x/2)$. Then,

$$\frac{x^2}{(0.20 - x/2)(0.20 - x/2)} = 2.5 \times 10^{-3}$$

Using the quadratic formula (Appendix A), we obtain $x = 9.8 \times 10^{-3}$ mol NO. The equilibrium concentrations are thus:

$$[NO] = 9.8 \times 10^{-3} \text{ mol}/5.00 \text{ L} = 2.0 \times 10^{-3} \text{ M}$$

$$[O_2] = [N_2] = (0.20 - 9.75 \times 10^{-3}/2)/5.00 \text{ L} = 3.9 \times 10^{-2} \text{ M}$$

14.22 We can see from the small value for K that most of the NO will have decomposed to form O_2 and N_2 at equilibrium. It is best to set up such a problem so that we solve directly for a small number, rather than try to find it as the difference between two larger numbers. Let's assume that nearly all the NO has reacted to form N_2 and O_2. The concentration of NO remaining we can call y. Then, from the balanced equation, the concentration of N_2 or O_2 is (0.250 mol/2.00 L - y) = (0.125 - y) M. Inserting into the expression for K_c we have

$$\frac{y^2}{(0.125 - y)^2} = 1.0 \times 10^{-5}$$

This equation is easily solved by taking the square root of both sides of the equation and solving for y. We obtain $y = 3.95 \times 10^{-4}$. Thus, $[NO] = 3.95 \times 10^{-4}$ M, $[O_2] = 0.125$ M, $[N_2] = 0.125$ M.

14.23 $K_c = [H_2S][NH_3] = 1.2 \times 10^{-4}$. The concentrations of H_2S and NH_3 will be the same; call this quantity y. Then, $y^2 = 1.2 \times 10^{-4}$, $y = 1.1 \times 10^{-2}$ M.

14.24 $K = \frac{[HI]^2}{[H_2][I_2]}$. Let us call the concentration of HI formed x. Then at equilibrium $[HI] = x$, $[I_2] = (0.0500 - x/2)$ M,

$[H_2] = (0.0700 - x/2)$ M; $\quad \frac{x^2}{(0.0500 - x/2)(0.0700 - x/2)} = 54.6$

The two solutions for x from the quadratic equation are approximately 0.173 and 0.0884. Only the smaller quantity is physically reasonable. Thus, at equilibrium we have $[HI] = 0.0884$ M; $[H_2] = 0.0258$ M; $[I_2] = 5.8 \times 10^{-3}$ M. Put back into the equilibrium-constant expression as a check, these values yield approximately the correct K.

14.25 (a) Shift equilibrium to the right; more CO will be formed. (b) No effect on equilibrium; so long as there is any C(s) present the amount is not involved in the equilibrium. (c) Equilibrium is shifted to the right, in the direction in which the reaction is endothermic. (d) An increase in pressure will cause a shift to the left; that is, some of the CO will be converted to CO_2(g) and C(s). (e) No effect on equilibrium. (f) Removal of CO will cause a shift to the right; that is, more CO_2(g) will react with C(s), forming CO(g).

14.26 (a) increase; (b) increase; (c) decrease; (d) no effect; (e) no effect

14.27 (a) no effect; (b) no effect; (c) increase equilibrium constant

14.28 The dissolving of $CuSO_4$ is an endothermic process. As temperature is increased, the equilibrium $CuSO_4(s) \rightleftarrows Cu^{2+}(aq) + SO_4^{2-}(aq)$ shifts in the direction that absorbs heat. We know from the data provided that the equilibrium shifts to the right. This is the direction in which the reaction is endothermic.

14.29 An increase in pressure will favor the phase that has the higher density; that is, the equilibrium will shift in a direction in which the system occupies a smaller space. The liquid phase is thus favored. The melting point of ice thus decreases with increasing pressure on the system.

14.30 $K = k_f/k_r = (1.38 \times 10^{-28}/M\text{-sec})/(9.3 \times 10^{10}/M\text{-sec}) = 1.48 \times 10^{-39}$

14.31

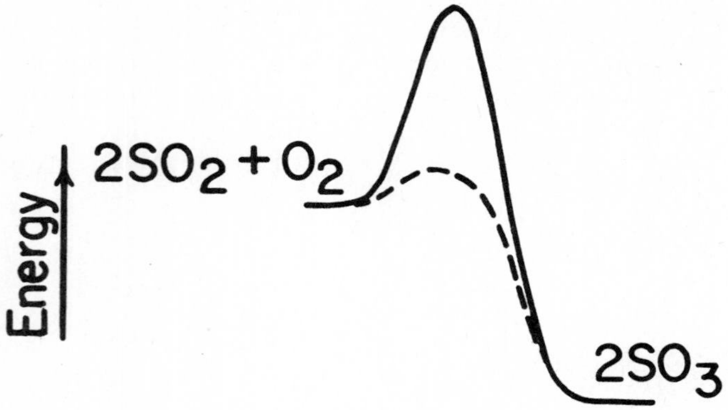

14.32 At equilibrium $r_f = r_r$. Thus, $k_f[NO][O_3] = k_r[NO_2][O_2]$

$$k_f/k_r = \frac{[NO_2][O_2]}{[NO][O_3]} = K_c$$

14.33 (a) Statement is correct. (b) False. A catalyst may change the rate at which a system approaches equilibrium, but it does not alter the position of equilibrium. (c) False. Homogeneous and heterogeneous catalysts must both obey the same laws of nature. A catalyst affects both forward and reverse reaction pathways to the same degree overall. (d) False. The speed of a reaction has nothing to do with the position of equilibrium. In Chapters 15 and 16, we will encounter many examples of aqueous equilibria in which the equilibrium is attained very rapidly, yet the equilibrium constant is small.

14.34 (a) $K = \frac{[PCl_3][Cl_2]}{[PCl_5]}$ (b) $K = [SO_2]$ (c) $K = \frac{1}{[CO][H_2]^2}$

(We assume CH_3OH is a pure liquid, so its concentration is constant, and not included in the expression for K.)

(d) $K = \frac{[H_2O]^2}{[H_2]^2[O_2]}$ (e) $K = \frac{[O_3]^2}{[O_2]^3}$

14.35 Reactions (a), (d) and (e) are homogeneous, (b) and (c) are heterogeneous.

14.36 (a) $K_c = [Pb^{2+}][Cl^-]^2$ (b) $K_c = \frac{[H^+][CN^-]}{[HCN]}$ (c) $K_c = \frac{[NH_4^+][OH^-]}{[NH_3]}$

(d) $K_c = \frac{[Ag(NH_3)_2^+]}{[Ag^+][NH_3]^2}$ (e) $K_c = \frac{[Zn^{2+}]}{[Cu^{2+}]}$ (f) $K_c = [Ba^{2+}][SO_4^{2-}]$

Note in these examples that we include in the expression for K any species whose concentration may vary. We exclude all pure substances, solid or liquid, for which the concentration is a fixed constant. In (c), for example, we leave $[NH_3]$ in, because the concentration of NH_3 in water can vary over a wide range. We exclude the $H_2O(\ell)$ because the concentration of water in pure water or dilute solution is essentially constant.

14.37

Reaction	Units of K_p	Units of K_c
(a)	atm	M
(b)	atm^{-2}	M^{-2}
(c)	dimensionless	dimensionless
(d)	atm^2	M^2

14.38 $$K_c = \frac{[CO][H_2]^3}{[CH_4][H_2O]} = \frac{(0.0616)(0.260)^3}{(0.538)^2} = 3.74 \times 10^{-3}$$

14.39 (a) If there is 1.0 mol SO_2 at equilibrium, then 2.0 mol SO_2 reacted. This must have consumed 2.0 mol NO_2, and formed 2.0 mol each of SO_3 and NO. Thus, at equilibrium:

$$[SO_2] = 1.0 \text{ mol}/2.0 \text{ L} = 0.50 \text{ M}$$
$$[NO_2] = (4.0 - 2.0) \text{ mol}/2.0 \text{ L} = 1.0 \text{ M}$$
$$[SO_3] = (1.0 + 2.0) \text{ mol}/2.0 \text{ L} = 1.5 \text{ M}$$
$$[NO] = (4.0 + 2.0) \text{ mol}/2.0 \text{ L} = 3.0 \text{ M}$$

(b) $$K_c = \frac{(3.0 \text{ M})(1.5 \text{ M})}{(1.0 \text{ M})(1.5 \text{ M})} = 3.0$$

14.40 We must first find the number of moles present in the container:

$$n = \frac{(1.30 \text{ atm})(1.00 \text{ L})}{\left(\frac{0.0821 \text{ L-atm}}{\text{mol-K}}\right)(1100 \text{ K})} = 0.0144 \text{ mol}$$

We began with $0.831 \text{ g } SO_3 \left(\frac{1 \text{ mol } SO_3}{80.1 \text{ g } SO_3}\right) = 0.0104 \text{ mol}$

Let the number of moles of SO_2 that decomposes be x. Then, $0.0104 - x + x + 0.5x = 0.0144$; $x = 0.0080$. Thus, the number of moles of SO_3 at equilibrium is 0.0024 mol. There are 0.0080 moles of SO_2 and 0.0040 moles of O_2. Since the volume is 1 L, the molar concentrations are just these same values.

$$K_c = \frac{[SO_2]^2[O_2]}{[SO_3]^2} = \frac{(0.0080)^2(0.0040)}{(0.0024)^2} = 4.4 \times 10^{-2}$$

To calculate K_p we use equation [14.7]:

$$K_p = K_c(RT)^{\Delta n} = 4.4 \times 10^{-2}(0.0821 \times 1100) = 4.0$$

14.41 Because we began with pure PCl_5, we know that the molar quantities of PCl_3 and Cl_2 present at equilibrium are equal. Thus, if the total pressure is 2.00 atm, and if 0.37 atm of this is PCl_5, the remaining 1.63 atm is equally divided between Cl_2 and PCl_3. Then,

$$K_p = \frac{P_{PCl_3} \times P_{Cl_2}}{P_{PCl_5}} = \frac{(0.815)(0.815)}{(0.37)} = 1.8$$

14.42 Formation of 0.371 mol of N_2 requires $(0.5)(0.371) = 0.186$ mol N_2 and $(1.5)(0.371) = 0.556$ mol H_2, based on the balanced equation. The equilibrium molarities are then $[N_2] = (1.000 - 0.186)$ mol/1 L = 0.814 M;

$[H_2] = (3.000 = 0.556)$ mol/1 L = 2.444 M; $[NH_3]$ = 0.371 mol/1 L = 0.371 M

$$K_c = \frac{(0.371)^2}{(0.814)(2.444)^3} = 1.16 \times 10^{-2}$$

14.43 (a) The initial concentrations of NO and O_2 are:
[NO] = 0.0400 mol/2.00 L = 0.0200 M
$[O_2]$ = 0.0600 mol/2.00 L = 0.0300 M

The change in NO concentration due to formation of 2.2×10^{-3} M NO_2 is 2.2×10^{-3} M, because the coefficients of the two substances in the balanced equation are the same. Thus at equilibrium, [NO] = (0.0200 - 0.0022) M $= 1.78 \times 10^{-2}$. Similarly, $[O_2] = (0.0300 - (0.5)(0.0022))$ M $= 2.89 \times 10^{-2}$.

$$\text{(b) } K_c = \frac{[NO_2]^2}{[NO]^2[O_2]} = \frac{(2.2 \times 10^{-3})^2}{(1.78 \times 10^{-2})^2(2.89 \times 10^{-2})} = 0.53$$

14.44 $P_{SO_3} = 0.328 \times 88.0 = 28.9$ atm

$P_{SO_2} = 0.566 \times 88.0 = 49.8$ atm

$P_{O_2} = 0.106 \times 88.0 = 9.3$ atm

$$K_p = \frac{P^2_{SO_3}}{P^2_{SO_2} \times P_{O_2}} = \frac{(28.9)^2}{(49.8)^2(9.3)} = 0.0362$$

14.45 $K_c = \dfrac{[CO]^2}{[CO_2]}$

To calculate K_c we need concentrations in units of moles/L. From the ideal gas law, n/V = P/RT. Since the total pressure is 1 atm in all cases, we can easily calculate n/V for CO_2 and CO at each temperature. For example, at 850°C (1123 K):

$$[CO_2] = \frac{(0.0623 \text{ atm})}{\left(\frac{0.0821 \text{ L-atm}}{\text{mol-K}}\right)(1123 \text{ K})} = 6.75 \times 10^{-4} \text{ M}$$

$$[CO] = \frac{(0.9377 \text{ atm})}{\left(\frac{0.0821 \text{ L-atm}}{\text{mol-K}}\right)(1123 \text{ K})} = 1.02 \times 10^{-2} \text{ M}$$

Temp (K)	$[CO_2]$	[CO]	K_c
1123	6.75×10^{-4}	1.02×10^{-2}	0.154
1223	1.31×10^{-4}	9.83×10^{-3}	0.735
1323	3.41×10^{-5}	9.18×10^{-3}	2.47
1473	5×10^{-6}	8.26×10^{-3}	14

From the fact that K_c grows larger with increasing temperature, we can infer that the reaction is endothermic in the forward direction.

14.46 In general, the reaction quotient is of the form

$$Q = \frac{[NOCl]^2}{[NO]^2[Cl_2]} \qquad \text{(a) } Q = \frac{(0.11)^2}{(0.15)^2(0.31)} = 1.7$$

$Q > K_P$; therefore, the reaction will shift toward reactants; i.e., to the left, in moving toward equilibrium.

$$\text{(b) } Q = \frac{(0.050)^2}{(0.12)^2(0.10)} = 1.7$$

$Q > K_P$; therefore, the reaction will shift toward reactants; i.e., to the left, in moving toward equilibrium.

$$\text{(c) } Q = \frac{(5.10 \times 10^{-3})^2}{(0.15)^2(0.20)} = 5.78 \times 10^{-3}$$

$Q < K_P$. Therefore, the reaction mixture will shift in the direction of more product, i.e., to the right, in moving toward equilibrium.

14.47 (a) To the left, to partially reduce the CO_2 concentration; (b) To the right, to form more CO_2; (c) To the left; a volume increase would produce a lower pressure, and the system will respond by a shift in equilibrium toward the side that produces more moles of gas. (d) Increase in pressure will cause a shift to the right, in the direction of fewer moles of gas. (e) Increase in temperature will cause a shift in equilibrium to the left, the direction in which the reaction is endothermic.

14.48 Increasing temperature causes a decrease in the value of K for reactions such as this, that are exothermic in the forward direction.

14.49 (a) The equilibrium as written is endothermic in the forward direction; that is, heat is required to drive the reaction to the right. Equilibrium is thus shifted to the right with increasing temperature. Of course, we already know this well; vapor pressure increases with increasing temperature (Section 11.3). (b) The equilibrium constant is not dependent on the quantity of liquid present, so long as there is any present.

14.50 $K = \frac{[CO_2]}{[CO]} = 600$

If P_{CO} is 150 mm Hg, P_{CO_2} can never exceed 760 - 150 = 610 mm Hg. Then $Q = 610/150 = 4.1$. Since this is far less than K, reaction will shift in the direction of more product. Reduction will therefore occur.

14.51 $CO(g) + 2H_2(g) \rightleftarrows CH_3OH(\ell)$; $\Delta H = -238.6 - (-110.5) = -128.1$ kJ

The process is exothermic; therefore, the equilibrium constant decreases with increasing temperature. All the gas components of the reaction are on the left. The extent of reaction would therefore increase with an increase in total pressure.

14.52 To solve this problem we must be careful about units. We should place all concentrations into units of atm. For example, using $P = (n/v)RT$, the hydrogen pressure is calculated to be:

$$P = \left(\frac{10^2 \text{ molecules}}{1 \text{ cm}^3}\right)\left(\frac{1 \text{ mol}}{6.02 \times 10^{23} \text{ molecules}}\right)\left(\frac{1000 \text{ cm}^3}{\text{L}}\right)\left(\frac{0.0821 \text{ L-atm}}{\text{mol-K}}\right)$$

$$\times (100 \text{ K}) = 1.36 \times 10^{-18} \text{ atm.}$$

Similarly, for N_2, $P = 1.36 \times 10^{-20}$ atm. $K = \frac{[NH_3]^2}{[H_2]^3[N_2]} = 6 \times 10^{37}$.

If we insert the above concentrations for H_2 and N_2 and solve for $[NH_3]$, we obtain $[NH_3] = 2 \times 10^{-18}$ atm. Since this is on the same order of magnitude as the pressure of the more abundant H_2, we can conclude that at equilibrium a significant fraction of the N_2 present would be converted to NH_3.

14.53 (a) The mixing of equal volumes reduces each concentration by a factor of 2. $Q = 1/[Ag^+][Cl^-] = 1/(1.0 \times 10^{-3}/2)(2.0 \times 10^{-4}/2) = 2.0 \times 10^7$. (b) Because Q is less than K, reaction will proceed to the right to attain equilibrium. That is, AgCl(s) will form until the concentrations of $Ag^+(aq)$ and $Cl^-(aq)$ are sufficiently low to meet the condition Q = K.

14.54 $K_c = [Br]^2/[Br_2] = 1.04 \times 10^{-3}$. Let the concentration of Br atoms at equilibrium be x. Then at equilibrium [Br] = x, $[Br_2] = 1.00 - x/2$

$$\frac{x^2}{1.00 - x/2} = 1.04 \times 10^{-3}$$

We use the quadratic formula to obtain $x = [Br] = 3.2 \times 10^{-2}$ M.
(b) The equilibrium $[Br_2]$ is 1.00 - 0.016 = 0.984 M. Thus, the fraction of the original Br_2 that dissociates is 0.016/1 = 0.016.

14.55 Let $P_{NH_3} = P_{HCl} = x$; $K_P = (P_{NH_3})(P_{HCl}) = 6.09 \times 10^{-9}$;

$$x = (6.09 \times 10^{-9})^{1/2} = 7.7 \times 10^{-5} \text{ atm}$$

14.56 $H_2(g) + I_2(g) \rightleftarrows 2HI(g)$

	$H_2(g)$	$I_2(g)$	$2HI(g)$
initial	0.055 M	0.055 M	0.39 M + 0.06 M
equilibrium	0.055 + x	0.055 + x	0.45 - x

$$K_c = 50.5 = \frac{[HI]^2}{[H_2][I_2]} = \frac{(0.45 - x)^2}{(0.055 + x)^2}$$

Take the square root of both sides:

$$\frac{0.45 - x}{0.055 + x} = (50.5)^{1/2} = 7.11$$

Solve for x: $x = 7.27 \times 10^{-3}$. New equilibrium concentrations are thus $[H_2] = [I_2] = 0.062$ M, [HI] = 0.44 M.

14.57 The patent claim is false. A catalyst does not alter the position of equilibrium in a system, only the rate of approach to the equilibrium condition.

14.58 $$K_p = \frac{P_{O_2} \times P_{CO}^2}{P_{CO_2}^2}$$

$$P_{O_2} = (0.03)(1 \text{ atm}) = 0.03 \text{ atm}$$

$$P_{CO} = (0.002)(1 \text{ atm}) = 0.002 \text{ atm}$$

$$P_{CO_2} = (0.12)(1 \text{ atm}) = 0.12 \text{ atm}$$

$$Q = \frac{(0.03)(0.002)^2}{(0.12)^2} = 8.3 \times 10^{-6}$$

Since $Q > K_P$, the system will shift to the left to attain equilibrium. Thus a catalyst that promoted the attainment of equilibrium would result in a lower CO content in the exhaust.

14.59 $K_p = k_f/k_r = 1.32 \times 10^{10}$

$$k_r = \frac{6.26 \times 10^{8}/\text{M-sec}}{1.32 \times 10^{10}} = 4.74 \times 10^{-2}/\text{M-sec}$$

15

Acids and bases

15.1 The HCl molecule is polar. There is thus little driving force for solution of HCl in the nonpolar benzene solvent. HCl reacts with water to form $H^+(aq)$ and $Cl^-(aq)$. It is this reaction that gives rise to the high solubility of HCl in water.

15.2 The proton is bound to one or more water molecules, forming species such as H_3O^+, $H_5O_2^+$ and $H_9O_4^+$ (Figure 15.3). The state of the proton is essentially independent of the anion from which it is derived, so long as the ionization is complete, or nearly so. This is the case for the strong acids HI and HCl.

15.3 It is probable that the cation is $H_5O_2^+$.

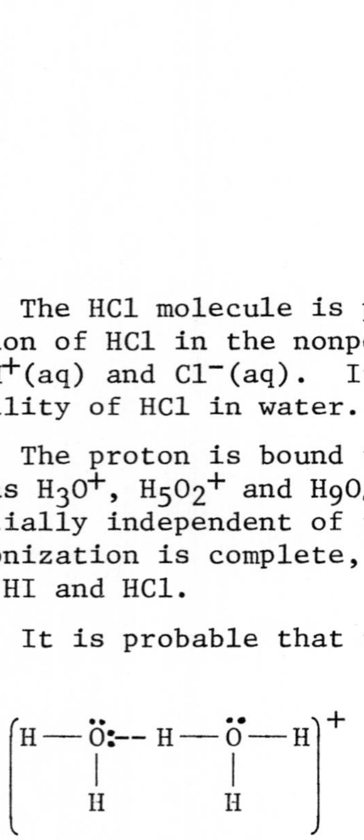

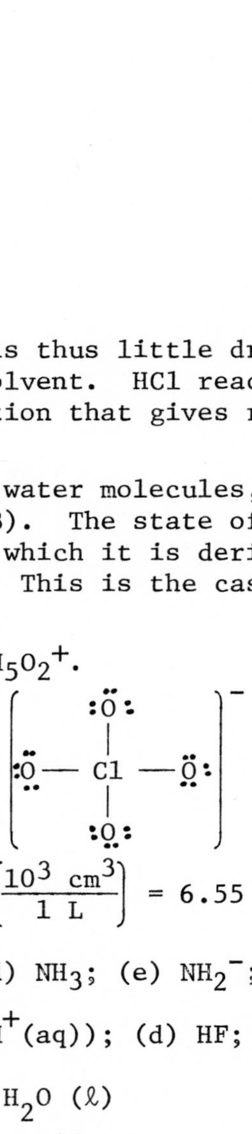

15.4 $$\left(\frac{524\ \text{g HI}}{1000\ \text{g Soln}}\right)\left(\frac{1\ \text{mol HI}}{128\ \text{g HI}}\right)\left(\frac{1.60\ \text{g HI}}{1\ \text{cm}^3}\right)\left(\frac{10^3\ \text{cm}^3}{1\ \text{L}}\right) = 6.55\ M$$

15.5 (a) CN^-; (b) HSO_4^-; (c) $C_2H_3O_2^-$; (d) NH_3; (e) NH_2^-; (f) SO_3^{2-}; (g) O^{2-}

15.6 (a) H_2O; (b) $H_2PO_4^-$; (c) H_3O^+ (or $H^+(aq)$); (d) HF; (e) NH_4^+

15.7 (a) $NH_4^+(aq) + OH^-(aq) \rightleftarrows NH_3(aq) + H_2O\ (\ell)$

acid base base acid

(b) $HCN(aq) + H_2O(\ell) \rightleftarrows H_3O^+(aq) + CN^-(aq)$

acid base acid base

(c) $O^{2-}(aq) + H_2O(\ell) \rightleftarrows OH^-(aq) + OH^-(aq)$

base acid base acid

In the reverse reaction, one OH^- group acts as an acid (proton donor) while the other acts as a base (proton acceptor). The reverse reaction does not occur to a significant extent, so the labels on the OH^- ions are hypothetical.

(d) $H^-(aq) + H_2O(\ell) \rightleftarrows H_2(g) + OH^-(aq)$

base acid acid base

(Again, the reverse reaction does not occur to a significant extent.)

(e) $HC_2O_4^-(aq) + CO_3^{2-}(aq) \rightleftarrows C_2O_4^{2-}(aq) + HCO_3^-(aq)$

acid base base acid

15.8 (a) F^-; (b) S^{2-}; (c) NO_2^-; (d) OH^-

15.9 In each case we must identify the two bases that compete for the proton. If the stronger base has the proton on the left, then reaction will not proceed far to the right. If the weaker base has the proton on the left side of the equation, then the reaction will proceed far to the right. (a) No. The two competing bases are CO_3^{2-} and F^-. Since CO_3^{2-} is a stronger base (Figure 15.4), reaction will not proceed far to the right. (b) yes; (c) no; (d) yes

15.10 (a) basic; (b) acidic; (c) neutral; (d) acidic; (e) basic (if $[OH^-] > 1 \times 10^{-7}$, $[H^+]$ is $< 1 \times 10^{-7}$); (f) acidic; (g) basic

15.11 $[H^+]^2 = 1.14 \times 10^{-15}$; $[H^+] = 3.38 \times 10^{-8}$ M

15.12 (a) 1.30; (b) 4.16; (c) 11.40; (d) 6.94. Solutions (a), (b) and (d) are acidic (pH < 7.0). Solution (c) is basic.

15.13 (a) 3.63×10^{-3} M; (b) 1.0×10^{-3} M; (c) 1.35×10^{-13} M; (d) 2.82×10^{-7} M; (e) 8.91×10^{-9} M; (f) 1.07×10^{-7} M. Solutions (a), (b), (d) and (f) are acidic (pH < 7, $[H^+] > 1.0 \times 10^{-7}$); solutions (c) and (e) are basic.

15.14 (a) 8×10^{-7} M; (b) 2.5×10^{-7} M; (c) 5×10^{-3} M; (d) 3×10^{-9} M; (e) 1×10^{-12}. Solutions (a), (b) and (c) are acidic; solutions (d) and (e) are basic.

15.15 (a) pOH = $-\log[6.8 \times 10^{-4}] = 3.17$; (b) $[OH^-] = 5 \times 10^{-5}$ M; (c) pOH = 14 - pH = 8.2; (d) pH = 14 - pOH = 11.38; (e) pKa = $-\log[K_a] = -\log[5.82 \times 10^{-4}] = 3.23$; (f) $K_a = 1.1 \times 10^{-6}$

15.16 pH is above 2.3, because methyl yellow is yellow, below 4.2 because it turns methyl red to red.

15.17 (a) 12.0; (b) 1.26; (c) 12.60 (Remember there are two OH^- groups per $Ca(OH)_2$ formula.) (d) 4.43

15.18 If pH = 12.00, $[OH^-] = 1.00 \times 10^{-2}$ M

$$\left(\frac{1.00 \times 10^{-2} \text{ mol NaOH}}{1 \text{ L}}\right)(0.500 \text{ L})\left(\frac{40.0 \text{ g NaOH}}{1 \text{ mol NaOH}}\right) = 0.200 \text{ g NaOH}$$

15.19 (a) $\left(\frac{0.640 \text{ g KOH}}{0.800 \text{ L Soln}}\right)\left(\frac{1 \text{ mol KOH}}{56.1 \text{ g KOH}}\right)\left(\frac{1 \text{ mol OH}^-}{1 \text{ mol KOH}}\right) = 1.43 \times 10^{-2}$ M OH^-

pH = 14 - pOH = 14 - 1.85 = 12.15

(b) $\left(\frac{0.807 \text{ g HI}}{0.200 \text{ L Soln}}\right)\left(\frac{1 \text{ mol HI}}{128 \text{ g HI}}\right)\left(\frac{1 \text{ mol H}^+}{1 \text{ mol HI}}\right) = 0.0315$ M H^+

pH = -log(0.0315) = 1.50

15.20 pH will be higher for the HF solution, because HF is a weak acid, whereas HCl is a strong acid. Thus $[H^+]$ is smaller in the HF solution, and pH is higher.

15.21 A strong acid in water is a substance that transfers a proton to the solvent in a reaction that proceeds essentially to completion. In the proton transfer reaction of a weak acid with water, the reaction proceeds only partly toward completion.

15.22 The common strong acids are $HClO_4$, HCl, HBr, HI, HNO_3, and H_2SO_4. The common strong bases are NaOH, KOH, $Mg(OH)_2$, $Ca(OH)_2$, $Ba(OH)_2$. The hydroxides of the other less familiar metals of groups 1A and 2A, and the oxides of all the metals of these two families (excepting Be) are also strong bases in water.

15.23 In pure water the pH is 7.0, because of the autoionization of the solvent. The concentration of H^+ from this source is greater than the concentration that derives from the acid when the acid concentration is so low as in this case. Even for strong acids the acid concentration must be about 10^{-6} M or greater in order that the concentration of H^+ from the acid be large in comparison with that deriving from the solvent itself. We are normally interested in solutions of much higher concentration, so this matter doesn't arise.

15.24 Phenol, with a K_a of only 1.3×10^{-10}, is the weakest acid in the group; ascorbic acid is the strongest.

15.25 (a) $HNO_2(aq) \rightleftarrows H^+(aq) + NO_2^-(aq)$

$$K_a = \frac{[H^+][NO_2^-]}{[HNO_2]}$$

(b) $HC_2O_4^-(aq) \rightleftarrows H^+(aq) + C_2O_4^{2-}(aq)$

$$K_a = \frac{[H^+][C_2O_4^{2-}]}{[HC_2O_4^-]}$$

15.26 In working these problems, you should follow the procedure outlined in Sample Exercise 15.8. The calculations are outlined briefly as follows:

(a) $\frac{x^2}{(0.01 - x)} \approx \frac{x^2}{0.01} = 1.9 \times 10^{-5}$

$x = 4.4 \times 10^{-4}$ M $= [H^+]$

(b) $\frac{x^2}{0.10 - x} \approx \frac{x^2}{0.10} = 6.5 \times 10^{-5}$

$x = 2.5 \times 10^{-3}$ M $= [H^+]$

(c) $\frac{x^2}{2.5 - x} = 3.5 \times 10^{-4} \approx \frac{x^2}{2.5}$

$x = 3.0 \times 10^{-2}$ M $= [H^+]$

15.27 First we must calculate $[H^+]$, then the concentration of $HC_7H_5O_2$ that can produce this concentration of $[H^+]$. We also know that

$[H^+] = [C_7H_5O_2^-]$. Because $-\log[H^+] = 2.80$, $[H^+] = 1.58 \times 10^{-3}$ M. Thus,

$$\frac{(1.58 \times 10^{-3})^2}{y} = K_a = 6.5 \times 10^{-5}; \; y = 3.8 \times 10^{-2} \text{ M}$$

The total amount of benzoic acid required is that which is present as unionized material plus that which has been ionized: 3.8×10^{-2} M $+ 1.6 \times 10^{-3}$ M $= 4.0 \times 10^{-2}$ M.

$$\left(\frac{4.0 \times 10^{-2} \text{ mol } HC_7H_5O_2}{1 \text{ L Soln}}\right)(3.00 \text{ L})\left(\frac{122 \text{ g } HC_7H_5O_2}{1 \text{ mol } HC_7H_5O_2}\right) = 14.6 \text{ g} \simeq 15 \text{ g } HC_7H_5O_2$$

15.28 We would need to consider the H^+ ions derived from water itself when the acidic substance present as solute produces H^+ ions in concentrations on the order of 10^{-7} M or less. This could occur because the substance is a very weak acid, $K_a \sim 10^{-14}$, or is present in such low concentration that only a low concentration of $H^+(aq)$ can result.

15.29 $$\left(\frac{500 \text{ mg } HC_6H_7O_6}{0.200 \text{ L}}\right)\left(\frac{1 \text{ g}}{1000 \text{ mg}}\right)\left(\frac{1 \text{ mol } HC_6H_7O_6}{176 \text{ g } HC_6H_7O_6}\right) = 1.42 \times 10^{-2} \text{ M}$$

$$\frac{x^2}{1.42 \times 10^{-2} - x} = 8.0 \times 10^{-5} \simeq \frac{x^2}{1.42 \times 10^{-2}}. \quad x = 1.06 \times 10^{-3} \text{ M}$$

Note that this is nearly 10% of the original concentration of acid. Our approximation that x is small in comparison with 1.42×10^{-2} is thus not very good. One easy way of getting a better estimate for x is to put the value we have just calculated for x in the demoninator, then solve for a new x, as follows:

$$\frac{x^2}{1.42 \times 10^{-2} - 1.06 \times 10^{-3}} = 8.0 \times 10^{-5}. \quad x = 1.03 \times 10^{-3} \text{ M} = [H^+]$$

$$pH = 2.99$$

15.30 We follow the general procedure outlined in Sample Exercise 15.9:

(a) $\frac{x^2}{1.00} \approx 1.31 \times 10^{-5}$; $x = 3.6 \times 10^{-3}$ M $= [H^+]$

percent ionized $= \left(\frac{3.6 \times 10^{-3}}{1.00}\right) 100 = 0.36\%$

(b) $\frac{x^2}{0.100} \approx 1.31 \times 10^{-5}$; $x = 1.14 \times 10^{-3} = [H^+]$

percent ionized $= \left(\frac{1.14 \times 10^{-3}}{0.100}\right) 100 = 1.1\%$

(c) $\frac{x^2}{0.0100} \approx 1.31 \times 10^{-5}$; $x = 3.6 \times 10^{-4} = [H^+]$

percent ionized $= \left(\frac{3.6 \times 10^{-4}}{0.0100}\right) 100 = 3.6\%$

15.31

	$HX(aq)$ $\rightleftarrows$	$H^+(aq)$ +	$X^-(aq)$
Equil. Concentration =	(0.815)(0.010)	(0.185)(0.010)	(0.185)(0.010)
=	8.15×10^{-3} M	1.85×10^{-3} M	1.85×10^{-3} M

$$K_a = (1.85 \times 10^{-3})^2/8.15 \times 10^{-3} = 4.20 \times 10^{-4}$$

15.32 $H_3C_6H_5O_7(aq) \rightleftharpoons H^+(aq) + H_2C_6H_5O_7^-(aq)$ (largest K_a)

$H_2C_6H_5O_7^-(aq) \rightleftharpoons H^+(aq) + HC_6H_5O_7^{2-}(aq)$

$HC_6H_5O_7^{2-}(aq) \rightleftharpoons H^+(aq) + C_6H_5O_7^{3-}(aq)$ (smallest K_a)

The dissociation constants should decrease in the order listed. As each proton is removed, it becomes increasingly difficult to remove the next positive charge.

15.33 HSO_3^- is the strongest acid, HPO_4^{2-} is the weakest.

15.34 $$\left(\frac{180 \text{ g Rhu}}{1.0 \text{ L}}\right)\left(\frac{2.4 \times 10^{-3} \text{ g } H_2C_2O_4}{1 \text{ g Rhu}}\right)\left(\frac{1 \text{ mol } H_2C_2O_4}{90.0 \text{ g } H_2C_2O_4}\right) = 4.8 \times 10^{-3} \text{ M}$$

Because K_{a1} for oxalic acid is fairly large, we must use the quadratic formula in solving for $[H^+]$.

$$\frac{x^2}{4.8 \times 10^{-3} - x} = 5.9 \times 10^{-2}; \quad x^2 + 5.9 \times 10^{-2}\, x - 2.83 \times 10^{-4} = 0$$

$$x = \frac{-5.9 \times 10^{-2} \pm [3.48 \times 10^{-3} + 1.13 \times 10^{-3}]^{1/2}}{2}$$

Only the solution with x positive, $x = 4.4 \times 10^{-3}$ M, is physically reasonable. Thus, $pH = -\log(4.4 \times 10^{-3}) = 2.4$. With only two significant figures, our calculation is not very precise, but it is adequate for the needs of the problem. Note that a large fraction of the oxalic acid is dissociated.

15.35 (a) $CH_3NH_2(aq) + H_2O(\ell) \rightleftharpoons CH_3NH_3^+(aq) + OH^-(aq)$

$$K_b = \frac{[CH_3NH_3^+][OH^-]}{[CH_3NH_2]}$$

(b) $CN^-(aq) + H_2O(\ell) \rightleftharpoons HCN(aq) + OH^-(aq)$

$$K_b = \frac{[HCN][OH^-]}{[CN^-]}$$

15.36 We follow the procedure spelled out in detail in Sample Exercise 15.11. The calculations are in brief as follows:

(a) $\frac{x^2}{0.010 - x} = 1.7 \times 10^{-9} \approx \frac{x^2}{0.010}$; $x^2 = 1.7 \times 10^{-11}$; $x = 4.1 \times 10^{-6}$

$x = [C_6H_5NH^+] = [OH^-]$. Thus, $pOH = 5.39$; $pH = 14 - 5.39 = 8.61$

(b) $\frac{x^2}{0.020 - x} = 4.4 \times 10^{-4} \approx \frac{x^2}{0.020}$; $x^2 = 8.8 \times 10^{-6}$; $x = 2.96 \times 10^{-3}$

At this point we note that the value of x obtained by the short-cut of ignoring x in the denominator is rather large in comparison with 0.020. We can go back and use the quadratic formula, which will give us $x = [OH^-] = 2.75 \times 10^{-3}$. A quick way to get an accurate answer. however, is simply to take our first, approximate value of x, 2.96×10^{-3}, and substitute it into the denominator, then solve for x^2 again:

$$\frac{x^2}{(0.020 - 2.96 \times 10^{-3})} = 4.4 \times 10^{-4}; \quad x^2 = 7.50 \times 10^{-6}$$

$$x = 2.74 \times 10^{-3} = [OH^-]$$

$pOH = -\log(2.74 \times 10^{-3}) = 2.56$. $pH = 14 - 2.56 = 11.44$

(c) $\frac{x^2}{0.10 - x} = 1.8 \times 10^{-5} \approx \frac{x^2}{0.10}$; $x^2 = 1.8 \times 10^{-6}$; $x = 1.3 \times 10^{-3}$

$x = [OH^-] = 1.3 \times 10^{-3}$ $\quad$ $pOH = -\log(1.3 \times 10^{-3}) = 2.87$

$pH = 14 - 2.87 = 11.13$

15.37 $pK_b = -\log(K_b) = 3.267$. If you don't have a log calculator, you need to find the antilog of -3.267. This is -4.000 + 0.733. In a log table you will find that the antilog of 0.733 is 5.41. Thus, $K_b = 5.41 \times 10^{-4}$. The equilibrium of interest is $(CH_3)_2NH(aq) + H_2O(\ell) \rightleftarrows (CH_3)_2NH_2^+(aq) + OH^-(aq)$. Proceeding as usual:

$$\frac{x^2}{0.010 - x} = 5.41 \times 10^{-4} \approx \frac{x^2}{0.010} ; \quad x^2 = 5.41 \times 10^{-6}$$

$x = 2.3 \times 10^{-3} = [OH^-]$; $pOH = -\log(2.3 \times 10^{-3}) = 2.63$

$pH = 14 - pOH = 11.37$. Using the quadratic formula we obtain $pH = 11.32$.

15.38 We employ the relationship that for a conjugate acid-base pair $K_a \times K_b = K_w = 1.0 \times 10^{-14}$.

(a) $K_b = 10^{-14}/1.8 \times 10^{-4} = 5.6 \times 10^{-11}$

(b) $K_b = 10^{-14}/4.2 \times 10^{-13} = 2.38 \times 10^{-2}$

(Note that we use K_{a3} here because it is the reaction $PO_4^{3-} + H_2O \rightleftarrows HPO_4^{2-} + OH^-$ that is of interest.)

(c) $K_b = 10^{-14}/4.9 \times 10^{-10} = 2.0 \times 10^{-5}$

(d) $K_b = 10^{-14}/1.4 \times 10^{-4} = 7.1 \times 10^{-11}$

15.39 As in Exercise 15.38, we employ the relationship that for a conjugate acid-base pair, $K_a \times K_b = K_w = 1 \times 10^{-14}$.

(a) $K_a = 10^{-14}/4.4 \times 10^{-4} = 2.3 \times 10^{-11}$

(b) $K_a = 10^{-14}/1.7 \times 10^{-9} = 5.9 \times 10^{-6}$

15.40 The reaction of interest is $PrnH^+(aq) \rightleftarrows Prn(aq) + H^+(aq)$. For this process $K_a = 10^{-14}/K_b$. Taking the negative log of both sides, $pK_a = 14 - pK_b$. We know that pK_b is 5.05; therefore, $pK_a = 14 - 5.05 = 8.95$. (Alternatively, we could have taken the antilog of pK_b to obtain $K_b = 8.91 \times 10^{-6}$, then substituted in above.) The antilog of -8.95 is $1.12 \times 10^{-9} = K_a$.

Proceeding as in the usual weak acid dissociation problem:

$$\frac{x^2}{0.010 - x} = 1.12 \times 10^{-9} \approx \frac{x^2}{0.010} ; \quad x^2 = 1.12 \times 10^{-11}$$

$x = [H^+] = 3.3 \times 10^{-6}$; $pH = -\log(3.3 \times 10^{-6}) = 5.48$

15.41 $K_{b_1} = 10^{-14}/K_{a_2} = 10^{-14}/6.4 \times 10^{-8} = 1.6 \times 10^{-7}$

$K_{b_2} = 10^{-14}/K_{a_1} = 10^{-14}/1.7 \times 10^{-2} = 5.9 \times 10^{-13}$

We can see from the small value of K_{b_2} that the second reaction will not proceed to a very large extent at all in comparison with the first. Thus, we need take account only of K_{b_1} when we ask about the pH of a solution of SO_3^{2-}.

15.42 The reaction of interest is:

$CO_3^{2-}(aq) + H_2O(\ell) \rightleftarrows HCO_3^-(aq) + OH^-(aq)$

$K_b = 10^{-14}/K_{a_2} = 10^{-14}/5.6 \times 10^{-11} = 1.8 \times 10^{-4}$

Let $[HCO_3^-] = [OH^-] = x$. Then,

$\frac{x^2}{0.020 - x} = 1.8 \times 10^{-4} \approx \frac{x^2}{0.02}$; $x^2 = 3.6 \times 10^{-6}$

$[OH^-] = x = 1.9 \times 10^{-3}$ M ; $pOH = -\log[OH^-] = -\log(1.9 \times 10^{-3}) = 2.72$

$pH = 14 - 2.72 = 11.28$

To more accurately assess the $[OH^-]$ we should use the quadratic formula, which gives us $[OH^-] = 1.8 \times 10^{-3}$, pOH = 2.74, pH = 11.26. We conclude that the 0.020 M Na_2CO_3 solution should contain a warning label.

15.43 $C_5H_5N(aq) + H_2O(\ell) \rightleftarrows C_5H_5NH^+(aq) + OH^-(aq)$

Let $[C_5H_5NH^+] = [OH^-] = x$; $\frac{x^2}{0.10 - x} = 1.7 \times 10^{-9} \approx \frac{x^2}{0.10}$

$x = 1.3 \times 10^{-5}$ M $= [C_5H_5NH^+]$

percentage ionized $= \left(\frac{1.3 \times 10^{-5}}{0.1}\right)100 = 1.3 \times 10^{-2}\%$

15.44 (a) $CN^-(aq) + H_2O(\ell) \rightleftarrows HCN(aq) + OH^-(aq)$

$K_b = \frac{[HCN][OH^-]}{[CN^-]} = K_w/K_a = 10^{-14}/4.9 \times 10^{-10} = 2.0 \times 10^{-5}$

Let $x = [HCN] = [OH^-]$. Then,

$\frac{x^2}{0.10 - x} \approx \frac{x^2}{0.10} = 2.0 \times 10^{-5}$; $x^2 = 2.0 \times 10^{-6}$

$x = 1.4 \times 10^{-3} = [OH^-]$; $pOH = -\log(1.4 \times 10^{-3}) = 2.85$

$pH = 14 - 2.85 = 11.15$

(b) $C_2H_3O_2^-(aq) + H_2O(\ell) \rightleftarrows HC_2H_3O_2(aq) + OH^-(aq)$

$K_b = \frac{[HC_2H_3O_2][OH^-]}{[C_2H_3O_2^-]} = K_w/K_a = 10^{-14}/1.8 \times 10^{-5} = 5.6 \times 10^{-10}$

Let $[HC_2H_3O_2] = [OH^-] = x$; $\frac{x^2}{0.30 - x} \approx \frac{x^2}{0.30} = 5.6 \times 10^{-10}$

(Note that two acetate ions are formed per mole of $Ca(C_2H_3O_2)_2$.)

$x^2 = 1.65 \times 10^{-10}$; $x = 1.3 \times 10^{-5} = [OH^-]$

$pOH = -\log(1.3 \times 10^{-5}) = 4.89$; $pH = 14 - 4.89 = 9.11$

(c) $CH_3NH_3^+(aq) \rightleftarrows CH_3NH_2(aq) + H^+(aq)$

$K_a = \frac{[CH_3NH_2][H^+]}{[CH_3NH_3^+]} = K_w/K_b = 10^{-14}/4.4 \times 10^{-4} = 2.3 \times 10^{-11}$

Let $[CH_3NH_2] = [H^+] = y$; $\frac{y^2}{0.10 - y} \approx \frac{y^2}{0.10} = 2.3 \times 10^{-11}$

$y^2 = 2.3 \times 10^{-12}$; $y = 1.5 \times 10^{-6}$

$[H^+] = 1.5 \times 10^{-6}$; $pH = -\log(1.5 \times 10^{-6}) = 5.82$

(d) $PO_4^{3-}(aq) + H_2O(\ell) \rightleftarrows HPO_4^{2-}(aq) + OH^-(aq)$

$$K_b = \frac{[HPO_4^{2-}][OH^-]}{[PO_4^{3-}]} = K_w/K_{a3} = 10^{-14}/4.2 \times 10^{-13} = 2.4 \times 10^{-2}$$

Let $[HPO_4^{2-}] = [OH^-] = y$

$$\frac{y^2}{0.020 - y} = 2.4 \times 10^{-2}$$

We must use the quadratic formula to solve for y in this case:

$$y^2 + 2.4 \times 10^{-2} y - 4.8 \times 10^{-4} = 0$$

$$y = \frac{-2.4 \times 10^{-2} \pm (5.76 \times 10^{-4} + 1.92 \times 10^{-3})^{1/2}}{2}$$

$y = 1.30 \times 10^{-2}$ M $= [OH^-]$; $pOH = -\log(1.30 \times 10^{-2}) = 1.89$

$pH = 14 - 1.89 = 12.11$

15.45 (a) neutral; (b) acidic; (c) acidic; (d) neutral or slightly acidic; (e) basic; (f) neutral

15.46 $Br^- < NO_2^- < C_2H_3O_2^- < CN^-$. We deduce this order from the order of acid strengths.

15.47 (a) To answer this question you must proceed as in Sample Exercise 15.14: $HCO_3^-(aq) \rightleftarrows H^+(aq) + CO_3^{2-}(aq)$

$K_a = K_{a2} = 5.6 \times 10^{-11}$

$HCO_3^-(aq) + H_2O(\ell) \rightleftarrows H_2CO_3(aq) + OH^-(aq)$

$K_b = K_w/K_{a_1} = 10^{-14}/4.3 \times 10^{-7} = 2.3 \times 10^{-8}$

We see that the second reaction, in which HCO_3^- acts as a base, has a larger equilibrium constant than the first one listed. We can thus conclude that the solution will be basic.

(b) acidic; NH_4^+ is a weak acid, I^- is a very weak base.
(c) basic; see Sample Exercise 15.14.
(d) acidic; compare K_{a_2} for oxalic acid with K_w/K_{a_1}.

15.48 (a) $SO_3^{2-}(aq) + H_2O(\ell) \rightleftarrows HSO_3^-(aq) + OH^-(aq)$

$K_b = K_w/K_{a_2} = 1 \times 10^{-14}/6.4 \times 10^{-8} = 1.6 \times 10^{-7}$

$\frac{x^2}{0.10} = 1.6 \times 10^{-7}$; $x^2 = 1.6 \times 10^{-8}$; $x = 1.2 \times 10^{-4} = [OH^-]$

$pOH = -\log(1.2 \times 10^{-4}) = 3.90$; $pH = 14 - 3.90 = 10.10$

(b) $NH_4^+(aq) \rightleftarrows NH_3(aq) + H^+(aq)$

$K_a = K_w/K_b = 10^{-14}/1.8 \times 10^{-5} = 5.6 \times 10^{-10}$

$\frac{x^2}{0.30} = 5.6 \times 10^{-10}$; $x^2 = 1.68 \times 10^{-10}$; $x = [H^+] = 1.3 \times 10^{-5}$ M

$$pH = -\log(1.3 \times 10^{-5}) = 4.89$$

(c) $BrO^-(aq) + H_2O(\ell) \rightleftarrows HBrO(aq) + OH^-(aq)$

$$K_b = K_w/K_a = 10^{-14}/2.1 \times 10^{-9} = 4.8 \times 10^{-6}$$

$$\frac{x^2}{0.10} = 4.8 \times 10^{-6}; \quad x^2 = 4.8 \times 10^{-7}; \quad x = 6.9 \times 10^{-4} \text{ M} = [OH^-]$$

$$pOH = 3.16; \quad pH = 14 - 3.16 = 10.84$$

<u>15.49</u> (a) Fe^{3+}; the higher-charged, smaller ion will cause a larger electron shift toward the metal from coordinated water molecules (Figure 15.8), causing the protons to be more acidic. (b) The Co^{2+} ion has higher charge, smaller radius, and will thus polarize water molecules more strongly. (c) Al^{3+} is smaller than Ga^{3+} and will thus be more effective in polarizing coordinated water molecules.

<u>15.50</u> See Sample Exercise 15.14 for a worked-out analogous case.
(a) K_{a_2} is 6.8×10^{-8}; $K_w/K_{a_1} = 1.3 \times 10^{-12}$. The solution is <u>acidic</u>.
(b) <u>neutral</u>. ClO_4^- is the (very weak) conjugate base of a strong acid, $HClO_4$, thus has no tendency to undergo hydrolysis.
(c) $K_{a_2} = 1.3 \times 10^{-13}$; $K_w/K_{a_1} = 10^{-14}/5.7 \times 10^{-8} = 1.8 \times 10^{-7}$. The reaction with water, $HS^-(aq) + H_2O(\ell) \rightleftarrows H_2S(aq) + OH^-(aq)$, is the more important, and the solution is <u>basic</u>. (d) K_a for HSO_4^- is 1.1×10^{-2}, whereas the hydrolysis reaction is completely negligible. The solution is <u>acidic</u>.

<u>15.51</u> (a) Acidity increases with H-X bond polarity, because the hydrogen is more readily lost as H^+ when the H-X bond is already polarized in that direction. (b) Acidity will generally decrease as H-X bond strength increases, all other factors being equal.

<u>15.52</u> (a) H_3O^+ is stronger. The larger nuclear charge on O in H_3O^+ causes higher polarity in the O-H bonds, and thus promotes H^+ transfer.
(b) NH_4^+; higher nuclear charge on the central atom. (c) HCl. The H-F bond energy is higher than for the other hydrogen halides; this bond energy has to be given up in forming $H^+(aq) + X^-(aq)$.

<u>15.53</u>

$$H-\overset{\cdot\cdot}{\underset{\cdot\cdot}{O}}-\overset{\cdot\cdot}{N}=\overset{\cdot\cdot}{\underset{\cdot\cdot}{O}} \qquad\qquad H-\overset{\cdot\cdot}{\underset{\cdot\cdot}{O}}-\underset{\underset{:\overset{\cdot\cdot}{O}:}{|}}{N}=\overset{\cdot\cdot}{\underset{\cdot\cdot}{O}}$$

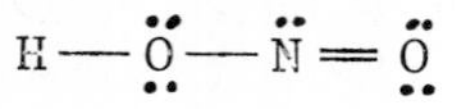

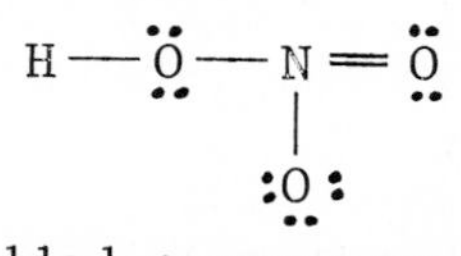

HNO_3 is the stronger acid; the added oxygen on nitrogen in HNO_3 produces an increased polarization of the HO-N bond, thus making the proton more positive, and more readily transferred as H^+.

<u>15.54</u> (a) H_2SO_4; higher charge on central sulfur; (b) H_2SO_4. Sulfur is a more electronegative element than Se. (c) HClO. Chlorine is more electro-negative than Br, and thus polarizes the OH bond a little more strongly.
(d) H_3PO_4. In $H_2PO_4^-$ the proton that departs must separate from a species that already has a -1 charge. This will occur less readily than proton loss from neutral H_3PO_4. (e) H_2SO_3. Sulfur is a more electronegative element than carbon. Thus, the OH bonds in H_2SO_3 should be more highly polarized.

15.55

Theory	acid	base
Arrhenius	forms H^+ ions in water	produces OH^- in water
Brønsted-Lowry	proton (H^+) donor	proton acceptor
Lewis	electron pair acceptor	electron pair donor

The Brønsted-Lowry Theory is more general than Arrhenius's definition, because it is based on a unified model for what processes are responsible for acidic or basic character, and it shows the relationship between these processes. The Lewis Theory is more general still because it is not restricted to only H^+ as the acidic species. Other substances that can be viewed as electron pair acceptors are encompassed under the definition of "acid."

15.56 (a) $BF_3(g)$ is the Lewis acid, $F^-(aq)$ is the Lewis base. (b) $Ag^+(aq)$ is the Lewis acid, $NH_3(aq)$ is the Lewis base. (c) HF(aq) is the Lewis acid; even though HF is dissociated eventually, we can view the initial reaction with H_2O as a Lewis acid-base process in which HF is the electron-pair acceptor toward a lone pair of electrons on H_2O, the Lewis base. (d) HF(aq) (or the proton, H^+) is the Lewis acid, $OH^-(aq)$ is the Lewis base. (e) $SO_3(aq)$ is the Lewis acid, H_2O is the Lewis base.

15.57 The Al^{3+} ion is a strong Lewis acid toward water molecules, which act as electron pair donors. The charge distribution within the water molecule is displaced toward Al^{3+}, thus making the hydrogens of the coordinated H_2O molecules more positively charged. They can more readily hydrogen bond to other water molecules, and are more readily transferred to others (see Figure 15.8 for a schematic view.)

15.58 (a) ZnI_2 (higher charge on cation); (b) $FeCl_3$ (higher charge on cation); (c) NH_4Cl (NH_4^+ is weakly acidic); (d) $FeCl_3$ (higher charge on cation).

15.59 Because pH = $-\log[H^+]$, a change in pH means $pH_1 - pH_2 = -\log\left(\frac{[H^+]_1}{[H^+]_2}\right)$. Thus when the change in pH is 2, the ratio of the hydrogen ion concentrations must be the antilog of 2, or 100.

15.60 HBr, strong acid; KOH, strong base; H_3PO_4, weak acid; CH_3NH_2, weak base; NH_4^+, weak acid; CN^-, weak base; Rb_2O, strong base.

15.61 KOH(pH = 13) > KCN > KBr(pH = 7) > NH_4Br > $HClO_4$(pH = 1) > H_2SO_4 (pH less than 1, because two protons are ionized).

15.62 The reaction rate, which depends on $[H^+]$, will decrease in the order 0.1 M H_3PO_4($K_{a_1} = 7.5 \times 10^{-3}$) > 0.1 M formic acid($K_a = 1.8 \times 10^{-4}$) > 0.1 M HCN($K_a = 4.9 \times 10^{-10}$).

15.63 (a) 12.18; (b) 2.98; (c) Proceed to solve the weak acid dissociation in the usual way; pH = 4.76. (d) CN^- is a weak base, $K_b = 10^{-14}/4.9 \times 10^{-10} = 2.0 \times 10^{-5}$; solve the weak base problem in the usual way (Sample Exercise 15.12); pH = 11.24. (e) Follow the procedure of Sample Exercise 15.11; pH = 11.97. (f) $K_a = 10^{-14}/1.8 \times 10^{-5} = 5.6 \times 10^{-10}$; pH = 5.04. (g) Convert to molarity; pH = 14.17. (h) pH = -0.05.

15.64 In all cases $[H^+] = 10^{-3}$ M. (a) Since HCl is a strong acid, 0.500 L of 1×10^{-3} M acid requires 5.00×10^{-4} mol HCl. (b) Let the concentration of nonionized acetic acid be z. Then, assuming we can ignore the small fraction of the acid that is ionized, z is the concentration we would make up. $[H^+] = [C_2H_3O_2^-] = 1 \times 10^{-3}$

$$\frac{(1 \times 10^{-3})^2}{z} = 1.8 \times 10^{-5}; \quad z = [HC_2H_3O_2] = 5.6 \times 10^{-2}\ M$$

$$\left(\frac{5.6 \times 10^{-2}\ \text{mol}\ HC_2H_3O_2}{1\ L}\right)(0.500\ L) = 2.8 \times 10^{-2}\ \text{mol}\ HC_2H_3O_2$$

(c) Proceeding as in (b), we let z = [HF]. We know that $[H^+] = [F^-] = 1 \times 10^{-3}$ M. Then,

$$\frac{(1 \times 10^{-3})^2}{z} = 6.8 \times 10^{-4}; \quad z = [HF] = 1.5 \times 10^{-3}\ M$$

Notice that in this case the equilibrium concentration of acid is not much greater than the equilibrium concentrations of H^+ and F^-. To get the correct total moles of HF we must add to the 1.5×10^{-3} M the concentration of H^+, since it requires 1 mol of HF to form 1 mol of H^+. Thus, the total concentration we want is $1.5 \times 10^{-3} + 1.0 \times 10^{-3} = 2.5 \times 10^{-3}$ M.

$$\left(\frac{2.5 \times 10^{-3}\ \text{mol HF}}{1\ L}\right)(0.500\ L) = 1.25 \times 10^{-3}\ \text{mol HF}$$

15.65 We proceed as in Sample Exercise 15.9. (a) percent ionization = 2.2×10^{-2}%; (b) percent ionization = 4.2%.

15.66 Soaps should be basic. The anions are weak bases, inasmuch as they derive from weak acids. The slippery feeling of soaps in water is the most obvious indication of basic character.

15.67 By analogy with the water case, $[NH_2^-][NH_4^+] = 1.0 \times 10^{-33}$

$$[NH_2^-] = \frac{1.0 \times 10^{-4}\ \text{mol}}{0.500\ L} = 2.0 \times 10^{-4}\ M; \text{ thus, } [NH_4^+] = \frac{1.0 \times 10^{-33}}{2.0 \times 10^{-4}} = 5.0 \times 10^{-30}\ M$$

15.68 (a) A high O_2 concentration displaces protons from Hb, and thus produces a more acidic solution, with lower pH. (b) $[H^+]$ = antilog(-7.4) = 4.0×10^{-8} M. The blood is very slightly basic. (c) We can see from the equilibrium that a high $[H^+]$ shifts the equilibrium toward the proton-bound form, HbH^+, which means a lower concentration of HbO_2^+ in the blood. Thus the ability of hemoglobin to transport oxygen is impeded.

15.69 The equilibrium is shifted to the right by addition of OH^-, which acts to remove H^+. Thus the concentration of the conjugate base from, Bb^-, increases at the expense of HBb; the blue color of Bb^- is observed in NaOH solution.

15.70 (a) increases pH; F^- addition represses ionization of HF. (b) increases pH; neutralization of HF; (c) no effect; (d) decreases pH; increases $[H^+]$; (e) increases pH; CN^- is a weak base; it reacts with $H^+(aq)$ to form HCN, thus reduces $[H^+]$ to some extent.

15.71 (a) NH_3 can't be a proton donor in water. Suppose there were a base strong enough to remove a proton from NH_3. That base would preferentially react with H_2O, because H_2O is a stronger proton donor than NH_3. The rule is that one cannot have a stronger base in a solvent than the conjugate base characteristic of that solvent. In water the characteristic conjugate base is $OH^-(aq)$. (b) NH_2^- is a stronger base than OH^-. If NH_2^- is added to water the reaction that occurs is $NH_2^-(aq) + H_2O(\ell) \rightleftharpoons NH_3(aq) + OH^-$. NH_2^- is a strong base in water. That is, the reaction proceeds far to the right.

(c) $\left(\frac{0.30 \text{ g NaNH}_2}{0.500 \text{ L}}\right)\left(\frac{1 \text{ mol NaNH}_2}{39 \text{ g NaNH}_2}\right) = 0.015 \text{ M}$

Since all the NH_2^- reacts as in (b) above, $[OH^-] = 0.015$ M, pH = 12.18.

15.72 Let $[H^+] = [NC_7H_4SO_3^-] = z$. $K_a = -\text{antilog}(11.68) = 2.1 \times 10^{-12}$

$$\frac{z^2}{0.10 - z} \sim \frac{z^2}{0.10} = 2.1 \times 10^{-12} \qquad z = [H^+] = 4.6 \times 10^{-7} \text{ M; pH} = 6.34$$

Notice that the pH is not much lower than it is in pure water. We could make a correction for the contribution from ionization of water itself to the concentration of $[H^+]$, but it turns out that this contribution can be neglected even in this case where the source of acidity is not very great.

15.73 The first ionization of H_2SO_4 is complete. Then,

$$HSO_4^-(aq) \rightleftarrows H^+(aq) + SO_4^{2-}(aq)$$

Equil. Conc.: 0.100 − x 0.100 + x x

$$\frac{(0.100 + x)(x)}{0.100 - x} = K_a = 1.2 \times 10^{-2}; \quad x^2 + 0.112\, x - 1.2 \times 10^{-3} = 0$$

$$x = \frac{-0.112 + (1.23 \times 10^{-2} + 4.4 \times 10^{-3})^{1/2}}{2} = 9.76 \times 10^{-3} \text{ M}$$

Thus, $[HSO_4^-] = 0.100 - 9.8 \times 10^{-3} = 0.090$ M

$[SO_4^{2-}] = 9.8 \times 10^{-3}$ M $[H^+] = 0.110$ M

15.74 The initial concentration is 1×10^{-6} mol/1×10^{-6} L = 1 M

Let $x = [H^+] = [CHO_2^-]$. $\frac{x^2}{1 - x} \approx \frac{x^2}{1} = 1.8 \times 10^{-4}$

$x^2 = 2 \times 10^{-4}$; $x = 1.4 \times 10^{-2}$ M; pH = 1.8

15.75 (a) $H_2C_3H_2O_4(aq) \rightleftarrows H^+(aq) + HC_3H_2O_4^-(aq)$

Let $[H^+] = [HC_3H_2O_4^-] = y$

$$\frac{y^2}{0.500 - y} = 1.5 \times 10^{-3}$$

We need to use the full quadratic form here, to obtain $y = 2.67 \times 10^{-2}$ M. A small fraction of the $HC_3H_2O_4^-$ goes on to further ionize: $HC_3H_2O_4^-(aq) \rightleftarrows H^+(aq) + C_3H_2O_4^{2-}(aq)$. Because K_{a_2} is much smaller than K_{a_1} we can ignore this process as a source of $H^+(aq)$, or as a loss of $HC_3H_2O_4^-$, but it is the only source of $C_3H_2O_4^{2-}$. Let $[C_3H_2O_4^{2-}] = z$. Then, $[H^+] = (2.67 \times 10^{-2} + z)$, $[HC_3H_2O_4^-] = 2.67 \times 10^{-2} - z$.

$$\frac{(2.67 \times 10^{-2} + z)\, z}{(2.67 \times 10^{-2} - z)} = 2.0 \times 10^{-6}$$

Try ignoring z in comparison with 2.67×10^{-2} M. Then, we obtain $z = 2.0 \times 10^{-6}$ M. This is small in comparison with 2.67×10^{-2} M, so our approximation is satisfactory. As a summary, $[H^+] = 2.67 \times 10^{-2}$ M, $[HC_3H_2O_4^-] = 2.67 \times 10^{-2}$ M, $[C_3H_2O_4^{2-}] = 2.0 \times 10^{-6}$ M. pH = 1.57.

15.76 The first dissociation equilibrium is the major source of $H^+(aq)$. The others are important because they are the sources of HPO_4^{2-} and PO_4^{3-}. $H_3PO_4(aq) \rightleftarrows H^+(aq) + H_2PO_4^-(aq)$

$$K_{a_1} = 7.5 \times 10^{-3} = \frac{[H^+][H_2PO_4^-]}{[H_3PO_4]} = \frac{x^2}{0.1 - x}$$

Use the quadratic form to solve for x, obtain $x = 2.39 \times 10^{-2}$ M = $[H^+]$ = $[H_2PO_4^-]$. Now let us calculate the concentration of $[HPO_4^{2-}]$ present at equilibrium: $H_2PO_4^-(aq) \rightleftarrows HPO_4^{2-}(aq) + H^+(aq)$.

$$K_{a_2} = 6.2 \times 10^{-8} = \frac{y(2.39 \times 10^{-2} + y)}{(2.39 \times 10^{-2} - y)}$$

If y is small in comparison with 2.39×10^{-2}, drop it from the parenthesized terms, obtain $y = 6.2 \times 10^{-8}$ M = $[HPO_4^{2-}]$. $HPO_4^{2-}(aq) \rightleftarrows PO_4^{3-}(aq) + H^+(aq)$

$$K_{a_3} = 4.2 \times 10^{-13} = \frac{z(2.39 \times 10^{-2})}{(6.2 \times 10^{-8})}$$

We ignore the slight amount of HPO_4^{2-} that ionizes. Then $z = [PO_4^{3-}] = 1.1 \times 10^{-18}$ M. We see that there is very little $PO_4^{3-}(aq)$ present in a solution of H_3PO_4.

15.77 $Mor(aq) + H_2O(\ell) \rightleftarrows HMor^+(aq) + OH^-(aq)$

$$K_b = \frac{[HMor^+][OH^-]}{[Mor]} \qquad K_b = \text{antilog}(-6.1) = 7.9 \times 10^{-7}$$

(a) $$\frac{x^2}{1.0 \times 10^{3} - x} = 7.9 \times 10^{-7}$$

Ignore x in comparison with 1.0×10^{-3}. Then, $x^2 = 7.9 \times 10^{-10}$; $x = 2.8 \times 10^{-5}$ M = $[OH^-]$; pOH = 4.55; pH = 14 - 4.55 = 9.45.

(b) $HMor^+(aq) \rightleftarrows H^+(aq) + Mor(aq)$

$$K_a = \frac{[H^+][Mor]}{[HMor^+]} = K_w/K_b = 10^{-14}/7.9 \times 10^{-7} = 1.27 \times 10^{-8}$$

$$\frac{y^2}{1.0 \times 10^{-3} - y} = 1.27 \times 10^{-8}$$

Ignore y in comparison with 1.0×10^{-3}. $y^2 = 1.27 \times 10^{-11}$; $y = 3.6 \times 10^{-6} = [H^+]$; pH = 5.45.

15.78 Let us call each compound in the neutral form Q. Then, $Q(aq) + H_2O(\ell) \rightleftarrows QH^+(aq) + OH^-$.

$$K_b = \frac{[QH^+][OH^-]}{[Q]}$$

We want to know about the ratio $[QH^+]/[Q]$, which equals $K_b/[OH^-]$. At pH = 2.5, pOH = 11.5, $[OH^-]$ = antilog(-11.5) = 3.1×10^{-12} M. Now let us calculate $K_b/[OH^-]$ for each compound:

Nicotine $\frac{[QH^+]}{[Q]} = 7 \times 10^{-7}/3.1 \times 10^{-12} = 2 \times 10^{5}$

Caffeine $\frac{[QH^+]}{[Q]} = 4 \times 10^{-14}/3.1 \times 10^{-12} = 1 \times 10^{-2}$

Strychnine $\frac{[QH^+]}{[Q]} = 1 \times 10^{-6}/3.1 \times 10^{-12} = 3 \times 10^{5}$

Quinine $\frac{[QH^+]}{[Q]} = 1.1 \times 10^{-6}/3.1 \times 10^{-12} = 3.5 \times 10^5$

We see that for all the compounds except caffeine the protonated form is in much higher concentration than the neutral form. However, for caffeine, a very weak base, the neutral form dominates.

15.79 When glycine is dissolved in water, the following two reactions are possible for the two groups that have acid-base properties:

$$-COOH(aq) \rightleftarrows -CO_2^-(aq) + H^+(aq) \qquad K_a$$

$$-NH_2(aq) + H_2O(\ell) \rightleftarrows -NH_3^+(aq) + OH^-(aq) \qquad K_b$$

$$-COOH(aq) + -NH_2(aq) + H_2O(\ell) \rightleftarrows -CO_2^-(aq) + -NH_3^+(aq) + H^+(aq) + OH^-(aq)$$

$$K = K_a \times K_b$$

If we reverse the water ionization equilibrium (which means we invert the K), then add, we obtain:

$$-COOH(aq) + -NH_2(aq) \rightleftarrows -CO_2^-(aq) + -NH_3^+(aq)$$

$$K = K_a \times K_b/K_w = 4.3 \times 10^{-3} \times 6.0 \times 10^{-5}/1 \times 10^{-14} = 2.6 \times 10^7$$

This tells us that when glycine is placed in water it exists in form II. The pH is determined by the acid dissociation constant for the $-NH_3^+$ group, which is $K_a = K_w/K_b = 1 \times 10^{-14}/6.0 \times 10^{-5} = 1.67 \times 10^{-10}$. Let $(H^+) = (-NH_2) = x$. Then,

$$\frac{x^2}{0.1 - x} \approx \frac{x^2}{0.1} = 1.67 \times 10^{-10}; \quad x = 4.1 \times 10^{-6}; \quad pH = 5.39$$

(b) In a strongly acid solution the CO_2^- function would be protonated, so glycine would exist as $^+H_3NCH_2COOH$. In strongly basic solution the $-NH_3^+$ group could be deprotonated, so glycine would be in the form $H_2NCH_2CO_2^-$.

(c) The ionization equilibrium for the $-NH_3^+$ group is

$$-NH_3^+(aq) \rightleftarrows H^+(aq) + -NH_2(aq)$$

$$K_a = \frac{[H^+][-NH_2]}{[-NH_3^+]}$$

Let us solve for the ratio $\frac{[-NH_2]}{[-NH_3^+]}$; from part (a) we have

$$\frac{[-NH_2]}{[-NH_3^+]} = K_a/[H^+] = 1.67 \times 10^{-10}/[H^+].$$ At pH = 10, $[H^+] = 10^{-10}$, so

$$\frac{[-NH_2]}{[-NH_3^+]} = 1.67.$$

There is somewhat more $H_2NCH_2CO_2^-$ present than $^+H_3NCH_2CO_2^-$. In the acidic solution we must concern ourselves with the equilibrium:

$-COOH(aq) \rightleftarrows H^+(aq) + -CO_2^-(aq)$.

$$K_a = \frac{[H^+][-CO_2^-]}{[-COOH]}.$$ The ratio $\frac{[-CO_2^-]}{[-COOH]} = K_a/[H^+] = 4.3 \times 10^{-3}/[H^+]$.

At pH 2, $[H^+] = 1 \times 10^{-2}$, $[-CO_2^-]/[-COOH] = 0.43$. That is, about two-thirds of the glycine molecules are in the form $^+H_3NCH_2COOH$, about one-third are in the form $^+H_3NCH_2CO_2^-$.

15.80 The apparent molality of the solution is given by

$$1.90°C \left(\frac{1\ \underline{m}}{1.86°C}\right) = 1.02\ \underline{m}$$

The ionization equilibrium, and molalities of the various species are:

$$HF(aq) \rightleftarrows H^+(aq) + F^-(aq)$$

Equil. Conc.: 1.00 − x x x

The total molality is thus 1.00 + x = 1.02. Thus, x = 0.02. For such a dilute aqueous solution, molality and molarity are essentially identical (Section 12.1). Thus, we can write

$$\frac{(0.02)^2}{0.98} = K_a = 4 \times 10^{-4}$$

The accuracy of this number is not very good; to improve upon it, we would need to know the freezing point depression more accurately.

15.81 $Coc(aq) + H_2O(\ell) \rightleftarrows HCoc^+(aq) + OH^-(aq)$. When pH = 10.04, $[H^+]$ = antilog(−10.04) = 9.1×10^{-11}; $[OH^-] = 10^{-14}/9.1 \times 10^{-11} = 1.1 \times 10^{-4}$. We see from the above equilibrium that $[HCoc^+] = [OH^-]$. Thus,

$$\frac{(1.1 \times 10^{-4})^2}{(5.0 \times 10^{-3} - 1.1 \times 10^{-4})} = K_b;\ K_b = 2.5 \times 10^{-6}$$

16

Aqueous equilibria

16.1 You should proceed as in Sample Exercise 16.1. Briefly, $HC_7H_5O_2(aq) \rightleftarrows H^+(aq) + C_7H_5O_2^-(aq)$. Let $[H^+] = x$. Then

$$\frac{x(0.15 + x)}{(0.10 - x)} = K_a = 6.5 \times 10^{-5}; \quad \frac{x(0.15)}{(0.10)} = 6.5 \times 10^{-5}$$

$$x = 4.3 \times 10^{-5} = [H^+]; \quad pH = 4.36$$

16.2 (a) pH = 3.17; (b) $\frac{x(0.050)}{(0.035)} = 1.3 \times 10^{-5}$; $x = [H^+] = 9.1 \times 10^{-6}$; pH = 5.04

(c) If pH = 4.80, $[H^+] = 1.6 \times 10^{-5}$ M; $\frac{1.6 \times 10^{-5}[C_3H_5O_2^-]}{0.035} = 1.3 \times 10^{-5}$

$[C_3H_5O_2^-] = 2.8 \times 10^{-2}$ M

16.3 The methods for solving these problems are detailed in Sample Exercises 16.1, 16.2 and 16.3. Briefly: (a) $NaNO_3$ has no effect on the acid-base equilibrium involving $HNO_2(aq)$. Following the procedure for a simple weak acid dissociation, we obtain $x^2/(0.20 - x) = 4.5 \times 10^{-4}$. Using quadratic form, $x = [H^+] = 9.3 \times 10^{-3}$ M; pH = 2.03. (b) Because benzoic acid is a fairly weak acid, the concentration of $[H^+]$ that comes from its dissociation is entirely negligible in comparison with the 0.10 M that originates with the strong acid, HNO_3. Thus, pH = 1.0.

(c) $C_5H_5N(aq) + H_2O(\ell) \rightleftarrows C_5H_5NH^+(aq) + OH^-(aq)$

$$K_b = 1.7 \times 10^{-9} = \frac{[C_5H_5NH^+][OH^-]}{[C_5H_5N]} = \frac{(0.010 + x)x}{(0.010 - x)}$$

Ignore x relative to 0.010 M. $x = 1.7 \times 10^{-9}$ M $= [OH^-]$. pOH = 8.76; pH = 5.23.

(d) We can ignore the small contribution to the OH^- concentration that would come from $NH_3(aq)$ in comparison with NaOH. Thus, $[OH^-] = 0.10$, pOH = 1, pH = 13.

16.4 (a) $C_2H_3O_2^-(aq) + H^+(aq) \rightleftarrows HC_2H_3O_2(aq)$

(b) $$K = \frac{[HC_2H_3O_2]}{[C_2H_3O_2^-][H^+]} = 1/K_a = 1/1.8 \times 10^{-5} = 5.6 \times 10^4$$

(c) The total volume after mixing is 1.0 L. The concentration of Na^+ is 0.05 M, that of ClO_4^- is 0.10 M. As to the others, we know from the fact that K is large that the reaction will proceed far to the right. Let us assume complete reaction of the 0.05 mol of $C_2H_3O_2^-$ with an equal number of moles of H^+ to yield 0.05 mol $HC_2H_3O_2$, and leaving 0.05 mol $H^+(aq)$ unreacted. Then our "initial" situation, and equilibrium situations are as follows:

	$C_2H_3O_2^-(aq)$ +	$H^+(aq)$ $\rightleftarrows$	$HC_2H_3O_2(aq)$
initial concentrations	0	0.05 M	0.05 M
equilibrium concentrations	x	0.05 + x	0.05 − x

$$\frac{(0.05 - x)}{x(0.05 + x)} = 5.5 \times 10^4$$

Assume x is small relative to 0.05. Then $x = 1.8 \times 10^{-5}$ M = $[C_2H_3O_2^-]$. Since x is small relative to 0.05 we can say that $[H^+] = [HC_2H_3O_2] \cong 0.05$ M.

16.5 The reaction of interest is $F^-(aq) + H^+(aq) \rightleftarrows HF(aq)$. We can work this problem as we did 16.4, or we can go at it slightly differently, by writing the reverse of the above reaction, $HF(aq) \rightleftarrows H^+(aq) + F^-(aq)$, for which we know that $K_a = 6.8 \times 10^{-4}$. Let us first calculate how many moles we have of F^- and H^+ in the mixed solution:

$$\left(\frac{0.10 \text{ mol NaF}}{1 \text{ L}}\right)(0.050 \text{ L}) = 5.0 \times 10^{-3} \text{ mol NaF}$$

$$\left(\frac{0.20 \text{ mol HCl}}{1 \text{ L}}\right)(0.020 \text{ L}) = 4.0 \times 10^{-3} \text{ mol HCl}$$

When these react we form 4.0×10^{-3} mol HF, with 1.0×10^{-3} mol F^- remaining. The volume of solution is 0.070 L, so the respective molarities are 5.7×10^{-2} M and 1.4×10^{-2} M. Then we have the usual weak acid dissociation problem, with an added common ion. Let $[H^+] = x$.

$$\frac{x(1.42 \times 10^{-2} + x)}{(5.72 \times 10^{-2} - x)} = 6.8 \times 10^{-4}$$

Ignore x in the parenthetical terms, obtain $x = 2.7 \times 10^{-3}$ M. This is rather large in comparison with the concentrations in parentheses. To obtain a better estimate we could retain x in the parenthetical terms and solve the full quadratic equation, or we can use our first guess at x in the parenthetical terms, then solve for a new x:

$$\frac{x(1.42 \times 10^{-2} + 2.7 \times 10^{-3})}{(5.72 \times 10^{-2} - 2.7 \times 10^{-3})} = 6.8 \times 10^{-4}; \qquad x = 2.2 \times 10^{-3} \text{ M}$$

Thus, $[H^+] = 2.2 \times 10^{-3}$ M, $[F^-] = 1.64 \times 10^{-2}$ M, $[HF] = 5.5 \times 10^{-2}$ M; $[Cl^-] = 4.0 \times 10^{-3}$ mol/0.070 L = 5.7×10^{-2} M.

16.6 A buffer solution possesses a capacity for reacting with either an added acid or added base to maintain nearly constant pH. The buffer solution consists of an equilibrium mixture of a weak acid and its conjugate base. The pH of the mixture depends on the strength of the acid (that is, on pK_a). When an acid is added, there is a reaction with the conjugate base, thus consuming a large fraction of the added acid. Similarly, when a base is added, the weak acid reacts with it, consuming a large fraction of the added base.

16.7 Add a few drops of 1 M HCl solution to a sample from each. The pH of the buffer solution, which you can test by using pH paper, a pH meter or indicators, will change much less than the pH of the KBr solution.

16.8 The equilibrium established in the buffer solution is:

$$HC_2H_3O_2(aq) \rightleftarrows H^+(aq) + C_2H_3O_2^-(aq) \qquad (A)$$

There are roughly equal concentrations of $HC_2H_3O_2$ and $C_2H_3O_2^-$ present in the solution. When a strong acid is added, reaction occurs with the base:

$$H^+(aq) + C_2H_3O_2^-(aq) \rightarrow HC_2H_3O_2(aq)$$

That is, the added $H^+(aq)$ are largely consumed in reaction with $C_2H_3O_2^-(aq)$, thus maintaining relatively constant pH. When a strong base is added, reaction occurs with the proton: $H^+(aq) + OH^-(aq) \rightarrow H_2O(\ell)$. As $H^+(aq)$ is depleted in equilibrium (A), ionization of $HC_2H_3O_2(aq)$ occurs to restore $[H^+]$ (LeChatelier's principle), thus limiting the pH change that can occur.

16.9 The capacity of a buffer solution relates to the molar quantities of acid and conjugate base that are present to react with added $H^+(aq)$ or $OH^-(aq)$. Thus 0.500 L of a 0.10 M $HC_2H_3O_2/NaC_2H_3O_2$ buffer has a ten times greater capacity to absorb added acid or base than 0.500 L of a 0.010 M $HC_2H_3O_2/NaC_2H_3O_2$ buffer.

16.10 The greatest buffer capacity is possessed by solution (c), because the number of moles of acid or conjugate base per unit volume available to react with added base or added acid is largest. Solution (b) has the least capacity as a buffer, because there is no acid-conjugate base equilibrium present. The total quantity of acid available to react with added base is only 1.8×10^{-5} mol/L, and there is no buffering capacity toward added acid.

16.11 $HCO_3^-(aq) + H_2O(\ell) \rightleftarrows H_2CO_3(aq) + OH^-(aq)$

$$K_b = \frac{[H_2CO_3][OH^-]}{[HCO_3^-]} = 1 \times 10^{-14}/K_{a_1} = 2.3 \times 10^{-8}$$

$$\frac{[H_2CO_3]}{[HCO_3^-]} = 2.3 \times 10^{-8}/[OH^-]. \quad \text{Since pH} = 7.4,\ \text{pOH} = 6.6$$

$$[OH^-] = 2.5 \times 10^{-7}$$

$$\text{Thus} \quad \frac{[H_2CO_3]}{[HCO_3^-]} = \frac{2.3 \times 10^{-8}}{2.5 \times 10^{-7}} = 0.092.$$

$$\frac{[HCO_3^-]}{[H_2CO_3]} = 11$$

16.12 When the acid-base pair are present in equal concentrations, the capacity of the buffer to react with either an added acid or added base is the same. When one of the two components is present in lower concentration than the other, the buffer has a more limited ability to maintain near constant pH when the component present in lower concentration is called upon to react. For example, in a buffer solution that is 0.10 M in $NaC_2H_3O_2$ and 0.01 M in $HC_2H_3O_2$, the ability of the solution to control pH when base is added is weaker than its ability to control pH when acid is added.

16.13 The pH of this buffer initially is given by using Equation [16.12]. pH = 3.74 + log 1 = 3.74. We use Equation [16.12] in calculating pH following additions of acid or base.

(a) After addition of 1×10^{-3} mol HCl, $[HCHO_2]$ = 0.101 M, $[CHO_2^-]$ = 0.099 M.

$$pH = 3.74 + \log\frac{(0.099)}{(0.101)} = 3.73$$

(b) Similarly, $pH = 3.74 + \log\frac{(0.090)}{(0.110)} = 3.65$

(c) $pH = 3.74 + \log\frac{(0.101)}{(0.099)} = 3.75$

(d) $pH = 3.74 + \log\frac{(0.110)}{(0.090)} = 3.83$

16.14 (a) Using Equation [16.12], $pH = 4.74 + \log\frac{(0.10)}{(0.15)} = 4.56$

(b) After addition of 0.01 mol HNO_3, $[HC_2H_3O_2]$ = 0.16 M, $[C_2H_3O_2^-]$ = 0.090 M. Then

$$pH = 4.74 + \log\frac{(0.090)}{(0.16)} = 4.49$$

(c) Similarly, $pH = 4.74 + \log\frac{(0.11)}{(0.14)} = 4.64$

16.15 (a) The quantity of base required to reach the equivalence point is the same in the two titrations. (b) pH is higher initially in titration of weak acid. (c) It is higher at the equivalence point in titration of weak acid. (d) pH in excess base is essentially the same for the two cases. (e) In titrating a weak acid, one needs an indicator that changes at a higher pH than for the strong acid titration. The choice is more critical because the change in pH close to the equivalence point is smaller for the weak acid titration.

16.16 (a) We can proceed as follows:

$$\left(\frac{0.040 \text{ mol HBr}}{1 \text{ L Soln}}\right)(0.025 \text{ L Soln})\left(\frac{1 \text{ mol NaOH}}{1 \text{ mol HBr}}\right)\left(\frac{1 \text{ L NaOH Soln}}{0.025 \text{ mol NaOH}}\right)$$

$$= 0.040 \text{ L NaOH Soln}$$

(b) Proceeding similarly, 36.1 mL NaOH Soln.

16.17 (a) $\left(\frac{0.120 \text{ mol KOH}}{1 \text{ L}}\right)(0.0350 \text{ L})\left(\frac{1 \text{ mol HCl}}{1 \text{ mol KOH}}\right) = 4.20 \times 10^{-3} \text{ mol HCl}$

(b) $\left(\frac{0.120 \text{ mol NaOH}}{1 \text{ L}}\right)(0.0400 \text{ L})\left(\frac{1 \text{ mol HNO}_3}{1 \text{ mol NaOH}}\right)\left(\frac{1 \text{ L HNO}_3 \text{ Soln}}{0.100 \text{ mol HNO}_3}\right)\left(\frac{1000 \text{ mL}}{1 \text{ L}}\right)$

$= 48.0 \text{ mL HNO}_3 \text{ Soln}$

(c) $(0.510 \text{ g NaOH})\left(\frac{1 \text{ mol NaOH}}{40.0 \text{ g NaOH}}\right)\left(\frac{1 \text{ mol HNO}_3}{1 \text{ mol NaOH}}\right)\left(\frac{1 \text{ L HNO}_3 \text{ Soln}}{0.100 \text{ mol HNO}_3}\right)\left(\frac{1000 \text{ mL}}{1 \text{ L}}\right)$

$= 128 \text{ mL HNO}_3 \text{ Soln}$

(d) $\left(\frac{0.100 \text{ mol NaOH}}{1 \text{ L Soln}}\right)(0.0300 \text{ L Soln})\left(\frac{1 \text{ mol HCl}}{1 \text{ mol NaOH}}\right) = 3.00 \times 10^{-3} \text{ mol HCl}$

$$\text{molarity} = \frac{3.00 \times 10^{-3} \text{ mol HCl}}{0.0205 \text{ L}} = 0.146 \text{ M}$$

16.18 (a) Initially pH = 1.00. (b) volume of solution = 75.0 mL. Initially we had (0.050 L)(0.100 mol/1 L) = 5.00×10^{-3} mol HCl. After adding 25.0 mL of 0.100 M NaOH we have 2.5×10^{-3} mol HCl in 75.0 mL. Thus, $[H^+]$ = 2.5×10^{-3} mol/0.0750 L = 3.33×10^{-2} M. pH = 1.48. (c) We have added 4.99×10^{-3} mol NaOH; thus, there remain 1.0×10^{-5} mol HCl in 99.9 mL. $[H^+] = 1 \times 10^{-5}$ mol/0.0999 L = 1.0×10^{-4} M; pH = 4.00. (d) When exactly 50.0 mL NaOH have been added, neutralization is complete and pH = 7.00. (e) When 50.1 mL 0.100 M NaOH have been added, we have an excess of 1.0×10^{-5} mol OH^- in 100.1 mL solution. Thus, $[OH^-] = 1.0 \times 10^{-5}$ mol/ 0.1001 L = 1.00×10^{-4} M. pOH = 4.0, pH = 10.0. (f) We have an excess of 2.5×10^{-3} mol NaOH in 0.125 L solution. Thus $[OH^-] = 2.5 \times 10^{-3}$ mol/0.125 L = 2.00×10^{-2} M. pOH = 1.70, pH = 12.30.

16.19 (a) Solving the usual weak acid dissociation problem: $x^2 = (6.5 \times 10^{-5})(0.100) = 6.5 \times 10^{-6}$; $x = 2.5 \times 10^{-3} = [H^+]$; pH = 2.59. (b) When 25.0 mL base have been added, bringing the total volume to 75.0 mL, $[HC_7H_5O_2] = 3.33 \times 10^{-2}$ M, $[C_7H_5O_2^-] = 3.33 \times 10^{-2}$ M. Using Equation [16.12], pH = 4.19 + log(1) = 4.19. (c) At this point, $[HC_7H_5O_2] = 1.0 \times 10^{-4}$ M, $[C_7H_5O_2^-] = 4.99 \times 10^{-2}$ M. pH = 4.19 + log(4.99 x 10^{-2})/(1.0 x 10^{-5}) = 6.89. (d) At the equivalence point, we have precisely 5.00×10^{-3} mol $NaC_7H_5O_2$ in 100.0 mL solution, so $[C_7H_5O_2^-] = 5.00 \times 10^{-2}$ M.

$$C_7H_5O_2^-(aq) + H_2O(\ell) \rightleftarrows HC_7H_5O_2(aq) + OH^-(aq)$$

$$\frac{x^2}{5.00 \times 10^{-2}} = 1.54 \times 10^{-10}; \quad x = [OH^-] = 2.78 \times 10^{-6}; \quad pH = 8.44$$

(e) We now have essentially 5.00×10^{-2} M $C_7H_5O_2^-$, and 1.00×10^{-4} M OH^-, from the excess NaOH. Since this amount of OH^- is already substantially greater than that provided by hydrolysis of $C_7H_5O_2^-$, we may say that $[OH^-]$ is approximately 1.0×10^{-4} M, so pH = 10.0. (f) When we have added 75 mL NaOH, we have 2.5×10^{-3} mol NaOH in 125 mL, so $[OH^-] = 2.00 \times 10^{-2}$ M, pH = 12.3.

You will find it instructive to compare the answer to problem 16.18, which involves titration of a strong acid, to that for 16.19, which involves a titration of a weak acid.

16.20 (a) pH = 7.0, since we have just $KClO_4$ in solution at the equivalence point. (Remember that $HClO_4$ is one of the strong acids.) (b) At the equivalence point we have a 0.050 M solution of $HNC_5H_5^+$, which dissociates as a weak acid:

$$HNC_5H_5^+(aq) \rightleftarrows H^+(aq) + NC_5H_5(aq)$$

$$K_a = K_w/K_b = 5.9 \times 10^{-6} = x^2/(0.050); \; x = 5.4 \times 10^{-4} \text{ M} = [H^+]; \; pH = 3.27$$

(c) $C_2H_5NH_3^+(aq) \rightleftarrows C_2H_5NH_2(aq) + H^+(aq)$; $K_a = K_w/K_b = 1.56 \times 10^{-11} = x^2/0.050$

$$x = 8.84 \times 10^{-7} = [H^+]; \quad pH = 6.05$$

16.21 (a) We could use methyl red, bromthymol blue, thymol blue or phenolphthalein, because in a strong acid-strong base titration a large pH change occurs over a small range of volume change in the region about the equivalence point. (b) Methyl orange would be the best choice in this case. (c) Methyl red or bromthymol blue would be best choices.

16.22 (a) $K_{sp} = [Ag^+][Br^-]$; (b) $K_{sp} = [Mg^{2+}][C_2O_4^{2-}]$; (c) $K_{sp} = [Cd^{2+}][IO_3^-]^2$; (d) $K_{sp} = [Ag^+]^2[SO_4^{2-}]$

16.23 The solubility equilibrium is $Mg(OH)_2(s) \rightleftarrows Mg^{2+}(aq) + 2OH^-(aq)$. $K_{sp} = [Mg^{2+}][OH^-]^2$. We see from the balanced equation that two moles of $OH^-(aq)$ are formed for each $Mg^{2+}(aq)$. Thus, $[Mg^{2+}] = 1.8 \times 10^{-4}$ M, $[OH^-] = 3.6 \times 10^{-4}$ M. $K_{sp} = (1.8 \times 10^{-4})(3.6 \times 10^{-4})^2 = 2.3 \times 10^{-11}$.

16.24 (a) $$\left(\frac{0.0283 \text{ g AgIO}_3}{1 \text{ L}}\right)\left(\frac{1 \text{ mol AgIO}_3}{283 \text{ g AgIO}_3}\right) = \frac{1.0 \times 10^{-4} \text{ mol AgIO}_3}{\text{L}}$$

$K_{sp} = [Ag^+][IO_3^-] = (1.0 \times 10^{-4})^2 = 1.0 \times 10^{-8}$

(b) $$\left(\frac{0.0729 \text{ g MgF}_2}{1 \text{ L}}\right)\left(\frac{1 \text{ mol MgF}_2}{62.3 \text{ g MgF}_2}\right) = \frac{1.17 \times 10^{-3} \text{ mol MgF}_2}{\text{L}}$$

$K_{sp} = [Mg^{2+}][F^-]^2 = (1.17 \times 10^{-3})(2.34 \times 10^{-3})^2 = 6.4 \times 10^{-9}$

(c) $$\left(\frac{3.51 \times 10^{-2} \text{ g Ag}_2\text{CO}_3}{1 \text{ L}}\right)\left(\frac{1 \text{ mol Ag}_2\text{CO}_3}{276 \text{ g Ag}_2\text{CO}_3}\right) = 1.27 \times 10^{-4} \frac{\text{mol Ag}_2\text{CO}_3}{1 \text{ L}}$$

$K_{sp} = [Ag^+]^2[CO_3^{2-}] = (2.54 \times 10^{-4})^2(1.27 \times 10^{-4}) = 8.2 \times 10^{-12}$

16.25 $NiCO_3(s) \rightleftarrows Ni^{2+}(aq) + CO_3^{2-}(aq)$. $K_{sp} = [Ni^{2+}][CO_3^{2-}]$. In a solution containing only Ni^{2+} and CO_3^{2-} in equilibrium with solid $NiCO_3$, and neglecting hydrolysis of CO_3^{2-}, we can write $[Ni^{2+}] = [CO_3^{2-}]$. Thus, $x^2 = 6.6 \times 10^{-9}$; $x = [Ni^{2+}] = [CO_3^{2-}] = 8.1 \times 10^{-5}$ M.

16.26 (a) $SrCO_3(s) \rightleftarrows Sr^{2+}(aq) + CO_3^{2-}(aq)$; $K_{sp} = [Sr^{2+}][CO_3^{2-}] = 1.1 \times 10^{-10}$. Let $[Sr^{2+}] = [CO_3^{2-}] = x$; $x^2 = 1.1 \times 10^{-10}$; $x = 1.0 \times 10^{-5}$ M = solubility of $SrCO_3$.

$$\left(\frac{1.0 \times 10^{-5} \text{ mol SrCO}_3}{1}\right)\left(\frac{148 \text{ g SrCO}_3}{1 \text{ mol SrCO}_3}\right) = 1.5 \times 10^{-3} \text{ g SrCO}_3\text{/L}$$

(b) $Cd(OH)_2(s) \rightleftarrows Cd^{2+}(aq) + 2OH^-(aq)$; $K_{sp} = [Cd^{2+}][OH^-]^2 = 2.5 \times 10^{-14}$ Let $[Cd^{2+}] = y$. Then $[OH^-] = 2y$; $[y][2y]^2 = 2.5 \times 10^{-14}$. $4y^3 = 2.5 \times 10^{-14}$; $y = 1.8 \times 10^{-5}$

$$\left(\frac{1.8 \times 10^{-5} \text{ mol Cd(OH)}_2}{\text{L}}\right)\left(\frac{146 \text{ g Cd(OH)}_2}{1 \text{ mol Cd(OH)}_2}\right) = 2.6 \times 10^{-3} \text{ g Cd(OH)}_2\text{/L}$$

(c) $Cu_2S(s) \rightleftarrows 2Cu^+(aq) + S^{2-}(aq)$; $K_{sp} = [Cu^+]^2[S^{2-}]$. We must take account of the sulfide ion hydrolysis: $S^{2-}(aq) + H_2O(\ell) \rightleftarrows HS^-(aq) + OH^-(aq)$, for which the equilibrium constant is $K_w/K_{a2} = 1 \times 10^{-14}/1.3 \times 10^{-13} = 8 \times 10^{-2}$. Because K_{sp} for Cu_2S is very small, we can assume that $[OH^-]$ remains 1×10^{-7} M. Thus, $[HS^-]/[S^{2-}] = 8 \times 10^{-2}/[OH^-] = 8 \times 10^5$. Most of the S^{2-} is converted to HS^-. Let the number of moles of Cu_2S that dissolves per liter equal y. Then $[Cu^+] = 2y$, $[S^{2-}] = y/8 \times 10^5$. $(2y)^2(1.3 \times 10^{-6}\ y) = 2.5 \times 10^{-48}$; $y = 7.9 \times 10^{-15}$ M.

$$\left(\frac{7.9 \times 10^{-15} \text{ mol Cu}_2\text{S}}{\text{L}}\right)\left(\frac{159 \text{ g Cu}_2\text{S}}{1 \text{ mol Cu}_2\text{S}}\right) = 1.3 \times 10^{-12} \text{ Cu}_2\text{S g/L}$$

16.27 $K_{sp} = [Mg^{2+}][OH^-]^2 = 1.8 \times 10^{-11}$. Let $[Mg^{2+}] = x$. Then $[OH^-] = 2x$. $(x)(2x)^2 = 1.8 \times 10^{-11}$. $x = 1.6 \times 10^{-4}$ M; $[OH^-] = 2x = 3.2 \times 10^{-4}$ M. pOH = 3.49; pH = 10.51.

16.28 (a) $K_{sp} = [Ca^{2+}][OH^-]^2 = 5.5 \times 10^{-6}$. Let $[Ca^{2+}] = y$. Then $[OH^-] = 2y$. $4y^3 = 5.5 \times 10^{-6}$; $y = 1.1 \times 10^{-2}$ M.

$$\left(\frac{1.1 \times 10^{-2} \text{ mol Ca(OH)}_2}{1 \text{ L}}\right)\left(\frac{0.10 \text{ L}}{100 \text{ mL}}\right)\left(\frac{74 \text{ g Ca(OH)}_2}{1 \text{ mol Ca(OH)}_2}\right) = \frac{8.1 \times 10^{-2} \text{ g Ca(OH)}_2}{100 \text{ mL}}$$

(b) $[OH^-] = 2y = 2.2 \times 10^{-2}$ M; pOH = 1.66; pH = 12.34

16.29 (a) $K_{sp} = [Pb^{2+}][SO_4^{2-}] = 1.6 \times 10^{-8}$. Let $x = [Pb^{2+}] = [SO_4^{2-}]$. $x^2 = 1.6 \times 10^{-8}$; $x = 1.3 \times 10^{-4}$ M. This is the molar solubility. (b) Assume that the SO_4^{2-} concentration is determined by the added SO_4^{2-}. Then $[Pb^{2+}](0.010) = 1.6 \times 10^{-8}$; $[Pb^{2+}] = 1.6 \times 10^{-6}$ M. This is the molar solubility of $PbSO_4$ in 0.010 M SO_4^{2-} solution. (c) Here we have $[Pb^{2+}] = 0.010$ M. $(0.010)[SO_4^{2-}] = 1.6 \times 10^{-8}$. $[SO_4^{2-}] = 1.6 \times 10^{-6}$ M. This is the molar solubility of $PbSO_4$ in a solution containing 0.010 M Pb^{2+}. Note that the common-ion effect can be exerted by either the anion or cation of the salt to limit solubility.

16.30 (a) $K_{sp} = [Ag^+]^2[CrO_4^{2-}] = 1.1 \times 10^{-12}$. Let $[CrO_4^{2-}] = z$. Then $[Ag^+] = 2z$. $4z^3 = 1.1 \times 10^{-12}$. $z = 6.5 \times 10^{-5}$ M. This is the molar solubility of Ag_2CrO_4 in pure water. (b) The CrO_4^{2-} concentration from Na_2CrO_4 is much greater than that produced by Ag_2CrO_4. Thus, $[Ag^+]^2(0.010) = 1.1 \times 10^{-12}$; $[Ag^+] = 1.0 \times 10^{-5}$ M. Note that two Ag^+ ions are formed from each Ag_2CrO_4 formula unit that dissolves. Thus the molar solubility of Ag_2CrO_4 in 0.010 M Na_2CrO_4 solution is 5.0×10^{-6} M. (c) $(0.010)^2[CrO_4^{2-}] = 1.1 \times 10^{-12}$. $[CrO_4^{2-}] = 1.1 \times 10^{-8}$ M. This is the molar solubility of Ag_2CrO_4 in 0.010 M $AgNO_3$ solution.

16.31 After mixing the solutions the concentration of Pb^{2+} would be 0.050 M, that of Cl^- would be 0.025 M. Then $Q = [Pb^{2+}][Cl^-]^2 = (0.050)(0.025)^2 = 3.1 \times 10^{-5}$. Since this exceeds K_{sp}, 1.6×10^{-5}, a precipitate should form.

16.32 We know that $K_{sp} = [Ca^{2+}][OH^-]^2 = 5.5 \times 10^{-6}$. If we assume $[Ca^{2+}]$ remains constant at 0.10 M, then $[OH^-]^2 = 5.5 \times 10^{-6}/0.10 = 5.5 \times 10^{-5}$; $[OH^-] = 7.4 \times 10^{-3}$ M. Thus, pH = 14 − pOH = 11.87. Precipitation of $Ca(OH)_2$ will occur when pH is 11.87 or greater.

16.33 When pH = 10.0, pOH = 4, thus $[OH^-] = 1 \times 10^{-4}$ M. We have $Q = [Mg^{2+}][OH^-]^2 = (0.010)(1 \times 10^{-4})^2 = 1.0 \times 10^{-10}$. This is greater than K_{sp}, 1.8×10^{-11}. Thus, $Mg(OH)_2$ should begin to precipitate.

16.34 $$\left(\frac{0.010 \text{ mol CaCl}_2}{1 \text{ L}}\right)(0.0300 \text{ L}) = 3.00 \times 10^{-4} \text{ mol CaCl}_2$$

$$\left(\frac{1.0 \times 10^{-3} \text{ mol Na}_2\text{CO}_3}{1 \text{ L}}\right)(0.0700 \text{ L}) = 7.0 \times 10^{-5} \text{ mol Na}_2\text{CO}_3$$

Upon mixing the total solution, volume is 0.100 L. Thus, $[Ca^{2+}] = 3.00 \times 10^{-3}$ M, $[CO_3^{2-}] = 7.0 \times 10^{-4}$ M. In calculating $[CO_3^{2-}]$, we have neglected hydrolysis: $CO_3^{2-}(aq) + H_2O(\ell) \rightleftarrows HCO_3^-(aq) + OH^-(aq)$. $K = K_w/K_{a_2} = 1.8 \times 10^{-4}$.

$$\frac{x^2}{7.0 \times 10^{-4} - x} = 1.8 \times 10^{-4}$$

Solving for x, we obtain 2.76×10^{-4}. Thus, $[CO_3^{2-}] = 7.0 \times 10^{-4} - x = 7.0 \times 10^{-4} - 2.8 \times 10^{-4}$ M $= 4.2 \times 10^{-4}$ M. $[Ca^{2+}][CO_3^{2-}] = K_{sp} = 2.8 \times 10^{-9}$.

But $Q = [Ca^{2+}][CO_3^{2-}] = (3.0 \times 10^{-4})(4.2 \times 10^{-4}) = 1.3 \times 10^{-7}$. Because $Q > K_{sp}$ we can expect precipitation to occur.

16.35 The salts soluble in acid solution are those for which the anion is the conjugate base of a weak acid. These include (a) CdS, (d) $Co(OH)_2$ and (f) $MnCO_3$. $PbSO_4$ may be slightly more soluble in strongly acid solution; however, SO_4^{2-} is a pretty weak base.

16.36 We know that a saturated solution of H_2S is about 0.1 M in H_2S. Thus, (Equation [16.24]), we can write $[H^+]^2[S^{2-}] = 7.4 \times 10^{-21}(0.1) = 7 \times 10^{-22}$.

(a) When $[S^{2-}] = 7 \times 10^{-20}$ M, $[H^+]^2 = 1 \times 10^{-2}$ M, $[H^+] = 1 \times 10^{-1}$ M, pH = 1.0.

(b) When $[S^{2-}] = 7 \times 10^{-10}$ M, $[H^+]^2 = 1 \times 10^{-12}$ M, $[H^+] = 1 \times 10^{-6}$ M, pH = 6.0.

(c) When $[S^{2-}] = 7 \times 10^{-4}$ M, $[H^+]^2 = 1 \times 10^{-18}$ M, $[H^+] = 1 \times 10^{-9}$ M, pH = 9.0.

16.37 Using equation [16.24], we see that when $[H^+] = 0.010$ M, $[S^{2-}] = 7 \times 10^{-22}/(0.010)^2 = 7 \times 10^{-18}$ M. Then $Q = [Zn^{2+}][S^{2-}] = (0.10)(7 \times 10^{-18}) = 7 \times 10^{-19}$. $K_{sp} = 1.1 \times 10^{-21}$. Because $Q > K_{sp}$, precipitation of ZnS should occur. (In practice, one does not see precipitation of ZnS at this pH, because Q is only a little larger than K_{sp}. ZnS does form readily in more basic solutions.)

16.38 When $[H^+] = 0.10$ M, $[S^{2-}] = 7 \times 10^{-22}/(0.10)^2 = 7 \times 10^{-20}$ M. We know that $[Co^{2+}][S^{2-}] = K_{sp} = 4.0 \times 10^{-21}$. Substituting, $[Co^{2+}] = 4.0 \times 10^{-21}/7 \times 10^{-20} = 0.06$ M. This is the molar solubility of CoS in water at this pH.

16.39 The amphoteric metal hydroxides are those that are insoluble in neutral water, but soluble in the presence of either excess acid or base. The amphoteric ones are (b) $Cr(OH)_3$ and (e) $Al(OH)_3$.

16.40 (a) $Zn(OH)_2(s) + 2OH^-(aq) \rightleftarrows Zn(OH)_4^{2-}(aq)$
(b) $Zn(OH)_2(s) + 2H^+(aq) \rightleftarrows Zn^{2+}(aq) + 2H_2O(\ell)$
(c) $Zn(OH)_4^{2-}(aq) + 2H^+(aq) \rightleftarrows Zn(OH)_2(s) + 2H_2O(\ell)$
(d) $Zn(OH)_4^{2-}(aq) + 4H^+(aq) \rightleftarrows Zn^{2+}(aq) + 4H_2O(\ell)$
(e) $Zn^{2+}(aq) + 2OH^-(aq) \rightleftarrows Zn(OH)_2(s)$
(f) $Zn^{2+}(aq) + 4OH^-(aq) \rightleftarrows Zn(OH)_4^{2-}(aq)$

16.41 Ammonia has an unshared electron pair centered on the nitrogen atom. It satisfies the Lewis definition of a base as an electron pair donor. The Ag^+ ion possesses a capacity to accept one or more electron pair donors such as NH_3 because of its positive charge, and the presence of valence shell orbitals on the metal that are not occupied. It is thus a Lewis acid, an electron pair acceptor.

16.42 (a) $Co^{3+}(aq) + 6NH_3(aq) \rightleftarrows Co(NH_3)_6^{3+}(aq)$

$$K_f = \frac{[Co(NH_3)_6^{3+}]}{[Co^{3+}][NH_3]^6}$$

(b) $Ag^+(aq) + 2CN^-(aq) \rightleftarrows Ag(CN)_2^-(aq)$

$$K_f = \frac{[Ag(CN)_2^-]}{[Ag^+][CN^-]^2}$$

16.43 The first two experiments eliminate Group 1 and 2 ions (Figure 16.9). The fact that no insoluble carbonates form in the filtrate from the third experiment rules out Group 4 ions. The ions which might be in the sample are those of Group 3, that is, Al^{3+}, Fe^{2+}, Zn^{2+}, Cr^{3+}, Ni^{2+}, Co^{2+}, or Mn^{2+}, and those of Group 5, NH_4^+, Na^+ or K^+.

16.44 The solubility rules (Section 12.6) tell us that all carbonates other than those of the alkali metal ions and $NH_4^+(aq)$ are insoluble. Thus, any of the ions in Groups 1, 2 or 3 of Figure 16.9 could have been responsible for the precipitate seen. You can also see this is true by looking at the solubilities of the carbonates of some of these metals such as Ni^{2+}, Co^{2+} and so forth.

16.45 $K_{sp} = [Pb^{2+}][Cl^-]^2 = 1.6 \times 10^{-5}$. When $[Cl^-] = 0.10$ M, $[Pb^{2+}] = 1.6 \times 10^{-3}$ M. Using Equation [16.24], we find that when pH = 1 (that is, $[H^+] = 0.10$ M), $[S^{2-}] = 7 \times 10^{-20}$ M. Then $Q = [Pb^{2+}][S^{2-}] = [1.6 \times 10^{-3}](7 \times 10^{-20}) = 1.1 \times 10^{-22}$. From Appendix E we note that $Q > K_{sp}$ for PbS. Thus, PbS will precipitate from the solution. Normally, rather concentrated Cl^- solutions are used to precipitate Pb^{2+} and the other metal ions of Group 1, Figure 16.9. Even though K_{sp} may be exceeded later when H_2S is added, the quantity of Pb^{2+} present is so slight that a precipitate would not be noticeable.

16.46 (a) Add a fairly concentrated (about 6 M) solution of NaOH(aq) to the solids. $Al(OH)_3$ dissolves as $Al(OH)_4^-(aq)$. Filter to remove $Fe(OH)_3(s)$, add acid to the filtrate to neutralize, thus re-precipitating $Al(OH)_3(s)$. (b) Add water, heat to boiling; $PbCl_2$ dissolves. Filter, cool filtrate to re-precipitate $PbCl_2$. (c) Adjust pH to about 0.5. Saturate solution with H_2S; HgS(s) forms, Zn^{2+} remains in solution. (d) Adjust pH to neutral or slightly basic. Add CO_2 to form $CaCO_3(s)$. (One could also precipitate Ca^{2+} as one of several other insoluble salts (Section 12.6).)

16.47 $K_{sp} = [Ag^+]^2[SO_4^{2-}] = 1.4 \times 10^{-5}$. When $[SO_4^{2-}] = 0.1$ M, $[Ag^+]^2 = 1.4 \times 10^{-4}$; $[Ag^+] = 1.2 \times 10^{-2}$ M. This represents the minimum concentration of Ag^+ below which no Ag_2SO_4 will be seen as precipitate with $[SO_4^{2-}] = 0.1$ M.

16.48 (a) $C_2H_3O_2^-(aq) + H^+(aq) \rightleftarrows HC_2H_3O_2(aq)$

(b) We have: $\left(\frac{0.100 \text{ mol HCl}}{\text{L}}\right)(0.0100 \text{ L}) = 1.00 \times 10^{-3}$ mol HCl

$$\left(\frac{0.200 \text{ mol } NaC_2H_3O_2}{1 \text{ L}}\right)(0.0200 \text{ L}) = 4.00 \times 10^{-3} \text{ mol } C_2H_3O_2^-$$

Reaction occurs to leave 1.00×10^{-3} mol $HC_2H_3O_2$, 3.00×10^{-3} mol $C_2H_3O_2^-$, in 0.300 L solution. Thus, we may write for equilibrium concentrations: $[C_2H_3O_2^-] = (0.100 + x)$ M; $[H^+] = x$; $[HC_2H_3O_2] = (3.33 \times 10^{-2} - x)$ M

$$K = 1/K_a = 1/1.8 \times 10^{-5} = 5.6 \times 10^4 = \frac{(3.33 \times 10^{-2} - x)}{(0.100 + x)\,x}$$

Neglecting the x in parentheses, $x = \dfrac{(3.33 \times 10^{-2})}{(0.100)(5.6 \times 10^{4})}$

$= 6.0 \times 10^{-6}\ M = [H^+]$; pH = 5.22

16.49 (a) $HCO_2^-(aq) + H^+(aq) \rightleftarrows HCO_2H(aq)$

(b) $2H^+(aq) + Ca(OH)_2(s) \rightleftarrows 2H_2O(\ell) + Ca^{2+}(aq)$

(c) $HF(aq) + OH^-(aq) \rightleftarrows H_2O(\ell) + F^-(aq)$

(d) $HNO_2(aq) + NH_3(aq) \rightleftarrows NH_4^+(aq) + NO_2^-(aq)$

16.50 $Mg(OH)_2(s) + 2H^+(aq) \rightleftarrows Mg^{2+}(aq) + 2H_2O(\ell)$ (A)
We can view this reaction as the sum of two other reactions, one involving the dissolving of $Mg(OH)_2(s)$ in water, the other the neutralization reaction:

$Mg(OH)_2(s) \rightleftarrows Mg^{2+}(aq) + 2OH^-(aq)$ (B)

$K_B = K_{sp} = 1.8 \times 10^{-11}$

$2OH^-(aq) + 2H^+(aq) \rightleftarrows 2H_2O(\ell)$ (C)

$K_C = (1/K_w)^2 = 1.0 \times 10^{28}$

(Note that the equilibrium constant is squared because we have two moles of H^+ and OH^- reacting, as is necessary to attain the desired balanced equation. K for reaction A above is simply the product of the K values for reactions B and C, which must be added to obtain (A). Thus, $K_A = K_B \times K_C = (1.8 \times 10^{-11})(1.0 \times 10^{28}) = 1.8 \times 10^{17}$. Clearly, reaction A proceeds far to the right in acidic solution.

16.51 All of these problems involve dissociation of the weak acid $HC_2H_3O_2$ or hydrolysis of the weak base $C_2H_3O_2^-$, with varying concentrations of common ions.

(a) Here we have simply mixed two acid solutions, one from a strong acid, the other from a weak acid. Essentially, all of the $[H^+]$ present comes from $HNO_3(aq)$:

$$\left(\frac{0.100 \text{ mol } HNO_3}{1 \text{ L}}\right)(0.050 \text{ L}) = 5.00 \times 10^{-3} \text{ mol } HNO_3$$

$$[H^+] = \frac{5.00 \times 10^{-3} \text{ mol } HNO_3}{0.100 \text{ L}} = 0.0500\ M; \quad pH = 1.30$$

(b) The concentrations of $HC_2H_3O_2$ and $C_2H_3O_2^-$ are each nearly 0.0500 M. Thus, using equation [16.12], pH = 4.74 + log(0.0500/0.0500) = 4.74.
(c) Total volume of solution is 0.075 L. The reactant solutions contain 5.00×10^{-3} mol NaOH, 2.50×10^{-3} mol $HC_2H_3O_2$. After reaction, there is 2.5×10^{-3} mol excess NaOH. Thus, $[OH^-] = 2.5 \times 10^{-3}$ mol/0.075 L $= 3.33 \times 10^{-2}\ M$; pH = 14 - pOH = 12.52. (d) Neutralization occurs, following which there is a 0.0500 M solution of $C_2H_3O_2^-$.

$C_2H_3O_2^-(aq) + H_2O(\ell) \rightleftarrows HC_2H_3O_2(aq) + OH^-(aq)$

$$K_b = K_w/K_a = 10^{-14}/1.8 \times 10^{-5} = 5.6 \times 10^{-10} = \frac{[OH^-][HC_2H_3O_2]}{[C_2H_3O_2^-]}$$

Proceeding in the usual manner (Sample Exercise 15.12):

$$\frac{x^2}{0.0500} = 5.6 \times 10^{-10}; \quad x = [OH^-] = 5.2 \times 10^{-6}\ M; \quad pH = 8.72$$

(e) In this case, half the acetic acid is neutralized. We have:

$$HC_2H_3O_2(aq) \rightleftarrows H^+(aq) + C_2H_3O_2^-(aq)$$

Equil. Conc: $(3.33 \times 10^{-2} - x)M$ $\quad x \quad$ $(3.33 \times 10^{-2} + x)M$

Proceeding as in Sample Exercise 16.1, we obtain:

$$\frac{x(3.33 \times 10^{-2} + x)}{(3.33 \times 10^{-2} - x)} = 1.8 \times 10^{-5}; \quad x = [H^+] = 1.8 \times 10^{-5}; \quad pH = 4.74$$

(f) In this case, we have precisely the same result as in (e) above. The added HCl reacts with $C_2H_3O_2^-$ to form $HC_2H_3O_2$. The molar quantities of $HC_2H_3O_2$ and $C_2H_3O_2^-$ at equilibrium, and the final volume are just as in part (e).

16.52 The reaction involved is $HA(aq) + OH^-(aq) \rightarrow A^-(aq) + H_2O(\ell)$. We thus have 0.05 mol A^- and 0.20 mol HA in a total volume of 1 L, so the "initial" molarities of A^- and HA are 0.05 M and 0.20 M, respectively. The weak acid equilibrium of interest is $HA(aq) \rightleftarrows H^+(aq) + A^-(aq)$.

$$K_a = \frac{[H^+][A^-]}{[HA]}$$

Using Equation [16.12], $6.0 = pK_a + \log \frac{[A^-]}{[HA]}$. $pK_a + \log\left(\frac{0.05}{0.20}\right)$.

$$pK_a = 6.0 + 0.60 = 6.60$$

16.53 Using Equation [16.12], $5.32 = pK_a + \log\left(\frac{2.0}{0.15}\right)$; $pK_a = 4.20$

16.54 The equilibrium involved can be written as:

$$(CH_3)_2NH_2^+(aq) \rightleftarrows (CH_3)_2NH + H^+(aq)$$

for which $K_a = 10^{-14}/K_b = 10^{-14}/5.4 \times 10^{-4} = 1.8 \times 10^{-11}$.

$$10.38 = 10.73 + \log\left(\frac{(0.25)}{[(CH_3)_2NH_2^+]}\right)$$

$$\log[(CH_3)_2NH_2^+] = -0.252; \quad [(CH_3)_2NH_2^+] = 0.56\ M$$

16.55 Use the value obtained for pH of the acid solution to determine K_a. $[H^+]$ = antilog of 4.35 = 4.47×10^{-5} M.

$$\frac{(4.47 \times 10^{-5})^2}{0.10} = K_a; \quad K_a = 2.0 \times 10^{-8}; \quad pK_a = 7.70$$

$$pH = 7.70 + \log\left(\frac{0.020}{0.10}\right) = 7.00$$

16.56 $6.77 = 7.52 + \log \frac{[ClO^-]}{(0.100)}$; $[ClO^-] = 0.018$ M

Moles required is $\left(\frac{0.018 \text{ mol NaClO}}{1 \text{ L}}\right)(0.500 \text{ L}) = 9.0 \times 10^{-3}$ mol

16.57 We look for an acid with pK_a near 4.2. Benzoic acid, with pK_a 4.19 is the only suitable choice.

16.58 We use Eq. [16.12] in each case; then pH = pK_a. (a) 10.25; (b) 3.86; (c) 10.64; (d) 12.38

16.59 Buffer solutions limit the change in pH upon addition of acids or bases in molar quantities that do not exceed perhaps 20 percent of the molar quantity of the acid or base component of the mixture present in lower molar quantity. When a huge molar excess is added, as in this example, the acid-base equilibrium in the buffer, which is the basis for the control, is overwhelmed. The acetate ion in the buffer is completely used up in the reaction with added HCl.

16.60 (a) $\text{pH} = 3.85 + \log\left(\frac{0.040}{0.050}\right) = 3.75$. Let HL = lactic acid.

(b) $\left(\frac{0.10 \text{ mol HCl}}{1 \text{ L}}\right)(1.0 \times 10^{-2} \text{ L}) = 1.0 \times 10^{-3} \text{ mol HCl}$.

The initial molar quantities of HL and L^- are 1.0×10^{-2} mol HL and 8.0×10^{-3} mol L^-, respectively. After addition of HCl we have 1.10×10^{-2} mol HL, 7.0×10^{-3} mol L^-, in 0.210 L. Thus,

$$\text{pH} = 3.85 + \log\left(\frac{3.3 \times 10^{-2}}{5.2 \times 10^{-2}}\right) = 3.65$$

(c) $$\text{pH} = 3.85 + \log\left(\frac{4.3 \times 10^{-2}}{4.3 \times 10^{-2}}\right) = 3.85$$

16.61 $$\frac{12 \text{ g } NaH_2PO_4}{0.500 \text{ L Soln}}\left(\frac{1 \text{ mol } NaH_2PO_4}{120 \text{ g } NaH_2PO_4}\right) = 0.20 \text{ M}$$

$$\left(\frac{12 \text{ g } Na_2HPO_4}{0.500 \text{ L Soln}}\right)\left(\frac{1 \text{ mol } Na_2HPO_4}{142 \text{ g } Na_2HPO_4}\right) = 0.17 \text{ M}; \quad \text{pH} = 7.21 + \log\left(\frac{0.17}{0.20}\right) = 7.14$$

16.62 $$\left(\frac{0.30 \text{ mol } HC_2H_3O_2}{1 \text{ L}}\right)(0.400 \text{ L}) = 0.12 \text{ mol } HC_2H_3O_2$$

$$(0.12 \text{ mol } HC_2H_3O_2)\left(\frac{60 \text{ g } HC_2H_3O_2}{1 \text{ mol } HC_2H_3O_2}\right)\left(\frac{1.00 \text{ g gl acetic acid}}{0.99 \text{ g } HC_2H_3O_2}\right)$$

$$\times \left(\frac{1.00 \text{ mL gl acetic acid}}{1.05 \text{ g gl acetic acid}}\right) = 6.9 \text{ mL glacial acetic acid}$$

$$4.44 = 4.74 + \log\frac{[C_2H_3O_2^-]}{(0.30)}; \quad [C_2H_3O_2^-] = 0.15 \text{ M}$$

$$\left(\frac{0.15 \text{ mol } NaC_2H_3O_2}{1 \text{ L}}\right)(0.400 \text{ L})\left(\frac{82 \text{ g } NaC_2H_3O_2}{1 \text{ mol } NaC_2H_3O_2}\right) = 4.9 \text{ g } NaC_2H_3O_2$$

16.63 (a) $[CH_3NH_3^+] = 0.050$ M at equivalence point. Solve weak acid dissociation problem, $K_a = K_w/K_b = 1 \times 10^{-14}/4.4 \times 10^{-4} = 2.3 \times 10\text{-}11$. $[H^+] = 1.1 \times 10^{-6}$; pH = 5.97.

(b) K_b of ascorbate ion is $1 \times 10^{-14}/8.0 \times 10^{-5} = 1.2 \times 10^{-10}$. Solve for $[OH^-]$ in weak base problem, then obtain

$$\frac{x^2}{0.05} = 1.2 \times 10^{-10}; \quad x = [OH^-] = 2.4 \times 10^{-6}; \text{ pH} = 14 - \text{pOH} = 8.39$$

(c) This solution represents exact neutralization of a strong acid by a strong base. Thus, pH = 7.0.

16.64 $$\left(\frac{0.100 \text{ mol NaOH}}{1 \text{ L}}\right)(0.025 \text{ L}) = 2.5 \times 10^{-3} \text{ mol } OH^-$$

$$\left(\frac{0.100 \text{ mol } HCHO_2}{1 \text{ L}}\right)(0.050 \text{ L}) = 5.0 \times 10^{-3} \text{ mol } HCHO_2$$

Following reaction we have 2.5×10^{-3} mol each of CHO_2^-, $HCHO_2$ and Na^+, and a small quantity of H^+. The concentrations are: $[CHO_2^-] = (2.5 \times 10^{-3}$ mol/0.075 L) = 3.33×10^{-2} M; $[Na^+] = 3.33 \times 10^{-2}$ M; $[HCHO_2] = 3.33 \times 10^{-2}$ M. We employ Equation [16.12] to obtain pH:

$$\text{pH} = 3.74 + \log \frac{[CHO_2^-]}{[HCHO_2]} = 3.74; \quad [H^+] = 1.8 \times 10^{-4}$$

16.65 (a) $CO_3^{2-}(aq) + H_2O(\ell) \rightleftarrows HCO_3^-(aq) + OH^-(aq)$

$$\frac{x^2}{0.10} = K_b = 1.8 \times 10^{-4}; \quad x = [OH^-] = 4.2 \times 10^{-3} \text{ M}; \quad \text{pH} = 11.63$$

(b) Since the concentrations of the two solutions are the same, it requires 50.0 mL of 0.10 M HCl to react with the CO_3^{2-} present in 50.0 mL of 0.10 M H_2CO_3, to form HCO_3^-. (c) An additional 50.0 mL of 0.10 M HCl solution is needed to react with HCO_3^- to form H_2CO_3 at the second equivalence point.

16.66 (a) $HA(aq) + B(aq) \rightleftarrows HB^+(aq) + A^-(aq)$ (i)

$$K_c = \frac{[HB^+][A^-]}{[HA][B]}$$

(b) If we look at the above reaction, we see that the solution is slightly basic because B is a stronger base than HA is an acid. (Or, equivalently, that A^- is a stronger base than HB^+ is an acid.) Thus, a little of the A^- is used up in reaction: $A^-(aq) + H_2O(\ell) \rightleftarrows HA(aq) + OH^-(aq)$. Since pH is not very far from neutral, we can assume that (a) reaction (i) has gone far to the right, and (b) that $[A^-] \approx [HB^+]$, $[HA] \approx [B]$. Then

$$K_a = \frac{[A^-][H^+]}{[HA]} = 5.0 \times 10^{-6}; \text{ when pH} = 8.8, [H^+] = 1.6 \times 10^{-9},$$

$$\frac{[A^-]}{[HA]} = 5.0 \times 10^{-6}/1.6 \times 10^{-9} = 3.1 \times 10^3$$

From our assumptions above, $\frac{[A^-]}{[HA]} \approx \frac{[HB^+]}{[B]}$, so $K_c \approx \frac{[A^-]^2}{[HA]^2}$

Thus, $K_c = 9.8 \times 10^6$. Incidentally, we can calculate K_b for the reaction $B(aq) + H_2O(\ell) \rightleftarrows BH^+(aq) + OH^-(aq)$ by noting that the equilibrium constant for reaction (i) can be written as $K_c = K_a^{HA} K_b^B/K_w$ (we'll leave it to you to see that this is so). Then,

$$K_b^B = \frac{(9.8 \times 10^6)(1 \times 10^{-14})}{5.0 \times 10^{-6}} = 2 \times 10^{-2}$$

K_b^B is larger than K_a^A, as it must be if the solution is to be basic.

16.67 The operative reaction is $C_6H_5OH(aq) + NO_2^-(aq) \rightleftarrows C_6H_5O^-(aq) + HNO_2(aq)$.

$$K = \frac{[C_6H_5O^-][HNO_2]}{[NO_2^-][C_6H_5OH]} = \frac{K_a^{phenol}}{K_a^{HNO_2}} = 2.9 \times 10^{-7}$$

We also know that we must continue to satisfy the acid dissociation equilibrium for C_6H_5OH: $C_6H_5OH(aq) \rightleftarrows C_6H_5O^-(aq) + H^+(aq)$.

$$K_a = 1.3 \times 10^{-10} = \frac{[C_6H_5O^-][H^+]}{[C_6H_5OH]}$$

Since $[H^+] = 1.0 \times 10^{-7}$, $[C_6H_5O^-]/[C_6H_5OH] = 1.3 \times 10^{-3}$. This means that $[C_6H_5O^-] = [1.3 \times 10^{-3}][0.1] = 1.3 \times 10^{-4}$. (We have used the small value for the concentration ratio to assume that $[C_6H_5OH] = 0.1$ M.) From our first equilibrium, we see that $[HNO_2] = [C_6H_5O^-]$. Thus,

$$2.9 \times 10^{-7} = \frac{(1.3 \times 10^{-4})^2}{(0.1)[NO_2^-]} \quad ; \quad [NO_2^-] = 0.58 \text{ M}$$

This is the concentration of added $NaNO_2$, since nearly all of the NO_2^- remains unreacted.

16.68 $K_{sp} = [Mg^{2+}][C_2O_4^{2-}] = 8.6 \times 10^{-5}$

(a) $[Mg^{2+}] = [C_2O_4^{2-}] = 9.3 \times 10^{-3}$ M

$$\left(\frac{9.3 \times 10^{-3} \text{ mol } MgC_2O_4}{1 \text{ L}}\right)(0.500 \text{ L})\left(\frac{112 \text{ g } MgC_2O_4}{1 \text{ mol } MgC_2O_4}\right) = 0.53 \text{ g } MgC_2O_4$$

(b) In this case, $[Mg^{2+}] \approx 0.10$ M, $[C_2O_4^{2-}] = x$; $x(0.10) = 8.6 \times 10^{-5}$; $x = 8.6 \times 10^{-4}$ M

$$\left(\frac{8.6 \times 10^{-4} \text{ mol } MgC_2O_4}{1 \text{ L}}\right)(0.500 \text{ L})\left(\frac{112 \text{ g } MgC_2O_4}{1 \text{ mol } MgC_2O_4}\right) = 4.8 \times 10^{-2} \text{ g } MgC_2O_4$$

16.69 The solubility equilibrium involved in this problem is $PbCO_3(s) \rightleftarrows Pb^{2+}(aq) + CO_3^{2-}(aq)$.

(a) The common ion effect of CO_3^{2-} from the Na_2CO_3 represses the solubility equilibrium, shifting it toward solid $PbCO_3$ in accord with LeChatelier's principle. (b) The solubility is enhanced in acid solution because CO_3^{2-} is removed via reaction with $H^+(aq)$, to form first HCO_3^-, then $CO_2(g) + H_2O(\ell)$.

16.70 (a) $Pb^{2+}(aq) + SO_4^{2-}(aq) \rightarrow PbSO_4(s)$

(b) $2Ag^+(aq) + H_2S(aq) \rightarrow Ag_2S + 2H^+(aq)$

(c) $CaCO_3(s) + 2H^+(aq) \rightarrow Ca^{2+}(aq) + CO_2(g) + H_2O(\ell)$

(d) $AgCl(s) + 2NH_3(aq) \rightarrow Ag(NH_3)_2^+(aq) + Cl^-(aq)$

16.71 $[Ca^{2+}][F^-]^2 = K_{sp} = 3.9 \times 10^{-11}$. When $[Ca^{2+}] = x$, $[F^-] = 2x$. Thus, $4x^3 = 3.9 \times 10^{-11}$, $x = 2.1 \times 10^{-4}$ M, $[F^-] = 2(2.1 \times 10^{-4} \text{ M}) = 4.2 \times 10^{-4}$ M.

$$\left(\frac{1 \text{ g } F^-}{10^6 \text{ g } H_2O}\right)\left(\frac{1 \text{ mol } F^-}{19 \text{ g } F^-}\right)\left(\frac{10^3 \text{ g } H_2O}{1 \text{ L } H_2O}\right) = 5.3 \times 10^{-5} \text{ M}$$

16.72 (a) We know that $[Fe^{2+}][S^{2-}] = 6.3 \times 10^{-18}$. If $[Fe^{2+}]$ is 1×10^{-4} M, then $[S^{2-}]$ can be at most 6.3×10^{-14} M. We also know that $[H^+]^2[S^{2-}] \approx 7.4 \times 10^{-22}$ (Equation [16.24]). Thus, $[H^+] = 1.1 \times 10^{-4}$ M or greater, to keep FeS from forming. Thus, maximum allowable pH is about 3.9 to 4.0. (b) Using Equation [16.12],

$$4.0 = 4.74 + \log \frac{[C_2H_3O_2^-]}{[HC_2H_3O_2]}\ ; \quad \frac{[C_2H_3O_2^-]}{[HC_2H_3O_2]} = 0.18, \text{ or } \frac{[HC_2H_3O_2]}{[C_2H_3O_2^-]} = 5.6$$

16.73 The concentration of the two ions in the mixed solution are as follows:

$$(2.0 \times 10^{-6}\ M\ Cl^-)\left(\frac{10\ mL}{50\ mL}\right) = 4.0 \times 10^{-7}\ M\ Cl^-$$

$$(1.0\ M\ Ag^+)\left(\frac{40\ mL}{50\ mL}\right) = 0.80\ M\ ag^+$$

$$Q = (0.80)(4.0 \times 10^{-7}) = 3.2 \times 10^{-7} > K_{sp} = 1.8 \times 10^{-10}$$

Thus a precipitate will form. However, it may be difficult to detect because of the small quantity of chloride present.

16.74 After mixing the "initial" concentrations of the components are:

$$[Ca^{2+}] = (0.24\ M)\left(\frac{10.0\ mL}{20.0\ mL}\right) = 0.12\ M$$

$$[NH_3] = (0.20\ M)\left(\frac{10.0\ mL}{20.0\ mL}\right) = 0.10\ M$$

$$[NH_4^+] = (0.050\ M)\left(\frac{10.0\ mL}{20.0\ mL}\right) = 0.025\ M$$

First find pH. Using Equation [16.12], in which pK_a is pK_a for NH_4^+, pH = 9.26 + log(0.10/0.025) = 9.86. Thus, $[OH^-] = 7.3 \times 10^{-5}$. $Q = [Ca^{2+}][OH^-]^2 = (0.12)(7.3 \times 10^{-5})^2 = 6.4 \times 10^{-10}$. This is less than K_{sp} for $Ca(OH)_2$, 5.5×10^{-6}. Thus no precipitation will occur.

16.75 We use Equation [16.24], in which the 0.1 M concentration of H_2S in a saturated solution is included.

(a) $[S^{2-}] = 7 \times 10^{-22}/(10^{-2})^2 = 7 \times 10^{-18}$ M

(b) $[S^{2-}] = 7 \times 10^{-22}/(10^{-7})^2 = 7 \times 10^{-8}$ M

(c) $[S^{2-}] = 7 \times 10^{-22}/(10^{-11}) = 7$ M

16.76 $(1\ g\ HgS)\left(\frac{1\ mol\ HgS}{233\ g\ HgS}\right) = 4.3 \times 10^{-3}$ mol HgS; $K_{sp} = [Hg^{2+}][S^{2-}] = 4 \times 10^{-53}$

To solve this problem properly, we need to take into account the hydrolysis of S^{2-} ion, as discussed in the solution of problem 16.26(c). Using the same approach, we have for a saturated solution of the salt, $[Hg^{2+}] = y$, $[S^{2-}] = 1.3 \times 10^{-6}\ y$. Then, $y(1.3 \times 10^{-6}\ y) = 4 \times 10^{-53}$; $y = 5.4 \times 10^{-24}$.

$$(4.3 \times 10^{-3}\ mol\ HgS)\left(\frac{1\ L}{5.4 \times 10^{-24}\ mol\ HgS}\right) = 8 \times 10^{20}\ L$$

(Only one significant figure is justified by the data given.)

16.77 At pH = 2.0, $[S^{2-}] = (7 \times 10^{-22})/(1 \times 10^{-2})^2 = 7 \times 10^{-18}$ M
For HgS, $Q = (7 \times 10^{-18})(0.020) = 1.4 \times 10^{-19} > K_{sp}$

For NiS, $Q = (7 \times 10^{-18})(0.020) = 1.4 \times 10^{-19} < K_{sp}$

For PbS, $Q = (7 \times 10^{-18})(0.020) = 1.4 \times 10^{-19} > K_{sp}$

Thus, we expect HgS and PbS to precipitate, but not NiS.

16.78 $$\left(\frac{0.042\ g\ MgC_2O_4\cdot 2H_2O}{0.10\ L}\right)\left(\frac{1\ mol\ MgC_2O_4\cdot 2H_2O}{148\ g}\right) = 2.8 \times 10^{-3}\ M$$

The acid-base reaction of importance is hydrolysis of $C_2O_4^{2-}$: $C_2O_4^{2-}(aq) + H_2O(\ell) \rightleftarrows HC_2O_4^{-}(aq) + OH^-$; $K_b = K_w/K_{a_2} = 1 \times 10^{-14}/6.4 \times 10^{-5} = 1.6 \times 10^{-10}$. Proceeding in the usual way, we have:

$$\frac{x^2}{2.8 \times 10^{-3}} = 1.6 \times 10^{-10}; \quad x = 6.6 \times 10^{-7} = [OH^-];$$

$$pH = 14 - pOH = 7.82$$

16.79 The equilibrium of interest is $Fe(OH)_3(s) \rightleftarrows Fe^{3+}(aq) + 3OH^-(aq)$. If we begin by ignoring the contribution to $[OH^-]$ from the ionization of water itself, we can proceed as follows: Let $[Fe^{3+}] = x$. Then $[OH^-] = 3x$, and $(x)(3x)^3 = 4 \times 10^{-38}$; $x = 2 \times 10^{-10}$ M. But this would mean that $[OH^-] = 3x = 6 \times 10^{-10}$ M. We know that $(H^+)(OH^-) = 1 \times 10^{-14}$. In neutral water $[OH^-] = 1 \times 10^{-7}$. Thus, it appears that the added concentration of $[OH^-]$ due to dissolving of $Fe(OH)_3$ is negligible in comparison with the OH^- already present in water. We should thus use 1×10^{-7} M as the appropriate value for $[OH^-]$. Then, $[Fe^{3+}](1 \times 10^{-7})^3 = 4 \times 10^{-38}$. $[Fe^{3+}] = 4 \times 10^{-17}$ M. This value represents the molar solubility of $Fe(OH)_3$ in pure water.

16.80 We have $[Pb^{2+}][S^{2-}] = 8.0 \times 10^{-28}$; $[Mn^{2+}][S^{2-}] = 1.0 \times 10^{-13}$. If $[Mn^{2+}] = 1.0 \times 10^{-2}$ M, then $[S^{2-}]$ may be as high as 10^{-11} M before precipitation occurs. Using Equation [16.24], we see that this corresponds to $[H^+] = 8 \times 10^{-6}$ M, or pH = 5.1.

For PbS to precipitate when $[Pb^{2+}] = 1 \times 10^{-2}$ M, $[S^{2-}]$ must be 8×10^{-26} M. Using Equation [16.24], we see that this corresponds to $[H^+]$ about 100 M. Thus, for all practical purposes PbS will precipitate from any acidic aqueous solution. We can separate Pb^{2+} from Mn^{2+} simply by adjusting the pH to lower than about 5, then adding H_2S. PbS precipitates, MnS does not.

16.81 We have $Ag^+(aq) + 2NH_3(aq) \rightleftarrows Ag(NH_3)_2^+(aq)$

$$K_f = \frac{[Ag(NH_3)_2^+]}{[Ag^+][NH_3]^2} = 1.7 \times 10^7$$

We can assume that essentially all the Ag^+ is complexed, so that the original 0.010 M Ag^+ is now converted to 0.010 M $Ag(NH_3)_2^+$. The NH_3 concentration we are given as 0.20 M. Let us call the small remaining concentration of Ag^+ in the solution y. Then,

$$1.7 \times 10^7 = \frac{(0.010)}{(0.20)^2(y)} \quad ; \quad y = 1.5 \times 10^{-8}\ M = [Ag^+]$$

17

Chemistry of natural waters

17.1 The dissolved salts constitute 3.5 percent by weight, and the volume of the oceans is 1.35 x 10^9 km^3. Thus, we have in the entire world ocean:

$$(1.35 \times 10^9\ km^3)\left(\frac{1000\ m}{1\ km}\right)^3\left(\frac{100\ cm}{1\ m}\right)^3\left(\frac{1\ g}{1\ cm^3}\right)\left(\frac{0.035\ g\ salts}{g\ water}\right)$$

$$= 4.7 \times 10^{22}\ g\ salts$$

The amount that is added each year, 4 x 10^{15} g, is 4 x 10^{15}/4.7 x 10^{22} ≃ 1 x 10^{-7}, a tiny fraction of the quantity already present. Thus, it is easy to see that the composition of the ocean is not subject to rapid change from this source.

17.2 A salinity of 5 denotes that there are 5 g of dry salt per kg of water.

$$\left(\frac{5\ g\ NaCl}{1\ kg\ soln}\right)\left(\frac{1\ kg\ soln}{1\ L\ soln}\right)\left(\frac{1\ mol\ NaCl}{58\ g\ NaCl}\right)\left(\frac{1\ mol\ Na^+}{1\ mol\ NaCl}\right) = 0.09\ M$$

17.3 If the phosphorus is present as phosphate, there is a 1:1 ratio between the molarity of phosphorus and molarity of phosphate. Thus, we can calculate the molarity based on the given mass of P.

$$\left(\frac{0.07\ g\ P}{10^6\ g\ H_2O}\right)\left(\frac{1\ mol\ P}{31\ g\ P}\right)\left(\frac{1\ mol\ PO_4{}^{3-}}{1\ mol\ P}\right)\left(\frac{10^3\ g\ H_2O}{1\ L\ H_2O}\right) = 2.3 \times 10^{-6}\ \frac{mol\ PO_4{}^{3-}}{L}$$

17.4 (a) $K_{sp} = 4 \times 10^{-38} = [Fe^{3+}][OH^-]^3$. If pH = 8.0, pOH = 6.0 $[OH^-] = 1 \times 10^{-6}$. $[Fe^{3+}][1 \times 10^{-6}]^3 = 4 \times 10^{-38}$; $[Fe^{3+}] = 4 \times 10^{-20}$ M. This is the molar solubility of $Fe(OH)_3(s)$ in seawater.

(b) $\left(\frac{1 \text{ g } Fe^{3+}}{10^6 \text{ g } H_2O}\right)\left(\frac{1 \text{ mol } Fe^{3+}}{56 \text{ g } Fe^{3+}}\right)\left(\frac{10^3 \text{ g } H_2O}{1 \text{ L } H_2O}\right) = 2 \times 10^{-5} \text{ M } Fe^{3+}$

This is far greater than the molar solubility of $Fe(OH)_3$, as calculated in (a). Therefore, $Fe(OH)_3(s)$ will form.

17.5 (a) Seawater is a solution in which the vapor pressure of water (the solvent) is reduced by the presence of solutes, in accordance with Raoult's law (Section 12.7). (b) The effect of interionic attractive forces and the slightly alkaline character of seawater (pH~8) combine to shift the $H_2CO_3 \rightleftarrows HCO_3^- \rightleftarrows CO_3^{2-}$ equilibria in seawater further to the right. The result is that CO_3^{2-} is a more abundant component in seawater than in fresh water, and HCO_3^- is correspondingly low in concentration. (c) Near the surface in the photosynthetic zone nutrients such as nitrate are used up in forming phytoplankton. Thus, the concentration of nitrate is low near the surface. At greater depths, the light intensity is insufficient to support photosynthesis. Furthermore, dead plant and animal matter decomposes via oxidation to yield nitrate ion. Thus, the nitrate ion concentrations from about 1 km on down tend to be rather constant, since these waters mix with one another over a long period of time.

17.6 $(10^{11} \text{ g Br})\left(\frac{10^3 \text{ g } H_2O}{0.067 \text{ g Br}}\right)\left(\frac{1 \text{ L } H_2O}{10^3 \text{ g } H_2O}\right) = 1.5 \times 10^{12} \text{ L } H_2O$

Because the process is only 10% efficient, ten times this much, or 1.5×10^{13} L H_2O, must be processed.

17.7 The electrolysis of molten $MgCl_2$ represents an oxidation-reduction process: $MgCl_2(\ell) \rightarrow Mg(\ell) + Cl_2(g)$. Magnesium is reduced from the +2 to the 0 oxidation state; chlorine is oxidized from the −1 to 0 oxidation state.

17.8 (a) $2Br^-(aq) + Cl_2(aq) \rightarrow Br_2(\ell) + 2Cl^-(aq)$
(b) $CaCO_3(s) \rightarrow CaO(s) + CO_2(g)$
(c) $Mg(OH)_2(s) + 2H^+(aq) \rightarrow Mg^{2+}(aq) + 2H_2O(\ell)$

17.9 $(5.0 \times 10^6 \text{ g } Mg(OH)_2)\left(\frac{1 \text{ mol } Mg(OH)_2}{58 \text{ g } Mg(OH)_2}\right)\left(\frac{1 \text{ mol CaO}}{1 \text{ mol } Mg(OH)_2}\right)\left(\frac{56 \text{ g CaO}}{1 \text{ mol CaO}}\right)$

$= 4.8 \times 10^6 \text{ g CaO}$

17.10 A salinity of 7 means that there are 7 g dry salts per kg of seawater. If these 7g are considered to be all NaCl, the molarity is approximately:

$$\left(\frac{7 \text{ g NaCl}}{1 \times 10^3 \text{ g } H_2O}\right)\left(\frac{1 \text{ mol NaCl}}{58 \text{ g NaCl}}\right)\left(\frac{1 \times 10^3 \text{ g } H_2O}{1 \text{ L soln}}\right) = 0.12 \text{ M}$$

Recall that each mole of NaCl(aq) produces two moles of solute particles. Then,

$$\pi = MRT = \left(\frac{0.24 \text{ mol}}{\text{L}}\right)\left(\frac{0.082 \text{ L-atm}}{\text{mol-K}}\right)(293 \text{ K}) = 5.8 \text{ atm}$$

17.11 We have:

$(4 \times 10^6 \text{ L water})\left(\frac{1000 \text{ g water}}{1 \text{ L water}}\right)\left(\frac{2500 \text{ J}}{1 \text{ g water}}\right)\left(\frac{1 \text{ g oil}}{46{,}000 \text{ J}}\right)\left(\frac{1}{0.20}\right)\left(\frac{1}{0.20}\right)$

$= 5 \times 10^9 \text{ g oil}$

Notice that the two factors regarding efficiency, (1/0.20), each result in factors larger than one, that is, they result in the need for a larger quantity of oil. It is also interesting to note that just slightly more than one gram of oil is required for each gram of water produced.

17.12 (a) Scale formation is a major problem. Large heat requirement is another. (b) The major problems in reverse osmosis are the development of adequate membranes, and the prevention of fouling due to growth of slimes and so forth, or to sediment impurities in the water.

17.13 Osmosis (Section 12.7) is the net movement of a solvent through a membrane that separates a solution from pure solvent. When the pressures applied to the liquids on the two sides of the membrane are equal, solvent moves through the membrane into the solution. By applying a sufficiently large pressure to the solution side, one can reverse this process to cause the solvent to pass from the solution into the pure solvent. This process is termed reverse osmosis.

17.14 Naturally occurring water contains dissolved gases such as O_2, N_2 and especially CO_2. It also contains ions leached from mineral formations through which the water may have passed, e.g., Na^+, Mg^{2+}, Fe^{2+}, HCO_3^-, Cl^-, and SO_4^{2-}. It may also contain colloidal matter (Section 12.8).

17.15 (a) Water hardness is due to Ca^{2+}, Mg^{2+} and Fe^{2+}. (b) These ions react with soaps to form insoluble materials. Further, precipitation of carbonate may occur when "hard" water is heated, as described by Equation [17.12]. This precipitation reduces heat transfer and water flow in pipes, and may eventually lead to complete blockage.

17.16 (a) A biodegradable material is one that is subject to bacterial degradation in the natural environment. That is, it is a substance that naturally occurring bacteria can feed on. (b) Aerobic decay occurs in the presence of air. The bacteria involved in aerobic decay thrive in an oxygen-containing environment. (c) Anaerobic decay occurs in the absence of air. The bacteria involved in anaerobic decomposition cannot tolerate an oxygen-containing environment. (d) BOD, biological or biochemical oxygen demand, is a measure of the amount of oxygen required to decompose all of the biodegradable organic wastes in water. (e) The so-called 5-day BOD test, BOD_5, measures the consumption of dissolved oxygen by bacteria over a five-day period in a sample of natural water prepared in a standardized way.

17.17 (a) The products of aerobic decomposition are mainly CO_2, H_2O, NO_3^-, SO_2 and SO_4^{2-} and phosphates, e.g., $H_2PO_4^-$. (b) Anaerobic decomposition leads to hydrocarbons such as CH_4, and other hydrides, e.g., NH_3, PH_3, H_2S, H_2O.

17.18 We must calculate the total mass of O_2 per gram of water required for oxidation of the dissolved detergent:

$$\left(\frac{1\text{ g } C_{18}H_{29}O_3S^-}{100\text{ L } H_2O}\right)\left(\frac{1\text{ mol } C_{18}H_{29}O_3S^-}{325\text{ g } C_{18}H_{29}O_3S^-}\right)\left(\frac{51\text{ mol } O_2}{2\text{ mol } C_{18}H_{29}O_3S^-}\right)\left(\frac{32\text{ g } O_2}{1\text{ mol } O_2}\right)$$

$$\times\left(\frac{1\text{ L } H_2O}{10^3\text{ g } H_2O}\right) = \frac{25.6\times10^{-6}\text{ g oxygen}}{1\text{ g } H_2O} = 26\text{ ppm } O_2.$$

17.19 The measured BOD_5 is 7.90 - 1.40 ppm = 6.50 ppm. However, we must scale this by the amount of dilution, which represents a factor of 30. Thus the true BOD_5 is 30 x 6.50 = 195 ppm O_2.

17.20 CaO is used in the so-called lime-soda process for water softening. The lime is added to increase pH, by reaction with HCO_3^- that is present, as described in Equation [17.13]. The $Al_2(SO_4)_3$ is added in the lime-soda process to help remove precipitates ($CaCO_3$ and $Mg(OH)_2$) that do not settle out very well. The Al^{3+} ion forms $Al(OH)_3(s)$ in basic solution. The gelatinous $Al(OH)_3$ precipitate carries finely divided precipitates and other suspended matter down with it as it settles.

17.21 Primary treatment consists of little more than a filtering operation. The sewage is passed into settling tanks where the suspended solids settle out. The water over these solids is then returned to the environment, perhaps after chlorination, or more usually passes on to secondary treatment. In the secondary stage the sewage is aerated and stirred to develop aerobic decomposition in large tanks. The bacteria form a mass called activated sludge, which settles out. The liquid over this sludge is then returned to the environment, perhaps after chlorination. Tertiary treatment involves removal of substances such as phosphates or metal ions from the water that has passed through secondary treatment. Tertiary treatment plants are relatively expensive, and there are not many of them.

17.22 $Ca(OH)_2$ is added to remove Ca^{2+} as $CaCO_3(s)$ according to Equation [17.13]: $Ca^{2+}(aq) + 2HCO_3^-(aq) + (Ca^{2+} + 2OH^-) \rightarrow 2CaCO_3(s) + 2H_2O(\ell)$ We therefore need to add 1 mol $Ca(OH)_2$ for each 2 mol of $HCO_3^-(aq)$ present.

$$\left(\frac{1.3 \times 10^{-3} \text{ mol } HCO_3^-}{1 \text{ L } H_2O}\right)(1 \times 10^7 \text{ L} H_2O)\left(\frac{1 \text{ mol } Ca(OH)_2}{2 \text{ mol } HCO_3^-}\right)\left(\frac{74 \text{ g } Ca(OH)_2}{1 \text{ mol } Ca(OH)_2}\right)$$
$$= 4.8 \times 10^{-5} \text{ g } Ca(OH)_2$$

This operation reduces the Ca^{2+} concentration from 2.2 x 10^{-3} M to (2.2 x 10^{-3} - 1.3 x 10^{-3}) M = 0.9 x 10^{-3} M. To achieve the reduction in $[Ca^{2+}]$ to the desired level, we must add sufficient Na_2CO_3 to reduce $[Ca^{2+}]$ to 0.55 x 10^{-3} M. Let us round this off to 0.5 x 10^{-3} M to be on the safe side. We thus need to reduce $[Ca^{2+}]$ by (0.9 x 10^{-3} - 0.5 x 10^{-3}) M = 0.4 x 10^{-3} M. $Ca^{2+}(aq) + CO_3^{2-}(aq) \rightarrow CaCO_3(s)$.

$$\left(\frac{0.4 \times 10^{-3} \text{ M } Ca^{2+}}{1 \text{ L } H_2O}\right)(1 \times 10^7 \text{ L } H_2O)\left(\frac{1 \text{ mol } Na_2CO_3}{1 \text{ mol } Ca^{2+}}\right)\left(\frac{10^6 \text{ g } Na_2CO_3}{1 \text{ mol } Na_2CO_3}\right)$$
$$= 4.2 \times 10^5 \text{ g } Na_2CO_3$$

17.23 $$(3000 \text{ mi})\left(\frac{1 \text{ gal gas}}{22 \text{ mi}}\right)\left(\frac{25 \text{ gal } H_2O}{1 \text{ gal gas}}\right) = 3.4 \times 10^3 \text{ gal } H_2O$$

17.24 (a) $$\left(\frac{5 \text{ mg Zn}}{1 \text{ L } H_2O}\right)(1 \times 10^7 \text{ L}) = 5 \times 10^7 \text{ mg Zn} = 5 \times 10^4 \text{ g Zn}$$

(b) $$\left(\frac{0.01 \text{ mg Cd}}{1 \text{ L } H_2O}\right)(1 \times 10^7 \text{ L}) = 1 \times 10^5 \text{ mg Cd} = 100 \text{ g Cd}$$

(c) $$\left(\frac{0.05 \text{ mg Mn}}{1 \text{ L } H_2O}\right)(1 \times 10^7 \text{ L}) = 5 \times 10^5 \text{ mg Mn} = 500 \text{ g Mn}$$

17.25 The culprit in this case is cadmium, a congener of zinc in group 2B. Cadmium is enough like zinc so that it occurs in trace quantities in zinc ores. Because it has a similar chemistry, it passes through all the metallurgical operations with the zinc and is present in the metal from which the pipe is formed. In the presence of acidic water, the cadmium can be slowly oxidized to form Cd^{2+} ions in solution. Even though the Cd^{2+} is present in the water in extremely low concentration, it can become a source of chronic metal poisoning.

17.26 The three considerations of importance are: (1) The toxicity of a substance may depend on its chemical state. (2) A substance may undergo chemical transformation in the environment that converts it from a relatively harmless form to a toxic form. (3) A toxic substance may become concentrated at some point in the food chain as compared with its overall concentration in the environment.

17.27 Recall from the solubility rules (Section 12.6) that essentially all salts of the alkali metal ions are water-soluble. By contrast, many salts of the alkaline earths are not very water-soluble. Thus, Na^+ is, in general, more readily leached out of mineral deposits that come in contact with ground water than is Ca^{2+}. Of course, many Ca^{2+} salts, especially $CaCO_3$, also have limited solubility in seawater, whereas the alkali metal salts are all soluble.

17.28 Parts per million refers to the mass of a solute component per million mass units of the total solution (corresponds to 1 mg per L).

17.29 Blue-green algae are often found in waters that reach an advanced stage of eutrophication. These plants are highly consumptive of oxygen and produce foul odors and tastes. When the algae die they settle to the bottom of the lake. If the lake lacks sufficient dissolved oxygen, the decomposition that occurs on the lake bottom is anaerobic; noxious substances are produced under these circumstances.

17.30 Among the processes that may occur are: i) precipitation, e.g., to form HgS or $Hg(OH)_2$; (ii) reduction to Hg_2^{2+} or Hg; (iii) conversion to an organomercury compound, CH_3Hg^+ or $(CH_3)_2Hg$. Of these, the conversion to organomercury compounds is potentially most harmful to us. In these forms the mercury can become concentrated in fish, and thus enter our food supply.

17.31 Substances become concentrated in living systems for various reasons such as higher solubility in fatty tissue (an example is DDT in fish or human fat tissue), formation of stable compounds (for example, Sr^{2+} may form stable complexes in plant cells) or conversion into other chemical substances that are incorporated into biological systems in some way (for example, selenium is incorporated into certain weeds). When animals ingest plant matter, they may selectively concentrate certain toxic impurities in certain organs or types of tissue. When we consume these animals in our diet, we in turn may selectively incorporate the trace toxic substances.

17.32 Carbon, oxygen, hydrogen, nitrogen, phosphorus. Availability of the last two, nitrogen and phosphorus, normally limits growth.

17.33 $$(50{,}000 \text{ persons})\left(\frac{59 \text{ g } O_2}{1 \text{ person}}\right)\left(\frac{10^6 \text{ g } H_2O}{9 \text{ g } O_2}\right)\left(\frac{1 \text{ L } H_2O}{10^3 \text{ g } H_2O}\right) = 3.3 \times 10^8 \text{ L } H_2O$$

17.34 $$\left(\frac{3.0 \times 10^4 \text{ g } CH_4N_2O}{3.0 \times 10^6 \text{ L } H_2O}\right)\left(\frac{1 \text{ mol } CH_4N_2O}{60 \text{ g } CH_4N_2O}\right)\left(\frac{4 \text{ mol } O_2}{1 \text{ mol } CH_4N_2O}\right)\left(\frac{32 \text{ g } O_2}{1 \text{ mol } O_2}\right)$$

$$\times \left(\frac{1 \text{ L } H_2O}{10^3 \text{ g } H_2O}\right) = \frac{2.1 \times 10^{-5} \text{ g } O_2}{1 \text{ g } H_2O} = 21 \text{ ppm } O_2$$

17.35 We need add only as much $Ca(OH)_2$ as is needed to remove bicarbonate in accordance with Equation [17.13]. If there are 7.0×10^{-4} mol $HCO_3^-(aq)$ per liter, we must add 3.5×10^{-4} mol $Ca(OH)_2$ per liter, or a total of 0.35 mol $Ca(OH)_2$ for 10^3 L. This reaction removes 3.5×10^{-4} mol of the original Ca^{2+} from each liter of solution, leaving 1.5×10^{-4} M $Ca^{2+}(aq)$. To remove this $Ca^{2+}(aq)$, we add 1.5×10^{-4} mol Na_2CO_3 per liter, or a total of 0.15 mol Na_2CO_3, forming $CaCO_3(s)$.

17.36 $$\pi = MRT = \left(\frac{0.200 \text{ mol}}{\text{L}}\right)\left(\frac{0.082 \text{ L-atm}}{\text{mol-K}}\right)(300 \text{ K}) = 4.92 \text{ atm}$$

Note that the molarity employed is the total molarity that results from dissociation of NaCl into Na^+ and Cl^- particles.

17.37 To produce PO_4^{3-} ions, we require a basic medium. Addition of CaO gives us this, via the reactions:

$$CaO(s) + H_2O(\ell) \rightarrow Ca^{2+}(aq) + 2OH^-(aq)$$

$$H_2PO_4^-(aq) + 2OH^-(aq) \rightarrow 2H_2O(\ell) + PO_4^{3-}(aq)$$

Precipitation of $Ca_3(PO_4)_2$, $K_{sp} = 2.0 \times 10^{-29}$ can then result in removal of the phosphate.

17.38 We have $K_{sp} = [Ca^{2+}][CO_3^{2-}] = 6.0 \times 10^{-7}$; $[CO_3^{2-}] = 6 \times 10^{-4}$ M. Thus, $[Ca^{2+}] = 6.0 \times 10^{-7}/6 \times 10^{-4} = 1 \times 10^{-3}$ M.

17.39 The solubility product differs because of the effects of all the other ions in the solution on the ions involved in the equilibrium. Interionic attractions between oppositely charged ions become significant at the high ionic concentrations characteristic of seawater. These interionic forces result in a lower energy (more stable state) for the ions in solution as compared with a highly dilute solution. Thus, solubility increases.

17.40 (a) $Cl_2(aq) + H_2O(\ell) \rightarrow HClO(aq) + H^+(aq) + Cl^-(aq)$

(b) $HClO(aq) + NH_3(aq) \rightarrow NH_2Cl(aq) + H_2O(\ell)$

(c) $Al^{3+}(aq) + 3OH^-(aq) \rightarrow Al(OH)_3(s)$

(d) $CaO(s) + H_2O(\ell) \rightarrow Ca^{2+}(aq) + 2OH^-(aq)$

(e) $Hg(aq) + Hg^{2+}(aq) \rightarrow Hg_2^{2+}(aq)$

(f) $Ca^{2+}(aq) + 2HCO_3^- \rightarrow CaCO_3(s) + CO_2(g) + H_2O(\ell)$

17.41 Let's first calculate the total mass of nitrogen present in the volume of water we are concerned with. This volume is $1 \text{ km}^2 \times 0.1 \text{ km} = 0.1 \text{ km}^3$. Converting this to liters, we have

$$0.1\ km^3 \left(\frac{1000\ m}{1\ km}\right)^3 \left(\frac{100\ cm}{1\ m}\right)^3 \left(\frac{1\ L}{1000\ cm^3}\right) = 1 \times 10^{11}\ L$$

The mass of nitrogen is thus:

$$\left(\frac{1 \times 10^{-6}\ mol\ N}{liter}\right)\left(\frac{14\ g\ N}{1\ mol\ N}\right) \times (1 \times 10^{11}\ L) = 14 \times 10^5\ g\ N$$

The formula weight of phytoplankton is obtained by adding up all the atomic weights of the elements present, each multiplied by its coefficient in the formula. This comes to 3524 g. The formula mass of nitrogen in this formula weight is 224 g. Thus, to obtain the total mass of phytoplankton formed, we have

$$\frac{1}{2}\left(\frac{14 \times 10^5\ g\ N}{0.1\ km^3\ water}\right)\left(\frac{3524\ g\ C_{108}H_{226}N_{16}O_{109}P}{224\ g\ N}\right)$$

$$= \frac{1 \times 10^7\ g\ C_{108}H_{226}N_{16}O_{109}P}{0.1\ km^3\ water}$$

17.42 $K_{sp} = [Hg^{2+}][OH^-]^2 = 1.6 \times 10^{-23}$; $pH = 7.8$;
Thus, $pOH = 6.2$; $[OH^-] = 6.3 \times 10^{-7}$; $[Hg^{2+}] = 1.6 \times 10^{-23}/(6.3 \times 10^{-7})^2$
$= 4.0 \times 10^{-11}\ M$

$$\left(\frac{4.0 \times 10^{-11}\ mol\ Hg^{2+}}{1\ L\ H_2O}\right)\left(\frac{201\ g\ Hg^{2+}}{1\ mol\ Hg^{2+}}\right)\left(\frac{1\ L\ H_2O}{10^3\ g\ H_2O}\right) = \frac{8.0 \times 10^{-12}\ g\ Hg^{2+}}{1\ g\ H_2O}$$

$$= \frac{8 \times 10^{-6}\ g\ Hg^{2+}}{10^6\ g\ H_2O} = 8 \times 10^{-6}\ ppm\ Hg^{2+}.$$

Thus, precipitation occurs at a Hg^{2+} concentration of about 10^{-5} ppm.

18

Free energy, entropy, and equilibrium

18.1 Processes (b) and (e) are spontaneous. That is, there is a "driving force" that promotes the process as described. On the other hand, processes (a), (c), (d) and (f) all require the input of work to cause them to proceed. This indicates that they are non-spontaneous. Processes (c), (d) and (f) are not simple processes, but rather are complex, consisting of many separate processes. Some of the individual processes that occur in the overall process may be spontaneous, but the overall process is one that involves making order from disorder, so that work is required.

18.2 (a) 1 mol He at 25°C; more kinetic energy of motion. (b) 2 mol H at 25°C; more particles, thus more randomness in the possible arrangements. (c) 1 mol $H_2O(g)$ at 100°C; more volume in gaseous phase, thus more randomness in locations of H_2O molecules. (d) 1 mol HCl(g) at 25°C; larger volume for gaseous molecules, thus more randomness in locations of molecules. (e) 1 mol $C_2H_6(g)$ at 25°C; larger molecule, thus more ways in which energy may be distributed in vibrations within molecule.

18.3 (a) lower (more rigid, ordered); (b) larger (more complex molecule, more ways of storing vibrational and rotational energy); (c) lower (single atom lacks rotational or vibrational energy); (d) lower (solid lattice restricts motion).

18.4 (a) S° for O_2 is 205 J/mol-K, for O_3 is 238 J/mol-K. It is higher for O_3 because the more complex molecule has more ways of storing rotational and vibrational energies. (b) S° for diamond is 2.43 J/mol-K; for graphite 5.69 J/mol-K. Both are small values; the diamond value is lower because it is a more rigid lattice than graphite. (c) S° is 245 J/mol-K for $Br_2(g)$, 152 J/mol-K for $Br_2(\ell)$. The higher value for the gaseous phase is due to the larger volume in which the $Br_2(g)$ molecules can move. (d) S° for NaCl(s) is 72 J/mol-K, for $MgCl_2$ is 90 J/mol-K. The value for $MgCl_2$ is larger because there are more particles per formula weight in the solid, thus more ways of distributing the energy at 298 K.

18.5 (a) ΔS positive; larger volume; (b) ΔS positive; larger volume for each component, increased randomness in mixing of two different substances; (c) ΔS negative; the entropy content of AgCl(s) is much smaller than that of the separated aqueous ions, which can move freely throughout the solution.

18.6 (a) ΔS positive; a gas is formed; (b) ΔS positive; a gas is formed; (c) ΔS negative; a gas is consumed; (d) ΔS positive; the number of gaseous particles is doubled; (e) ΔS positive; in aqueous NaCl the separated ions are able to move throughout the solution volume; (f) ΔS positive; dilution means that the ions in HCl(aq) are free to move throughout a larger volume.

18.7 (a) $\Delta S^\circ = 237.6 - 197.9 - 2(130.6) = -221.5$ J/K
(b) $\Delta S^\circ = 2(198.5) + 223.0 - 2(186.7) - 152.3 = 94.3$ J/K
(c) $\Delta S^\circ = 2(256.2) - 2(248.5) - 205.0 = -189.6$ J/K
(d) $\Delta S^\circ = 304.3 - 2(240.4) = -176.5$ J/K

ΔS° is negative in (a), (c) and (d) because a smaller number of gas molecules is formed in the products. In (b), the number of gas molecules is greater on the right, so ΔS° is positive.

18.8 The application of DDT over large areas represents an increase in the entropy of the DDT system. As the substance moves through the environment, it may become even more dispersed, in water bodies and so forth. These changes further increase the entropy of the system. However, biological systems may concentrate DDT, as in the example of its accumulation in the fatty tissues of fish. Such a process is associated with a negative ΔS in our "system."

Cleanup of DDT from the environment means gathering it up somehow; that is, it means a process which has associated with it a large, negative ΔS. Such processes are, of course, not spontaneous in general. This means that man would need to consume energy to bring them about. It is difficult to imagine any way in which this could be done. Our only "solution" to such environmental problems is not to allow the chemical to disperse in the first place.

18.9 We must distinguish the system from the surroundings. In the freezing of water the entropy of the water does indeed decrease. However, this decrease in the entropy of the system is more than compensated for by an accompanying increase in the entropy of the surroundings. The net overall entropy change is positive.

18.10 The spontaneity of any process is determined by two factors: the change in entropy, which measures the change in randomness of the system, and the enthalpy change, which measures the change in potential energy, as it were. Formation of $H_2O(\ell)$ from $H_2(g)$ and $O_2(g)$ represents a large decrease in potential energy (that is, the product is more stable, or lower in energy). The enthalpy change here is more important than the entropy decrease.

18.11 ΔG is negative for spontaneous processes, positive for non-spontaneous processes, and zero for systems at equilibrium. We cannot infer anything about reaction speed from the magnitude of ΔG, because reaction speed is related to the energy barrier between reactants and products (Section 13.4), and not to the final energy difference between reactants and products. When ΔG for the reaction is negative, this means that reaction in the forward direction is spontaneous. Thus, the system approaches equilibrium from the direction of reactants. When ΔG is positive, this means that the reaction proceeds spontaneously in the <u>reverse</u> direction as written. Thus, the system approaches equilibrium from the direction of excess products.

<u>18.12</u> $\Delta G° = -228.61 - (-236.81) = 8.2$ kJ
$\Delta G°$ is positive because at 298 K the energy required to overcome the intermolecular forces to form $H_2O(g)$ from $H_2O(\ell)$ is not overcome by the larger entropy of the gaseous molecules. At 373 K, 1 atm pressure, these two processes make equal contributions, and $\Delta G = 0$.

<u>18.13</u> (a) $\Delta G° = -394.4 - (-318.2) - (-137.3) = 61.1$ kJ
(b) $\Delta G° = 209.2 - 2(-394.4) - (-228.6) = 1226.6$ kJ
(c) $\Delta G° = 2(1.30) = 2.60$ kJ/mol
None of the three reactions is spontaneous under standard conditions.

<u>18.14</u> $\Delta G° = \Delta H° - T\Delta S° = -1.06 \times 10^5 \text{ J} - 298 \text{ K}(58 \text{ J/K}) = -1.23 \times 10^5 \text{ J}$
The process is highly spontaneous. One does not expect that $H_2O_2(g)$ would be stable in the gas phase at 298 K.

<u>18.15</u> (a) $\Delta G° = 2(-370.4) - 2(-300.4) = -140.0$ kJ
$\Delta H° = 2(-395.2) - 2(-296.9) = -196.6$ kJ
$\Delta S° = 2(256.2) - 2(248.5) - 205.0 = -189.6$ J/K
$-140.0 \times 10^3 = -196.6 \times 10^3 - 298(-189.6)$

(b) $\Delta G° = 2(-236.8) - 394.4 - (-50.8) = -817.2$ kJ
$\Delta H° = 2(-285.8) - 393.5 - (-74.8) = -890.3$ kJ
$\Delta S° = 2(69.96) + 213.6 - 186.3 - 2(205.0) = -242.8$ J/K
$-817.2 \times 10^3 = -890.3 \times 10^3 - 298(-242.8)$

(c) $\Delta G° = -1128.8 - (-604.2) - (-394.4) = -130.2$ kJ
$\Delta H° = -1207.1 - (-635.5) - (-393.5) = -178.1$ kJ
$\Delta S° = 92.88 - 39.8 - 213.6 = -160.5$ J/K
$-130.2 \times 10^3 = -178.1 \times 10^3 - 298(-160.5)$

(d) $\Delta G° = 3(-394.4) - 3(-137.3) - (-741.0) = -30.3$ kJ
$\Delta H° = 3(-393.5) - 3(-110.5) - (-822.2) = -26.8$ kJ
$\Delta S° = 2(27.2) + 3(213.6) - 3(197.9) - 90.0 = 11.5$ J/K
$-30.3 \times 10^3 = -26.8 \times 10^3 - 298(11.5)$

<u>18.16</u> $\Delta G° = 0.00 - 2.84 \text{ kJ} = -2.84$ kJ/mol C. Although $\Delta G°$ is negative, the conversion of diamonds to graphite does not occur at a measurable rate because the activation energy for the reaction (Section 13.4) is very large.

<u>18.17</u> (a) ΔG is negative at low temperatures, positive at high temperature. That is, the reaction proceeds in the forward direction spontaneously at lower temperatures but spontaneously reverses at higher temperatures.
(b) ΔG is positive at all temperatures. The reaction is nonspontaneous in the forward direction at all temperatures. (c) ΔG is positive at low temperatures, negative at high temperatures. That is, the reaction will proceed spontaneously in the forward direction at high temperature.
(d) ΔG is positive at lower temperatures, negative at higher temperatures. From our knowledge of the stability of the water molecule, we expect that it will require a very high temperature to dissociate the H_2O molecule.

<u>18.18</u> (a) We need to evaluate $\Delta H°$ and $\Delta S°$:
$\Delta H° = -201.2 - (-110.5) = -90.7$ kJ
$\Delta S° = 237.6 - 197.9 - 2(130.6) = -221.5$ J/K
The reaction has a negative ΔG at lower temperature and a positive ΔG at higher temperature. Thus, if we wished to synthesize CH_3OH from CO and H_2 it would be important to find a catalyst to cause the reaction to proceed

at a reasonable rate at lower temperature, where the reaction is more spontaneous. (b) $\Delta G(500°C) = \Delta G(773\ K) = -90.7 \times 10^3 - 773(221.5) = +80.5$ kJ

18.19 We must calculate $\Delta H°$ and $\Delta S°$

$\Delta H° = 3(-393.5) - 3(-110.5) - (-822.2) = -26.8$ kJ

$\Delta S° = 2(27.2) + 3(213.6) - 3(197.9) - 90.0 = 11.5$ J/K

The reaction is spontaneous at all temperatures because ΔG will be negative at all temperatures so long as ΔH remains negative and ΔS positive.

18.20 (a) Using LeChatelier's principle, we can reason that an increase in the pressure of N_2 will cause the reaction mixture to shift toward more product. Thus, we can surmise that an increase in N_2 pressure causes ΔG to become more negative. We can also see this from application of equation [18.14]:

$$\Delta G = \Delta G° + 2.3RT\log\left(\frac{[NH_3]^2}{[N_2][H_2]^3}\right)$$

An increase in $[N_2]$ will reduce the value of the log term, thus making ΔG less positive, or more negative.

(b) $\Delta G = -33.32\ \text{kJ} + 2.3\ RT\log Q$

$= -33.32\ \text{kJ} + 2.3(8.314\ \text{J/K})(298\ \text{K})\log\left(\frac{(0.10)^2}{(5.0)(2.0)^3}\right)$

$= -53.85$ kJ

18.21 Using Equation [18.14]:

$$\Delta G = -118.4 \times 10^3\ \text{J} + 2.3\left(\frac{8.314\ \text{J}}{\text{K}}\right)(298\ \text{K})\ \log P_{O_2}^{3/2}$$

If ΔG is to be zero, then

$$\log P_{O_2}^{3/2} = \left(\frac{118.4 \times 10^3\ \text{J}}{2.3}\right)\left(\frac{\text{K}}{8.314\ \text{J}}\right)\left(\frac{1}{298\ \text{K}}\right) = 20.8$$

Thus, $P_{O_2}^{3/2} = 6 \times 10^{20}$, $P_{O_2} = 7.1 \times 10^{13}$ atm

18.22 We must first determine $\Delta G°$:

$\Delta G° = 2(-394.4) - 2(-137.3) = -514.3$ kJ

$$\Delta G = -514.3 \times 10^3 + 2.3\left(\frac{8.314\ \text{J}}{\text{K}}\right)(298\text{K})\log\left(\frac{(10.0)^2}{(2.0)^2(2.0)}\right) = -508.0\ \text{kJ}$$

18.23 We employ Equation [18.15], and must therefore calculate $\Delta G°$ in each case:

(a) $\Delta G° = 2(-95.27) - 2(1.30) = -193.1$ kJ

$\log K = 193.1 \times 10^3\ \text{J}/2.3(8.314\ \text{J/K})(298\ \text{K})$

$= 33.89;\ K = 7.8 \times 10^{33}$

(b) $\Delta G° = 2(1.30) = 2.60$ kJ

$\log K = -2.60 \times 10^3\ \text{J}/2.3(8.314\ \text{J/K})(298\ \text{K})$

$= -0.456;\ K = 0.35$

18.24 $\Delta G° = 0 + (-157.3) - (-236.8) = +79.5$ kJ

The fact that $\Delta G°$ is large and positive indicates that the equilibrium lies far to the left, as indeed we know it does.

18.25 $\Delta G° = -2.3(8.314\ \text{J/K})(460\ \text{K})\log\left(\frac{(0.50)}{(0.25)(0.25)}\right)$

$= -7.94$ kJ

18.26 We have that $HX(aq) \rightleftarrows H^+(aq) + X^-(aq)$. If pH = 5.83, $[H^+] = 1.5 \times 10^{-6} = [X^-]$.

$$\Delta G° = -2.3(8.314\ J/K)(298)\ \log\left(\frac{(1.5 \times 10^{-6})^2}{0.10}\right) = 60.7\ kJ$$

18.27 Reactions (a), (b), (d) should proceed spontaneously in the forward direction at room temperature; therefore, K > 1. (Remember that we are not referring to the rate of reaction, but only to the position of equilibrium if the reaction were able to proceed.) Reactions (c) and (e) are quite clearly spontaneous in the reverse direction. Thus, for these, K < 1.

Reaction	Predicted $\Delta H°$	Predicted $\Delta S°$	Comment on K
(a)	–	near zero	K > 1 all temperatures
(b)	–	–	K > 1 at low temperatures, decreases at high temps.
(c)	+	+	K < 1 at low temperatures, increases at higher temps.
(d)	–	–	K > 1 at low temperatures, decreases at higher temps.
(e)	+	near zero	K < 1 at all attainable temperatures.

18.28 (a) $\Delta G° = 2(-228.6) + 3(0) - 2(-33.0) - (-300.4) = -90.8$ kJ
$\log K = 90.8 \times 10^3/2.3(8.314)(298) = 15.9$; $K = 8.6 \times 10^{15}$

(b) The large value for K suggests that, in principle, the reaction could remove $SO_2(g)$ from stack gases.

(c) $$K = 8.6 \times 10^{15} = \frac{P^2_{H_2O}}{P^2_{H_2S} \times P_{SO_2}} = \frac{(25/760)^2}{x^3}$$

$x^3 = 1.26 \times 10^{-19}$; $x = P_{SO_2} = 5 \times 10^{-7}$ atm

(d) Note that there are two moles of gaseous product on the right, four moles of gaseous reactant on the left. We can thus assume that $\Delta S°$ is negative. At higher temperatures, ΔG will be more positive, K will be smaller. Thus, at some range of higher temperature the reaction will not proceed sufficiently far to the right to be a useful means of sulfur removal.

18.29 A state function is a thermodynamic quantity which has a value that is determined once the state of the system is completely specified. That is, it is related to the inherent properties of the system itself rather than to any change the system undergoes. Enthalpy, entropy and free energy are state functions. The heat evolved or absorbed in a process, or the work done in a process, are not state functions; the values for these quantities depend on the particular pathway by which the system changes from one state to another.

18.30 (a) $\Delta G = \Delta H - T\Delta S$; (b) $\Delta G = \Delta G° + 2.3\ RT \log Q$; (c) $\Delta G° = -2.3\ RT \log K$

18.31 Calculate $\Delta G°$ for the reaction: $N_2(g) + 2O_2(g) \rightleftarrows 2NO_2(g)$

$\Delta G° = 2(51.84) - 0 = 103.0$ kJ.

With such a large positive $\Delta G°$, we can immediately conclude that the equilibrium constant for the reaction will be small:

$$\log K = \left(\frac{-103.0 \times 10^3\ J}{2.3\left(\frac{8.314\ J}{K}\right)298\ K}\right) = -18.1;\ K = 8.41 \times 10^{-19}$$

Now a catalyst can speed up the rate at which the system approaches equilibrium, but it cannot change the value for the equilibrium constant itself. There is no way in which a catalyst can circumvent the unfavorable value for K at 298 K.

18.32 The denaturation process is endothermic, because we must supply energy for rupture of the hydrogen bonds. Thus, ΔH is positive. The entropy change is also positive, because the denatured protein is less regular and ordered than the original protein. Because both ΔH and ΔS are positive, ΔG is positive at lower temperature, but becomes negative at higher temperature.

18.33 It requires energy to stretch the rubber band; we are forcing an unfavorable alignment on the rubber molecules. Thus, ΔH should be positive. The fact that the rubber molecules are increasingly aligned in the stretched state suggests that ΔS is negative for the stretch process. Thus, ΔG is made positive by both the ΔH and $T\Delta S$ terms.

18.34 As a means of establishing a reference point, the entropy of a pure crystalline solid at 0 K is defined as zero. As the temperature increases above 0 K, the system gains energy from its surroundings. This energy is distributed among the various forms of energy that the system can have. The entropy measures the degree to which this energy is randomized among the various energy states. This measure is inherently positive.

(It should be noted, however, that by defining the entropy of a pure crystalline substance as zero at 0 K, it turns out that for some species in a solution at 298 K the entropy can actually be negative. For example, note that S° for $OH^-(aq)$ is negative. This negative value arises from the extensive ordering of the water molecules around the ion.)

18.35 In all these comparisons we should ask which is the more "ordered" system with respect to some ideal. In (a) the shuffled deck is more random with respect to the ordering required to have the deck arranged by suits, so it has the higher entropy, or probability. There are almost an infinity of shuffled arrangements, only one arrangement in which each suit is arranged in order, suit by suit. (b) The distribution of toys throughout the house has a higher entropy than the distribution in a single room. (c) An assembled TV set represents a high degree of organization and order; it therefore has a lower entropy content than a collection of all the separate parts of the set.

18.36 We have learned that spontaneous processes proceed with a net increase in the entropy of the universe. Thus, no matter how a process is carried out, if it occurs in a finite time, it results in an increase in entropy. Spontaneous processes occur in a certain direction; thus, in a sense the direction of time that points toward the future is the direction in which spontaneous processes occur. One way of seeing this is to think how it feels to watch a movie run backwards. You know the movie is running in reverse when you see processes that you know are non-spontaneous happening as time moves "forward." For example, an egg spattered on the sidewalk reforms into a perfect egg in the thrower's hand.

18.37 (a) This process occurs spontaneously to yield a saturated $MgCl_2$ solution. When $[MgCl_2]$ = 6 M, $\Delta G = 0$, and equilibrium is attained. (b) This process cannot occur spontaneously; the reverse reaction is spontaneous. (c) This reaction occurs spontaneously.

18.38 (a) False. Some exothermic processes ($\Delta H < 0$) are non-spontaneous because the entropy change for the reaction is unfavorable. That is, $\Delta S < 0$, and sufficiently large so that the $-T\Delta S$ term in the formula $\Delta G = \Delta H - T\Delta S$ overrides the enthalpy term to make ΔG positive. (b) False. The entropy of the universe increases in all spontaneous processes. (c) The statement is true, but could be more general. For any process that has a positive ΔS, ΔG will grow more negative at higher temperature because of the $-T\Delta S$ term in $\Delta G = \Delta H - T\Delta S$.

18.39 Energy has value to us when we can use it to obtain useful work. We can use the chemical potential energy of gasoline to generate a great deal of heat in a piston, and thus propel a car. However, the heat that is eventually released to the surroundings as the car runs cannot be tapped for useful work because it is widely diffused, and not in a sufficiently hot system to permit extracting useful work. The energy crisis arises when not enough energy in a highly useful form, such as oil, is available for use.

18.40 (a) ΔH positive; ΔS positive; $\Delta G = 0$; (b) ΔH positive; ΔS positive; ΔG negative; (c) ΔH positive; ΔS positive; ΔG positive.

18.41 (a) $\Delta S° = 2(240.4) - 205.0 - 2(210.6) = -145.4$ J/K
$\Delta G° = 2(51.84) - 0 - 2(86.71) = -69.74$ kJ
(b) $\Delta S° = 229.5 - 130.6 - 219.4 = -120.5$ J/K
$\Delta G° = -32.89 - 0 - 68.11 = -101.0$ kJ
(c) $\Delta S° = 188.7 + 213.6 + 136.0 - 2(102.1) = 334.1$ J/K
$\Delta G° = -228.6 - 394.4 - 1047.7 - 2(-851.8) = 32.9$ kJ

18.42 If $\Delta G = 0$, then $0 = \Delta H - T\Delta S$, $\Delta S = \Delta H/T$.
$\Delta S = 40.7 \times 10^3$ J/373 K = 109.1 J/K

18.43 (a) This suggests that ΔH is positive. That is, the process is endothermic. (b) If ΔG is negative when ΔH is positive, ΔS must be positive, so that the $-T\Delta S$ term in $\Delta G = \Delta H - T\Delta S$ is larger than the ΔH term. The magnitude of ΔS can be obtained from a knowledge of ΔG and ΔH at 60°C (333 K): $\Delta S = (\Delta G - \Delta H)/333$.

The positive value for ΔS suggests that the higher temperature form (denatured state) is more disordered, more random, than the native state. Since ΔH is positive (that is, the process is endothermic), it would appear that bonds are broken in the transition from native to denatured states.

18.44

Reaction	$\Delta G°_{298}$	Spontaneous at 25°C?	Spontaneous at Higher Temp.?
(a)	-62 kJ	yes	yes
(b)	-304 kJ	yes	no
(c)	20.2 kJ	no	yes

18.45 We should estimate $\Delta G°$ at 1000°C (1273 K) from the $\Delta H°$ and $\Delta S°$ values at 298 K.

$\Delta H° = 2(-46.19) - 2(90.37) - 3(-241.8) = +452.2$ kJ
$\Delta S° = 2(192.5) + 5/2(205.0) - 2(210.6) - 3(188.7) = -89.8$ J/K

We need not proceed any further. Note that ΔH is positive, ΔS is negative. ΔG must therefore be large and positive at all temperatures. The equilibrium constant will be much too small for this reaction to be a feasible means of NO removal.

18.46 (a) $\Delta G° = -2.3\ RT \log K = 2.3\left(\frac{8.314\ J}{K}\right)(298)\log(1.8 \times 10^{-5})$

$= +27.0\ kJ$

(b) At equilibrium, ΔG is, by definition, zero.

(c) $\Delta G = \Delta G° + 2.3\left(\frac{8.314\ J}{K}\right)(298)\log\left(\frac{(1.0 \times 10^{-7})(1.0 \times 10^{-7})}{(2.0)}\right)$

$= +27.0\ kJ - 81.5\ kJ = -54.5\ kJ$

18.47 $\Delta G° = 2\Delta G°_f(NH_3) + \Delta G°_f(CO_2) - \Delta G°_f(H_2O) - \Delta G°(CO(NH_2)_2)$

$= 2(-16.7) + (-394.4) - (-228.6) - (-197.2) = -2.0\ kJ$

$\Delta G° = 2.3\ RT \log K_p$; $\log K_p = 2000/2.3(8.314)(298) = 0.351$

$K_p = 2.24$

18.48 Using $\Delta G° = 2.3\ RT \log K$, we see that when $K = 1$, $\Delta G° = 0$. Now $\Delta G°$ will be zero when $\Delta H° - T\Delta S° = 0$. So $T = \Delta H°/\Delta S°$ for the reaction $O_2(g) \rightleftarrows 2O(g)$, $\Delta H° = 2(247.5) = 495\ kJ$, $\Delta S = 2(161.0) - 205 = 117\ J/K$. Then $T = 495 \times 10^3/117 = 4230\ K$. Extrapolation of the thermodynamic quantities for 298 K to such a high temperature is not really valid, so the temperature we calculate is only approximate.

18.49 We must calculate $\Delta G°$:

$\Delta G° = -203.0 - (95.27) - (-16.7) = -91.0\ kJ$

$\Delta G° = -2.3\ RT \log K_p$

$\log K_p = -91.0 \times 10^3/2.3(8.314)\ 298 = 15.97$

$K_p = 9.3 \times 10^{15}$

18.50 We must calculate $\Delta G°$ for each reaction:

for $C_6H_{12}O_6(s) + 6O_2(g) \rightleftarrows 6CO_2(g) + 6H_2O(\ell)$ (A)

$\Delta G° = 6(-236.8) + 6(-394.4) - (-912) = -2875\ kJ$

for $C_6H_{12}O_6(s) \rightleftarrows 2C_2H_5OH(\ell) + 2CO_2(g)$ (B)

$\Delta G° = 2(-394.4) + 2(-174.8) - (-912) = -226\ kJ$

For Reaction (a), $\log K = 2875 \times 10^3/2.3(8.314)(298) = 504.5$;
$K = 3 \times 10^{504}$.

For Reaction (b), $\log K = 226 \times 10^3/2.3(8.314)(298) = 39.73$;
$K = 5 \times 10^{39}$.

Both these values for K are unimaginably large. However, K for reaction (A) is larger, because $\Delta G°$ is more negative. The magnitude of the work that can be accomplished by coupling a reaction to its surroundings is measured by ΔG. We can see from the above that considerably more work can in principle be obtained from reaction (A), because $\Delta G°$ is more negative.

18.51 (a) To obtain $\Delta H°$ from the equilibrium constant data, graph log K at various temperatures vs. 1/T, being sure to employ absolute temperature. The slope of the linear relationship that should result is $-\Delta H°/2.3\ R$; thus, $\Delta H°$ is easily calculated. (b) We use $\Delta G° = \Delta H° - T\Delta S°$, and $\Delta G° = -2.3\ RT \log K$. Substituting this second expression into the first, we obtain

$$-2.3\ RT \log K = \Delta H° - T\Delta S°; \quad \log K = \frac{-\Delta H°}{2.3\ RT} - \frac{-\Delta S°}{2.3\ R}$$

We see from this that the constant in the equation given in the exercise is $\Delta S°/2.3\ R$.

18.52 (a) The equilibrium of interest here can be written as:

$$K^+(\text{plasma}) \rightleftarrows K^+(\text{muscle})$$

Since an aqueous solution is involved in both cases, we can assume that the equilibrium constant for the above process is exactly 1. That is, $\Delta G° = 0$. However, ΔG is not zero because the concentrations are not the same on both sides of the membrane. Let's use equation [18.14] to calculate ΔG:

$$\Delta G = \Delta G° + 2.3\ RT \log \frac{[K^+(\text{muscle})]}{[K^+(\text{plasma})]}$$

$$= 0 + 2.3(8.314)(310)\log\left(\frac{(0.15)}{(5.0 \times 10^{-3})}\right) = 8.75\ \text{kJ}$$

(b) Note that ΔG is positive. This means that work must be done on the system (blood plasma plus muscle cells) to move the K^+ ions "uphill," as it were. The minimum amount of work possible is given by the value for ΔG. This value represents the minimum amount of work required to transfer one mole of K^+ ions from the blood plasma at 5×10^{-3} M to muscle cell fluids at 0.15 M, assuming constancy of concentrations. In practice, a larger than minimum amount of work is required.

19

Electrochemistry

19.1 (a) $SO_4^{2-}(aq) + 4H^+(aq) + 2e^- \rightarrow SO_2(g) + 2H_2O(\ell)$ reduction

(b) $Fe(s) \rightarrow Fe^{2+}(aq) + 2e^-$ oxidation

(c) $NO_2^-(aq) + H_2O(\ell) \rightarrow NO_3^-(aq) + 2H^+ + 2e^-$ oxidation

(d) $O_2(g) + 2H_2O(\ell) + 4e^- \rightarrow 4OH^-(aq)$ reduction

(e) $Cr(OH)_3(s) + 5OH^-(aq) \rightarrow CrO_4^{2-}(aq) + 4H_2O(\ell) + 3e^-$ oxidation

19.2 (a) no oxidation - reduction; (b) manganese is reduced from +4 to +2 oxidation state; chlorine is oxidized from -1 to zero oxidation state (that is, that part of the HCl that ends up as Cl_2). (c) no oxidation - reduction; (d) bromine is both oxidized (from zero to +5 oxidation state) and reduced (from zero to -1 oxidation state).

19.3 (a) $Fe(s) + H^+(aq) + SO_4^{2-}(aq) \rightarrow Fe^{3+}(aq) + SO_2(g) + H_2O(\ell)$

The half-reactions are:

$$2[Fe(s) \rightarrow Fe^{3+}(aq) + 3e^-] \quad \text{(oxidation)}$$

$$3[SO_4^{2-}(aq) + 4H^+(aq) + 2e^- \rightarrow SO_2(g) + 2H_2O(\ell)] \quad \text{(reduction)}$$

$$2Fe(s) + 12H^+(aq) + 3SO_4^{2-}(aq) \rightarrow 2Fe^{3+}(aq) + 3SO_2(g) + 6H_2O(\ell)$$

Fe is oxidized, SO_4^{2-} is reduced.

(b) $Cl_2(g) + OH^-(aq) \rightarrow Cl^-(aq) + ClO^-(aq) + H_2O(\ell)$

$$(1/2)Cl_2(g) + e^- \rightarrow Cl^-(aq) \quad \text{(reduction)}$$

$$(1/2)Cl_2(g) + 2OH^-(aq) \rightarrow ClO^-(aq) + H_2O(\ell) + e^- \quad \text{(oxidation)}$$

$$Cl_2(g) + 2OH^-(aq) \rightarrow Cl^-(aq) + ClO^-(aq) + H_2O(\ell)$$

Cl_2 is both reduced and oxidized.

For (c) and (d), we show just the final balanced equations:

(c) $4Zn(s) + NO_3^-(aq) + 10H^+(aq) \rightarrow 4Zn^{2+}(aq) + NH_4^+(aq) + 3H_2O(\ell)$

Zinc is oxidized, nitrate is reduced.

(d) $2MnO_4^-(aq) + 3S^{2-}(aq) + 4H_2O(\ell) \rightarrow 2MnO_2(s) + 3S(s) + 8OH^-(aq)$

Sulfide ion (S^{2-}) is oxidized to S(s); $MnO_4^-(aq)$ is reduced to $MnO_2(s)$.

(e) $H_2O_2(aq) + MnO_4^-(aq) + H^+(aq) \rightarrow O_2(g) + Mn^{2+}(aq) + H_2O(\ell)$

$$5[H_2O_2(aq) \rightarrow O_2(g) + 2e^- + 2H^+(aq)] \qquad \text{(oxidation)}$$

$$2[8H^+(aq) + MnO_4^-(aq) + 5e^- \rightarrow Mn^{2+}(aq) + 4H_2O(\ell)] \qquad \text{(reduction)}$$

$$5H_2O_2(aq) + 6H^+(aq) + 2MnO_4^-(aq) \rightarrow 2Mn^{2+}(aq) + 5O_2(g) + 8H_2O(\ell)$$

MnO_4^- is reduced, H_2O_2 is oxidized.

19.4 The balanced half reactions and balanced total reaction are shown in each case:

(a)
$$2[Cr_2O_7^{2-}(aq) + 14H^+(aq) + 6e^- \rightarrow 2Cr^{3+}(aq) + 7H_2O(\ell)]$$
$$3[CH_3OH(aq) + H_2O(\ell) \rightarrow HCO_2H(aq) + 4H^+(aq) + 4e^-]$$
$$2Cr_2O_7^{2-}(aq) + 3CH_3OH(aq) + 16H^+(aq) \rightarrow 4Cr^{3+}(aq) + 3HCO_2H(aq) + 11H_2O(\ell)$$

(b)
$$2[MnO_4^-(aq) + 2H_2O(\ell) + 3e^- \rightarrow MnO_2(s) + 4OH^-(aq)]$$
$$3[2I^-(aq) \rightarrow I_2(s) + 2e^-]$$
$$2MnO_4^-(aq) + 6I^-(aq) + 4H_2O(\ell) \rightarrow 2MnO_2(s) + 3I_2(s) + 8OH^-(aq)$$

(c)
$$2[Cr(OH)_4^-(aq) + 4OH^-(aq) \rightarrow CrO_4^{2-}(aq) + 4H_2O(\ell) + 3e^-]$$
$$3[H_2O_2(aq) + 2e^- \rightarrow 2OH^-(aq)]$$
$$2Cr(OH)_4^-(aq) + 3H_2O_2(aq) + 2OH^-(aq) \rightarrow 2CrO_4^{2-}(aq) + 8H_2O(\ell)$$

(d)
$$10[NO_3^-(aq) + 2H^+(aq) + e^- \rightarrow NO_2(g) + H_2O(\ell)]$$
$$I_2(s) + 6H_2O(\ell) \rightarrow 2IO_3^-(aq) + 12H^+(aq) + 10e^-$$
$$10NO_3^-(aq) + I_2(s) + 8H^+(aq) \rightarrow 10NO_2(g) + 2IO_3^-(aq) + 4H_2O(\ell)$$

(e)
$$3H_2O(\ell) + As(s) \rightarrow H_3AsO_3(aq) + 3H^+(aq) + 3e^-$$
$$NO_3^-(aq) + 4H^+(aq) + 3e^- \rightarrow NO(g) + 2H_2O(\ell)$$
$$As(s) + NO_3^-(aq) + H_2O(\ell) + H^+(aq) \rightarrow H_3AsO_3(aq) + NO(g)$$

(f) $H_2O_2(aq) + 2OH^-(aq) \rightarrow O_2(g) + 2H_2O(\ell) + 2e^-$

$2[ClO_2(aq) + e^- \rightarrow ClO_2^-(aq)]$

$$H_2O_2(aq) + 2ClO_2(aq) + 2OH^-(aq) \rightarrow O_2(g) + 2ClO_2^-(aq) + 2H_2O(\ell)$$

19.5 (a) $3N_2H_4(g) + N_2O_4(g) \rightarrow 4N_2(g) + 4H_2O(g) + 2H_2(g)$

(You may notice that there is no single unique set of coefficients that satisfies the observations we are given, because N_2H_4 can decompose on its own to form $N_2 + 2H_2$. Thus, any amount of excess N_2H_4 could be added in. We need to know the relative amounts of the products to write a uniquely balanced equation.) (b) N_2O_4 serves as the oxidant; it is itself reduced. N_2H_4 serves as the reducing agent; it is itself oxidized.

19.6 (a) $5[H_2C_2O_4(aq) \rightarrow 2CO_2(g) + 2H^+(aq) + 2e^-]$

$2[MnO_4^-(aq) + 8H^+(aq) + 5e^- \rightarrow Mn^{2+}(aq) + 4H_2O(\ell)]$

$$5H_2C_2O_4(aq) + 2MnO_4^-(aq) + 6H^+(aq) \rightarrow 2Mn^{2+}(aq) + 10CO_2 + 8H_2O(\ell)$$

(b)
$$\left(\frac{0.0500 \text{ mol } MnO_4^-}{1 \text{ L } MnO_4^- \text{ Soln}}\right)\left(\frac{0.022 \text{ L } MnO_4^- \text{ Soln}}{50.0 \text{ g sample}}\right)\left(\frac{5 \text{ mol } H_2C_2O_4}{2 \text{ mol } MnO_4^-}\right)$$
$$\times \left(\frac{90.0 \text{ g } H_2C_2O_4}{1 \text{ mol } H_2C_2O_4}\right) = \frac{4.95 \times 10^{-3} \text{ g } H_2C_2O_4}{1 \text{ g sample}}$$

Multiplying this by 100, we obtain 0.495 weight percent.

19.7

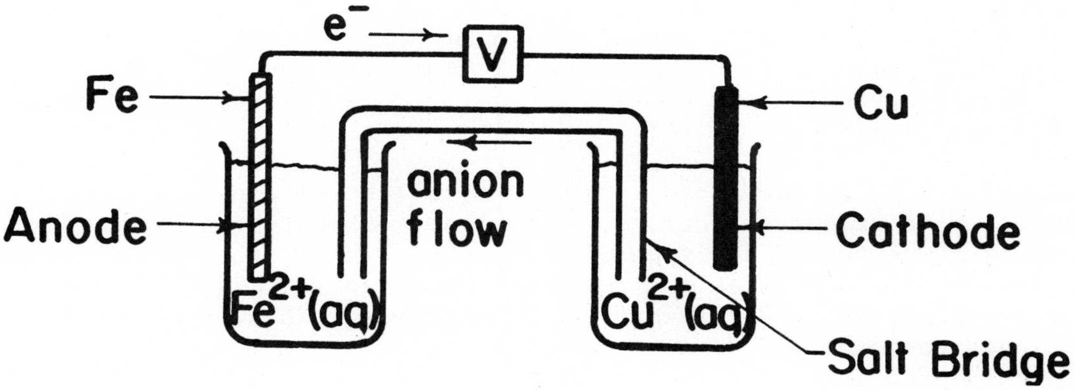

The standard emf for the cell is simply the sum of the two half-cell potentials:

$Fe(s) \rightarrow Fe^{2+}(aq) + 2e^-$	$E° = +0.44$ V
$2e^- + Cu^{2+}(aq) \rightarrow Cu(s)$	$E° = +0.34$ V
$Fe(s) + Cu^{2+}(aq) \rightarrow Fe^{2+}(aq) + Cu(s)$	$E° = +0.78$ V

19.8 (a) From the data in Table 19.1, we can calculate that for the reaction $Ni(s) + 2Ag^+(aq) \rightarrow Ni^{2+}(aq) + 2Ag(s)$, E° = 1.08 V. Nickel is oxidized; thus, Ni(s) serves as the anode. $Ag^+(aq)$ is reduced to Ag(s) at the Ag(s) cathode. (b) Ag gains mass as Ag^+ plates out; Ni(s) loses mass as nickel enters solution as $Ni^{2+}(aq)$. (c) The nickel electrode is negative; electrons move from it to the silver electrode, which appears positively charged. (d) As we saw above in (a), E° = 1.08 V.

19.9 (a) Reaction at anode: $5Fe^{2+}(aq) \rightarrow 5Fe^{3+}(aq) + 5e^-$. Reaction at cathode: $8H^+(aq) + MnO_4^-(aq) + 5e^- \rightarrow Mn^{2+}(aq) + 4H_2O(\ell)$. Electrons move from the Pt electrode in the iron solution to the Pt electrode in the $MnO_4^-(aq)$ solution. Anions migrate through the salt bridge from the cathode beaker to the anode beaker. The electrode in the iron-containing beaker has a negative sign, that in the MnO_4^- beaker has a positive sign. (b) Using Table 19.1, E° = (-0.77 V) + 1.51 V = 0.74 V.

19.10 (a) We look in Appendix F for the half-cell potentials:

$$Cr^{3+}(aq) + 3e^- \rightarrow Cr(s) \qquad E° = -0.74 \text{ V}$$

$$MnO_2(s) + 4H^+(aq) + 2e^- \rightarrow Mn^{2+}(aq) + 2H_2O(\ell) \qquad E° = +1.23 \text{ V}$$

We obtain the largest positive potential by reversing the first half-reaction and adding the two half-reactions (remember to balance the electron gain and loss).

$$2Cr(s) + 3MnO_2(s) + 12H^+(aq) \rightarrow 2Cr^{3+}(aq) + 3Mn^{2+}(aq) + 6H_2O(\ell)$$

$$E° = 1.97 \text{ V}$$

19.11 Na^+ and SO_4^{2-} are not readily oxidized. Oxidation of Na^+ would require removal of an electron from the closed inert-gas shell. SO_4^{2-} can be more easily oxidized, but such an oxidation does not occur readily, because SO_4^{2-} is a very stable anion. Chloride ion, Cl^-, can be oxidized moderately well. Oxidation of Cl_2 to form species such as $ClO_3^-(aq)$ occurs only with strong oxidizing agents, as can be seen from the E° values listed in Appendix F.

19.12 Fluoride ion, F^-, cannot be further reduced; it already possesses a completed octet of valence electrons, and carries a negative charge. There is no driving force toward addition of an electron to Na; this would form Na^-. (Although such a species has recently been prepared, it is formed only in the complete absence of water, and is very unstable.) Reduction of ClO_3^- leads to formation of a species with chlorine in a lower oxidation state. Such a process occurs very readily, as does reduction of Cl_2 (Appendix F).

19.13 (a) Arrange in order of increasing reduction potential:
$Cu^{2+}(aq) < O_2(g) < Cr_2O_7^{2-}(aq) < Cl_2(aq) < H_2O_2(aq)$

(b) Arrange in order of decreasing reduction potential:
$H_2O_2(aq) < I^-(aq) < Sn^{2+}(aq) < Zn(s) < Al(s)$

19.14 The $MnO_4^-(aq)$ reduction half-cell potential is about +1.5 V. Oxidation of all the other species involves half-cell potentials ranging from -1.47 for Cl_2 to -0.34 for Cu. Thus, all the species listed should be capable of oxidation by MnO_4^- in acidic solution.

19.15 We must look up the half-cell potentials for the oxidation and reduction half-cell reactions in each case. With a bit of practice, you will be able to discern the two half-cell reactions without separating the overall equation into two balanced half-cell reactions. You should look up the half-cell potential in each case, using Appendix F, and add to obtain the standard cell potential.

(a) $E° = +1.23\ V + (-1.36\ V) = -0.13\ V$ non-spontaneous
(b) $E° = +0.74\ V + (-1.66\ V) = -0.92\ V$ non-spontaneous
(c) $E° = +1.36\ V + (-1.06\ V) = +0.30\ V$ spontaneous
(d) $E° = -0.15\ V + 0.77\ V = +0.62\ V$ spontaneous
(e) $E° = -0.34\ V + 1.36\ V = +1.02\ V$ spontaneous

19.16 (a) $2Fe^{2+}(aq) + Br_2(aq) \rightarrow 2Fe^{3+}(aq) + 2Br^-(aq)$

$E° = +1.065\ V + (-0.771) = +0.294\ V$ (spontaneous)

(b) $6[MnO_4^-(aq) + 8H^+(aq) + 5e^- \rightarrow Mn^{2+}(aq) + 4H_2O(\ell)]$

$\underline{5[2Cr^{3+}(aq) + 7H_2O(\ell) \rightarrow Cr_2O_7^{2-}(aq) + 14H^+(aq) + 6e^-]}$

$6MnO_4^-(aq) + 10Cr^{3+}(aq) + 11H_2O(\ell) \rightarrow 5Cr_2O_7^{2-}(aq) + 6Mn^{2+}(aq) + 22H^+(aq)$

$E° = +1.51 + (-1.33) = +0.18\ V$ (spontaneous)

(c) $Cd^{2+}(aq) + Fe(s) \rightarrow Cd(s) + Fe^{2+}(aq)$

$E° = -0.403 + 0.440 = 0.037\ V$ (spontaneous)

19.17 We obtain E°, then employ Equation [19.23] to obtain ΔG°.

(a) $E° = -0.440 + 0.763 = 0.323\ V$ $\quad n = 2$

$$\Delta G° = -(2\ \text{mol } e^-)\left(\frac{96{,}500\ J}{V\text{-mol } e^-}\right)(0.323\ V) = -62.3\ kJ$$

(b) $E° = +1.685 + 0.356 = 2.041\ V$ $\quad n = 2$

$$\Delta G° = -(2\ \text{mol } e^-)\left(\frac{96{,}500\ J}{V\text{-mol } e^-}\right)(2.041\ V) = -393.9\ kJ$$

(c) $E° = +1.065 + (-1.359) = -0.294\ V$ $\quad n = 2$

$$\Delta G° = -(2\ \text{mol } e^-)\left(\frac{96{,}500\ J}{V\text{-mol } e^-}\right)(-0.294\ V) = +56.7\ kJ$$

(d) $E° = +2.87 + (-0.83) = 2.04\ V$ $\quad n = 2$

$$\Delta G° = -(2\ \text{mol } e^-)\left(\frac{96{,}500\ J}{V\text{-mol } e^-}\right)(2.04\ V) = -393.7\ kJ$$

19.18 We employ Equation [19.26], which relates E° to K:

$$E° = \frac{0.0591\ V}{n}\log K$$

(a) $\log K = \dfrac{1.00\ V}{0.0591\ V} = 16.92;\quad K = 8.3 \times 10^{16}$

(b) $\log K = \dfrac{0.10\ V}{0.0591\ V} = 1.69;\quad K = 49$

19.19 We must discern the two half-reactions in these equations, look up their potentials in Appendix F, then determine the value for n. We then employ Equation [19.26] to calculate K.

(a) $E° = 1.18 + (-0.763) = 0.42$ V; $\underline{n} = 2$

$$\log K = \frac{(2)(0.42\ V)}{0.0591\ V} = 14.2; \quad K = 2 \times 10^{14}$$

(b) $E° = +0.28 + 0 = +0.28$ V; $\underline{n} = 2$

$$\log K = \frac{(2)(0.28\ V)}{0.0591\ V} = 9.48; \quad K = 3 \times 10^{9}$$

(c) $E° = -1.359\ V + 1.51\ V = +0.15$ V; $\underline{n} = 10$

$$\log K = \frac{(10)(0.15\ V)}{0.0591\ V} = 25.4; \quad K = 2 \times 10^{25}$$

(d) $E° = 0.337 + (-0.536) = -0.199$ V; $\underline{n} = 2$

$$\log K = \frac{(2)(-0.199\ V)}{0.0591\ V} = -6.73; \quad K = 1.9 \times 10^{-7}$$

<u>19.20</u> Let us first balance the equation:

$$4CyFe^{2+}(aq) + O_2(g) + 4H^+(aq) \rightarrow CyFe^{3+}(aq) + 2H_2O(\ell)$$

$E = +0.60$ V; $\underline{n} = 4$

(a) From equation [19.21] we can relate ΔG for the process under the conditions specified to the measured potential E:

$$\Delta G = -n\mathcal{F}E = -(4\text{ mol } e^-)\left(\frac{96{,}500\ J}{V\text{-mol } e^-}\right)(0.60\ V) = -232\text{ kJ}$$

(b) The number of ATP molecules synthesized per O_2 molecule is given by:

$$\left(\frac{232\text{ kJ}}{O_2\text{ molecule}}\right)\left(\frac{1\text{ ATP formed}}{37.7\text{ kJ}}\right) = \text{approximately } \frac{6\text{ ATP}}{O_2}$$

<u>19.21</u> We can relate effects of concentration to emf by considering how concentration changes affect the position of equilibrium, then relating this to E through an expression such as Equation [19.29]:

$$E = E° - \frac{0.0591}{2}\log\left(\frac{[Cl^-]^2}{[I^-]^2}\right)$$

Increasing $[I^-]$ will clearly make the log term smaller (or make it more negative if it is already negative). Because of the negative sign before the term, the result is that E becomes more positive. This is what we expect. Le Chatelier's principle tells us that increasing $[I^-]$ will cause the reaction to shift to the right. E is a measure of the tendency of the reaction to proceed spontaneously toward products, so it thus should become more positive. The effect of increasing $[Cl^-]$ is exactly the opposite; E becomes smaller or more negative.

<u>19.22</u> $E° = +0.763 + 0.799 = 1.562$ V; $\underline{n} = 2$

$$E = 1.562 - \frac{0.0591}{2}\log\left(\frac{[Zn^{2+}]}{[Ag^+]^2}\right) = 1.562 - \frac{0.0591}{2}\log\left(\frac{0.01}{(0.5)^2}\right) = 1.603\ V$$

<u>19.23</u> $E° = -0.136\ V + 0.126\ V = -0.010$ V; $\underline{n} = 2$

$$0.22 = -0.010 - \frac{0.0591}{2}\log\left(\frac{[Pb^{2+}]}{[Sn^{2+}]}\right) = -0.010 - \frac{0.0591}{2}\log\frac{[Pb^{2+}]}{1.00}$$

$$\log [Pb^{2+}] = \frac{-0.23\ V(2)}{0.0591} = -7.78; \quad [Pb^{2+}] = 1.6 \times 10^{-8}$$

For $PbSO_4(s)$, $K_{sp} = [Pb^{2+}][SO_4^{2-}] = (1.0)(1.6 \times 10^{-8}) = 1.6 \times 10^{-8}$.

19.24 (a) increases emf; (b) decreases emf; (c) no effect; (d) increases emf; (e) decreases emf (see Equation [19.28]).

19.25 The cell reaction is $Zn(s) + 2H^+(aq) \rightarrow Zn^{2+}(aq) + H_2(g)$

$E° = 0.763 + 0 = 0.763$ V; $n = 2$

$$E = E° - \frac{0.0591}{2} \log \frac{[Zn^{2+}]P_{H_2}}{[H^+]^2}; \quad 0.702 = 0.763 - 0.0296 \log\left(\frac{1}{[H^+]^2}\right)$$

$$\log [H^+]^2 = \frac{-0.061}{0.0296} = -2.06; \quad [H^+] = 9.3 \times 10^{-2}\ M; \quad pH = 1.03$$

19.26 (a) A major advantage is that there is no need to separate the anode and cathode compartments via a salt bridge or some other similar device. Simple physical spacers are all that is needed. Mechanical ruggedness is improved by use of solids as reactants. (b) The concentration of a solid is effectively constant, so long as there is any solid present to contact the electrolyte. Thus, the voltage remains more constant as the cell discharges.

19.27 In the cell reaction, Equation [19.31], discharge of the cell is accompanied by removal of H_2SO_4 from the electrolyte, via formation of $PbSO_4(s)$ and $H_2O(\ell)$. As the concentration of H_2SO_4 in the electrolyte decreases, the lowering of the freezing point also decreases.

19.28 Loss of $PbSO_4(s)$ from the electrodes has two effects. First, the capacity of the battery is reduced; it delivers a smaller number of ampere-hours because some of the substance involved in the reaction has been lost. Secondly, the solid that accumulates in the base of the battery may eventually contact the plates, causing a short circuit. When this occurs, the cell may be completely inoperative.

19.29 The use of HNO_3 would cause problems. Since nitric acid is a good oxidizing agent, it is able to dissolve lead from the anode, thereby destroying the battery:

$$3[Pb \rightarrow Pb^{2+} + 2e^-] \qquad E° = -0.13\ V$$

$$2[4H^+ + NO_3^- + 3e^- \rightarrow NO + 2H_2O] \qquad E° = 0.96\ V$$

$$3Pb + 8H^+ + 2NO_3^- \rightarrow 3Pb^{2+} + 2NO + 4H_2O \qquad E° = 0.83\ V$$

Na_2SO_4 can be used in place of H_2SO_4, but it is less soluble than H_2SO_4, its density is greater than that of H_2SO_4, and evaporation of water from the solution leaves a salt residue.

19.30 Electrolysis of molten $AlCl_3$ can produce only Al(s) and $Cl_2(aq)$ as products: $2AlCl_3(\ell) \rightarrow 2Al(\ell) + 3Cl_2(g)$. When aqueous $AlCl_3$ is electrolyzed, water may be oxidized or reduced in preference to the solutes $Al^{3+}(aq)$ and $Cl^-(aq)$. In practice, water is reduced rather than $Al^{3+}(aq)$, but Cl^- would probably be oxidized. The overall cell reaction would probably be equation [19.39]. [Note: In practice, electrolysis of molten $AlCl_3$ is not employed to produce aluminum metal, for several reasons, including the fact that it sublimes at 181°C. The search for a proper electrolyte to use in preparing Al was the major aspect of Hall's research.]

19.31 (a) $Mg(\ell) + Cl_2(g)$; (b) $Cu(s) + O_2(g)$; (c) $Ag(s) + O_2(g)$; (d) $H_2(g) + Br_2(g)$.

19.32

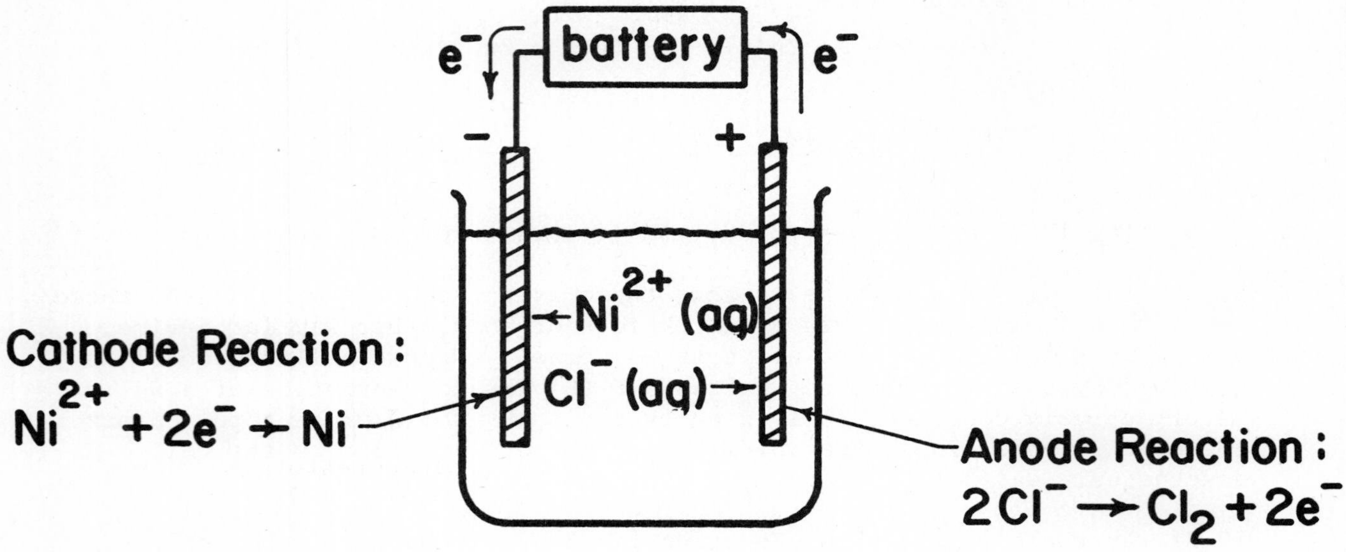

19.33

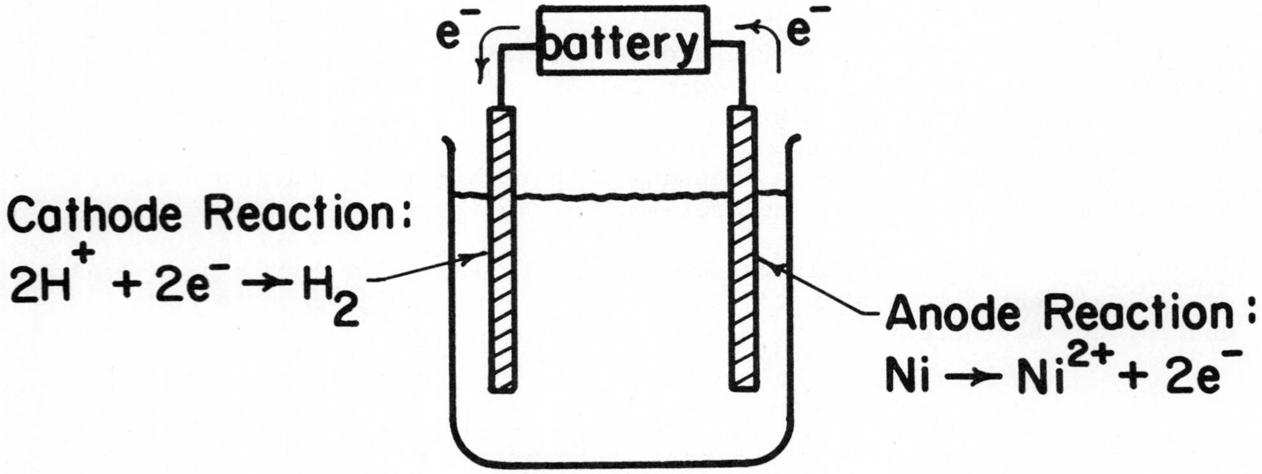

Ni is oxidized at lower potential ($E° = +0.28$ V) than $Br^-(aq)$ ($E° = -1.065$ V). H_2O is most readily reduced of any species present ($E° = -0.83$ V). Thus, the minimum applied potential required to initiate electrolysis would be 0.5 V. In practice, the hydrogen overvoltage on nickel ranges from about 0.2 to 0.9 volts, so something like 0.7 to 1.4 V would be required.

19.34 $$F = \left(\frac{96{,}500 \text{ coul}}{1 \text{ mol } e^-}\right)\left(\frac{1 \text{ mol } e^-}{6.022 \times 10^{23}\ e^-}\right) = 1.602 \times 10^{-19} \text{ coul}/e^-$$

19.35 $\left(\frac{3671.3 \text{ coul}}{4.8285 \text{ g } I_2}\right)\left(\frac{253.81 \text{ g } I_2}{1 \text{ mol } I_2}\right)\left(\frac{1 \text{ mol } I_2}{2 \text{ mol I}}\right)\left(\frac{1 \text{ mol I}}{1 \text{ mol } e^-}\right) = \frac{96491 \text{ coul}}{1 \text{ mol } e^-}$

19.36 (a) 5F; (b) 6F; (c) 1F; (d) 2F.

19.37 $(0.500 \text{ amp})(20.0 \text{ min})\left(\frac{60 \text{ sec}}{\text{min}}\right)\left(\frac{1 \text{ mol } e^-}{96{,}500 \text{ amp-sec}}\right)\left(\frac{1 \text{ mol Cd}}{2 \text{ mol } e^-}\right)\left(\frac{112 \text{ g Cd}}{1 \text{ mol Cd}}\right)$

$= 0.348$ g Cd

19.38 $(10.0 \text{ g Ni})\left(\frac{1 \text{ mol Ni}}{58.7 \text{ g Ni}}\right)\left(\frac{2 \text{ mol } e^-}{1 \text{ mol Ni}}\right)\left(\frac{96{,}500 \text{ amp-sec}}{1 \text{ mol } e^-}\right)\left(\frac{1}{1.50 \text{ amp}}\right)$

$= 2.19 \times 10^4$ sec = 6.08 hr

19.39 The half-cell reactions are Equations [19.40] and [19.41]:

$(1.50 \times 10^3 \text{ g C})\left(\frac{1 \text{ mol C}}{12.0 \text{ g C}}\right)\left(\frac{4 \text{ mol } e^-}{1 \text{ mol C}}\right)\left(\frac{1 \text{ mol Al}}{3 \text{ mol } e^-}\right)\left(\frac{27.0 \text{ g Al}}{1 \text{ mol Al}}\right) = 4.50 \text{ kg Al}$

19.40 $(2.00 \text{ amp})(1.00 \text{ hr})\left(\frac{3600 \text{ sec}}{1 \text{ hr}}\right)\left(\frac{1 \text{ mol } e^-}{96{,}500 \text{ amp-sec}}\right)\left(\frac{1 \text{ mol } Cl_2}{2 \text{ mol } e^-}\right)$

$= 0.0373 \text{ mol } Cl_2$

Using the ideal gas law, $V = \frac{(0.0373 \text{ mol})\left(\frac{0.082 \text{ L-atm}}{\text{mol-K}}\right) 273 \text{ K}}{1 \text{ atm}} = 0.835 \text{ L}$

19.41 The electrode reactions are: $Ni^{2+}(aq) + 2e^- \rightarrow Ni(s)$ and $2H_2O(\ell) \rightarrow 4H^+(aq) + O_2(g) + 4e^-$.

$(3.00 \text{ amp})(40.0 \text{ min})\left(\frac{60 \text{ sec}}{1 \text{ min}}\right)\left(\frac{1 \text{ mol } e^-}{96{,}500 \text{ amp-sec}}\right)\left(\frac{1 \text{ mol Ni}}{2 \text{ mol } e^-}\right) = 0.0373 \text{ mol Ni}$

There were originally $\left(\frac{0.500 \text{ mol } Ni^{2+}}{1 \text{ L}}\right)(0.250 \text{ L}) = 0.125 \text{ mol } Ni^{2+}$

in solution. After electrolysis, there are 0.0877 mol remaining, so $[Ni^{2+}] = 0.0877 \text{mol}/0.250 \text{ L} = 0.351$ M. $[SO_4^{2-}]$ remains unchanged.

$(3.00 \text{ amp})(40.0 \text{ min})\left(\frac{60 \text{ sec}}{1 \text{ min}}\right)\left(\frac{1 \text{ mol } e^-}{96{,}500 \text{ amp-sec}}\right)\left(\frac{1 \text{ mol } H^+}{1 \text{ mol } e^-}\right) = 0.0746 \text{ mol } H^+$

The concentration of $H^+(aq)$ is thus 0.0746 mol/0.250 L = 0.298 M.

19.42 (a) The record must be made the cathode, so that the metal ion in solution is reduced to metal on the surface. (b) The total area involved is $2\pi(15.0)^2 = 1414 \text{ cm}^2$. The total volume of Ni is $(1414 \text{ cm}^2)(1.0 \times 10^{-3} \text{ cm}) = 1.414 \text{ cm}^3$. The mass of nickel is $1.414 \text{ cm}^3(8.90 \text{ g/cm}^3) = 12.58$ g Ni.

$(12.58 \text{ g Ni})\left(\frac{1 \text{ mol Ni}}{58.71 \text{ g Ni}}\right)\left(\frac{2 \text{ mol } e^-}{1 \text{ mol Ni}}\right)\left(\frac{96{,}500 \text{ amp-sec}}{1 \text{ mol } e^-}\right)\left(\frac{1}{0.20 \text{ amp}}\right)$

$= 2.1 \times 10^5$ sec, about 57 hours.

19.43 (a) $\left(\frac{1.00 \times 10^3 \text{ g Al}}{1 \text{ hr}}\right)\left(\frac{1 \text{ hr}}{3600 \text{ sec}}\right)\left(\frac{1 \text{ mol Al}}{27.0 \text{ g Al}}\right)\left(\frac{3 \text{ mol } e^-}{1 \text{ mol Al}}\right)\left(\frac{96{,}500 \text{ amp-sec}}{1 \text{ mol } e^-}\right)$

$= 2.98 \times 10^3$ amp

(b) $\left(\frac{1.00 \times 10^3 \text{ g Cu}}{1 \text{ hr}}\right)\left(\frac{1 \text{ hr}}{3600 \text{ sec}}\right)\left(\frac{1 \text{ mol Cu}}{63.5 \text{ g Cu}}\right)\left(\frac{2 \text{ mol e}^-}{1 \text{ mol Cu}}\right)\left(\frac{96{,}500 \text{ amp-sec}}{1 \text{ mol e}^-}\right)$

$= 844$ amp

(c) $\left(\frac{1.00 \times 10^3 \text{ g Ag}}{1 \text{ hr}}\right)\left(\frac{1 \text{ hr}}{3600 \text{ sec}}\right)\left(\frac{1 \text{ mol Ag}}{108 \text{ g Ag}}\right)\left(\frac{1 \text{ mol e}^-}{1 \text{ mol Ag}}\right)\left(\frac{96{,}500 \text{ amp-sec}}{1 \text{ mol e}^-}\right)$

$= 248$ amp

19.44 The standard cell potential for this cell is 1.10 V (Equation [19.11]). To use Equation [19.47], we must calculate $n\mathcal{F}$.

$$(0.100 \text{ mol Zn})\left(\frac{2 \text{ mol e}^-}{1 \text{ mol Zn}}\right)\left(\frac{96{,}500 \text{ coul}}{1 \text{ mol e}^-}\right) = 1.93 \times 10^4 \text{ coul}$$

$$W_{max} = (1.93 \times 10^4 \text{ coul})(1.10 \text{ V}) = 2.12 \times 10^4 \text{ coul-V} = 2.12 \times 10^4 \text{ J}$$

19.45 Let us first calculate the energy change represented by the battery discharge. We must calculate $n\mathcal{F}$ from the quantity of Pb converted.

$$(15 \text{ g Pb})\left(\frac{1 \text{ mol Pb}}{207 \text{ g Pb}}\right)\left(\frac{2 \text{ mol e}^-}{1 \text{ mol Pb}}\right)\left(\frac{96{,}500 \text{ coul}}{1 \text{ mol e}^-}\right) = 1.40 \times 10^4 \text{ coul}$$

$$\Delta G = -n\mathcal{F}E = (1.4 \times 10^4 \text{ coul})(2.0 \text{ V}) = 2.8 \times 10^4 \text{ coul-V} = 2.8 \times 10^4 \text{ J}$$

$$(2.8 \times 10^4 \text{ J})\left(\frac{1 \text{ watt-sec}}{1 \text{ J}}\right)\left(\frac{1}{25 \text{ watt}}\right) = 1.12 \times 10^3 \text{ sec} = 0.3 \text{ hr}$$

19.46 The potential for oxidation of Zn to Zn^{2+} is much more positive than for oxidation of Fe to Fe^{2+}. Therefore, when a situation arises in which electrochemical oxidation could occur, any zinc in contact with an iron object will be oxidized in preference to the iron. Note that the potential for the process $Cr(s) \rightarrow Cr^{3+}(aq) + 3e^-$, $E° = +0.74$ V, is more positive than for the process $Fe(s) \rightarrow Fe^{2+}(aq) + 2e^-$, $E° = +0.44$ V. Thus, Cr should also act as a sacrificial anode when in contact with iron.

19.47 E° for the half reaction $Ni(s) \rightarrow Ni^{2+}(aq) + 2e^-$ is 0.28 V, lower than E° for oxidation of Fe(s) to $Fe^{2+}(aq)$. Nickel does not therefore act to protect iron cathodically. The nickel layer serves to completely cover the iron, and to provide a tight bond between the chrome plate layer and the metal underneath. Chromium does not stick very well on a pure iron surface.

19.48 The pH of an aqueous medium is an important factor in corrosion. Note from Equation [19.51] that an increased $[H^+]$ shifts the equilibrium to the right. When $[H^+]$ is depressed by the presence of a base, the reduction of O_2 is less favorable, and corrosion slows down. The added amines serve the function of keeping $[H^+]$ low.

19.49 E° for oxidation of copper to copper(II) is -0.34 V; that for oxidation of iron to iron(II) is $+0.44$ V. Thus, the iron will protect the copper from corrosion by corroding in preference to the copper. The fact that the steel pipe is galvanized (coated with Zn) means that the oxidation of Zn to Zn^{2+} ($E° = +0.77$ V) will occur in preference to oxidation of either metal, as long as there is any Zn left.

19.50 (a)

$$I_2(s) + 2e^- \rightarrow 2I(aq)$$
$$H_3AsO_3(aq) + H_2O(\ell) \rightarrow H_3AsO_4(aq) + 2H^+(aq) + 2e^-$$
$$\overline{I_2(s) + H_3AsO_3(aq) + H_2O \rightarrow 2I^-(aq) + H_3AsO_4(aq) + 2H^+(aq)}$$

(b)

$$5[(1/2)Br_2(\ell) + e^- \rightarrow Br^-(aq)]$$
$$(1/2)Br_2(\ell) + 6OH^-(aq) \rightarrow BrO_3^-(aq) + 3H_2O(\ell) + 5e^-$$
$$\overline{3Br_2(\ell) + 6OH^-(aq) \rightarrow BrO_3^-(aq) + 5Br^-(aq) + 3H_2O(\ell)}$$

(c)

$$2[Fe^{3+}(aq) + e^- \rightarrow Fe^{2+}(aq)]$$
$$2I^-(aq) \rightarrow I_2(s) + 2e^-$$
$$\overline{2Fe^{3+}(aq) + 2I^-(aq) \rightarrow 2Fe^{2+}(aq) + I_2(s)}$$

(d)

$$5[ClO_3^-(aq) + 6H^+(aq) + 6e^- \rightarrow Cl^-(aq) + 3H_2O(\ell)]$$
$$3[I_2(s) + 6H_2O(\ell) \rightarrow 2IO_3^-(aq) + 12H^+(aq) + 10e^-]$$
$$\overline{5ClO_3^-(aq) + 3I_2(s) + 3H_2O(\ell) \rightarrow 5Cl^-(aq) + 6IO_3^-(aq) + 6H^+(aq)}$$

(e)

$$6[Fe^{2+}(aq) \rightarrow Fe^{3+}(aq) + e^-]$$
$$Cr_2O_7^{2-}(aq) + 14H^+(aq) + 6e^- \rightarrow 2Cr^{3+}(aq) + 7H_2O(\ell)$$
$$\overline{6Fe^{2+}(aq) + Cr_2O_7^{2-}(aq) + 14H^+(aq) \rightarrow 6Fe^{3+}(aq) + 2Cr^{3+}(aq) + 7H_2O(\ell)}$$

(f)

$$2[CrO_4^{2-}(aq) + 2H_2O(\ell) + 3e^- \rightarrow CrO_2^-(aq) + 4OH^-(aq)]$$
$$3[HSnO_2^-(aq) + 2OH^-(aq) \rightarrow HSnO_3^-(aq) + H_2O(\ell) + 2e^-]$$
$$\overline{2CrO_4^{2-}(aq) + H_2O(\ell) + 3HSnO_2^-(aq) \rightarrow 2CrO_2^-(aq) + 3HSnO_3^-(aq) + 2OH^-(aq)}$$

19.51 (a)

$$5[C_6H_{12}O_6(aq) + 6H_2O(\ell) \rightarrow 6CO_2(g) + 24H^+(aq) + 24e^-]$$
$$12[2NO_3^-(aq) + 12H^+(aq) + 10e^- \rightarrow N_2(g) + 6H_2O(\ell)]$$
$$\overline{5C_6H_{12}O_6(aq) + 24NO_3^-(aq) + 24H^+(aq) \rightarrow 30CO_2(g) + 12N_2(g) + 42H_2O(\ell)}$$

(b)

$$Au(s) + 4Cl^-(aq) \rightarrow AuCl_4^-(aq) + 3e^-$$
$$3[NO_3^-(aq) + 2H^+(aq) + e^- \rightarrow NO_2(g) + H_2O(\ell)]$$
$$\overline{Au(s) + 3NO_3^-(aq) + 4Cl^-(aq) + 6H^+(aq) \rightarrow AuCl_4^-(aq) + 3NO_2(g) + 3H_2O(\ell)}$$

(c)

$$2Al(s) + 6H^+(aq) \rightarrow 2Al^{3+}(aq) + 3H_2(g)$$

(d)

$$Cu(s) \rightarrow Cu^{2+}(aq) + 2e^-$$
$$HSO_4^-(aq) + 3H^+(aq) + 2e^- \rightarrow SO_2(g) + 2H_2O(\ell)$$
$$\overline{Cu(s) + HSO_4^-(aq) + 3H^+(aq) \rightarrow Cu^{2+}(aq) + SO_2(g) + 2H_2O(\ell)}$$

(e)

$$\begin{array}{l} CH_3OH(g) + H_2O(g) \rightarrow CO_2(g) + 6H^+(aq) + 6e^- \\ 3[H_2O_2(g) + 2H^+(aq) + 2e^- \rightarrow 2H_2O(g)] \\ \hline CH_3OH(g) + 3H_2O_2(g) \rightarrow CO_2(g) + 5H_2O(g) \end{array}$$

$$\begin{array}{l} N_2H_4(g) \rightarrow N_2(g) + 4H^+(aq) + 4e^- \\ 2[H_2O_2(g) + 2H^+(aq) + 2e^- \rightarrow 2H_2O(g)] \\ \hline N_2H_4(g) + 2H_2O_2(g) \rightarrow N_2(g) + 4H_2O(g) \end{array}$$

19.52 (a)

$$\begin{array}{l} Cu(s) \rightarrow Cu^{2+}(aq) + 2e^- \\ 2[NO_3^-(aq) + 2H^+(aq) + e^- \rightarrow NO_2(g) + H_2O(\ell)] \\ \hline Cu(s) + 2NO_3^-(aq) + 4H^+(aq) \rightarrow 2NO_2(g) + Cu^{2+}(aq) + 2H_2O(\ell) \end{array}$$

(b) Copper is a reducing agent; it reduces NO_3^-. Nitrate is the oxidizing agent; it oxidizes Cu. (c) Chloride ion cannot be readily reduced, whereas NO_3^- is a strong oxidizing agent in acidic solution. The $H^+(aq)$ is also a possible oxidizing agent; it is itself reduced to H_2. However, copper is not an active metal; that is, it is not readily oxidized, and H^+ does not oxidize the copper even in strongly acidic medium.

19.53

$$\begin{array}{l} I_2(s) + 6H_2O(\ell) \rightarrow 2IO_3^-(aq) + 12H^+(aq) + 10e^- \\ 10[NO_3^-(aq) + 2H^+(aq) + e^- \rightarrow NO_2(g) + H_2O(\ell)] \\ \hline I_2(s) + 10NO_3^-(aq) + 8H^+(aq) \rightarrow 2IO_3^-(aq) + 10NO_2(g) + 4H_2O(\ell) \end{array}$$

Note that HIO_3 is a fairly strong acid; we could also have written $HIO_3(aq)$.

$$(10.0 \text{ g } HIO_3)\left(\frac{1 \text{ mol } HIO_3}{176 \text{ g } HIO_3}\right)\left(\frac{10 \text{ mol } NO_2}{2 \text{ mol } IO_3^-}\right)\left(\frac{46 \text{ g } NO_2}{1 \text{ mol } NO_2}\right) = 13.1 \text{ g } NO_2$$

19.54

$$\begin{array}{l} 2CN^-(aq) \rightarrow (CN)_2(aq) + 2e^- \\ 2[CN^-(aq) + Cu^{2+}(aq) + e^- \rightarrow CuCN(s)] \\ \hline 4CN^-(aq) + 2Cu^{2+}(aq) \rightarrow 2CuCN(s) + (CN)_2(aq) \end{array}$$

$$(5.0 \text{ g } CuSO_4)\left(\frac{1 \text{ mol } CuSO_4}{160 \text{ g } CuSO_4}\right)\left(\frac{1 \text{ mol} (CN)_2}{2 \text{ mol } CuSO_4}\right)\left(\frac{52 \text{ g } (CN)_2}{1 \text{ mol} (CN)_2}\right) = 0.81 \text{ g } (CN)_2$$

19.55 (a) Iron serves as anode, because E° for $Fe \rightarrow Fe^{2+} + 2e^-$ is +0.440 V, whereas for $Sn \rightarrow Sn^{2+} + 2e^-$ it is +0.136 V.

(b)

$$\begin{array}{ll} Fe(s) \rightarrow Fe^{2+}(aq) + 2e^- & E° = +0.440 \text{ V} \\ Sn^{2+}(aq) + 2e^- \rightarrow Sn(s) & E° = -0.136 \text{ V} \\ \hline Fe(s) + Sn^{2+}(aq) \rightarrow Fe^{2+}(aq) + Sn(s) & E° = +0.304 \text{ V} \end{array}$$

19.56 (a)

$$\begin{array}{l} Fe(s) \rightarrow Fe^{2+}(aq) + 2e^- \\ 2Ag^+(aq) + 2e^- \rightarrow 2Ag(s) \\ \hline Fe(s) + 2Ag^+(aq) \rightarrow Fe^{2+}(aq) + 2Ag(s) \end{array}$$

(b)

$$\begin{array}{l} Mg(s) \rightarrow Mg^{2+}(aq) + 2e^- \\ 2H^+(aq) + 2e^- \rightarrow H_2(g) \\ \hline Mg(s) + 2H^+(aq) \rightarrow Mg^{2+}(aq) + H_2(g) \end{array}$$

(c) $Zn|Zn^{2+}||Cu^{2+}|Cu$

19.57 (a) Reversal of the first half-reaction, then adding, leads to a large, positive overall cell potential. Zinc is oxidized to Zn^{2+}, silver oxide is reduced to metallic silver. (The element silver actually undergoes the reduction.) (b) Electrons come out of the anode ($Zn \rightarrow Zn^{2+} + 2e^-$) during cell operation. Thus, the anode is negatively charged. The cathode is positively charged; it is here that electrons are drawn to reduce the Ag_2O during cell operation.

(c)
$$Zn(s) \rightarrow Zn^{2+}(aq) + 2e^- \qquad E° = +0.763\ V$$
$$Ag_2O(s) + H_2O(\ell) \rightarrow 2Ag(s) + 2OH^-(aq) \qquad E° = +0.34\ V$$
$$Zn(s) + Ag_2O(s) + H_2O(\ell) \rightarrow Zn^{2+}(aq) + 2Ag(s) + 2OH^-(aq) \qquad E° = +1.10\ V$$

19.58 (a) H_2O; MnO_4^-; (b) Fe; H_2

19.59 E_{ox} must be between -0.54 V and -0.80 V. Thus, the potential choices are: (a) $Fe^{2+}(aq)$; $H_2O_2(aq)$ in acidic solution; $MnO_2(s)$ in basic solution. (b) $MnO_4^-(aq)$ in acid is the only choice with E_{red} between 1.36 and 2.87 V.

19.60 (a) yes; sum of E° values for the two half-cell reactions is positive. (b) yes.

19.61 (a) For this cell, E° = +0.14 V; $\underline{n} = 2$

$$E = E° - \frac{2.30\ RT}{n\mathcal{F}} \log Q = 0.14 - \frac{0.0591}{2} \log\left(\frac{[Ni^{2+}]}{[Sn^{2+}]}\right)$$
$$= 0.14 - 0.0295 \log\left(\frac{10^{-3}}{1.0}\right) = 0.22\ V.$$

(b) E° = +0.440 V; $\underline{n} = 2$

$$E = 0.440 - \frac{0.0591}{2} \log\left(\frac{(0.020)(0.10)}{(0.1)^2}\right) = 0.461\ V$$

19.62 $E = 0 - \frac{0.0591}{3} \log\left(\frac{5.0 \times 10^{-3}}{2.0}\right) = 0.051\ V$

The anode in this concentration cell is the electrode immersed in the less concentrated solution, 5×10^{-3} M.

19.63 (a) E° = 0.14 V; $\Delta G° = -n\mathcal{F}E° = -2(0.14)(96{,}500) = -27$ kJ;

$$E° = \frac{0.0591}{n} \log K; \quad \log K = \frac{2(0.14)}{0.0591} = 4.74; \quad K = 5.5 \times 10^4$$

(b) E° = 0.440 V; $\Delta G° = -n\mathcal{F}E° = -2(0.440)(96{,}500) = -84.9$ kJ;

$$E° = \frac{0.0591 \log K}{n}; \quad \log K = \frac{2(0.440)}{0.0591} = 14.9; \quad K = 8 \times 10^{14}$$

19.64 The cell reaction is given in Equation [19.31] and the half-cell reactions are written just above it. From Appendix F, we obtain the half-cell potentials: E° = 0.356 + 1.685 = 2.041 V;

$$E = E° - \frac{0.0591}{2} \log\left(\frac{1}{[H^+]^2[HSO_4^-]^2}\right)$$

$$\left(\frac{0.38 \text{ g } H_2SO_4}{1 \text{ g Soln}}\right)\left(\frac{1.286 \text{ g Soln}}{1 \text{ mL}}\right)\left(\frac{1000 \text{ mL}}{1 \text{ L}}\right)\left(\frac{1 \text{ mol } H_2SO_4}{98 \text{ g } H_2SO_4}\right) = 5.0 \text{ M } H_2SO_4$$

If we assume that the H_2SO_4 is completely ionized to form $H^+ + HSO_4^-$, and neglect further ionization of HSO_4^-, then $[H^+] = [HSO_4^-] = 5.0$ M.

$$E = 2.04 - 0.0295 \log\left(\frac{1}{(5.0)^2(5.0)^2}\right) = 2.12 \text{ V}$$

[Note: This is a rather approximate correction to E°, because at a concentration level of 5 M the solution behaves non-ideally. The main point is that the cell voltage is actually higher than E° in a fully charged lead-acid storage battery.]

(b) $$\left(\frac{0.05 \text{ g } H_2SO_4}{1 \text{ g Soln}}\right)\left(\frac{1.025 \text{ g Soln}}{1 \text{ mL}}\right)\left(\frac{1000 \text{ mL}}{1 \text{ L}}\right)\left(\frac{1 \text{ mol } H_2SO_4}{98 \text{ g } H_2SO_4}\right)$$

$$= 0.52 \text{ M } H_2SO_4$$

Let us again assume complete ionization of H_2SO_4 to form H^+ and HSO_4^-; ionization of HSO_4^- could now be important, but if you work through the weak-acid-dissociation problem, you will find that only about 2% of the HSO_4^- goes on to dissociate further. Thus,

$$E = 2.04 - 0.0295 \log\left(\frac{1}{[0.52]^2[0.52]^2}\right) = 2.04 - 0.03 = 2.01 \text{ V}$$

Note that the change in electrolyte concentration from part (a) to part (b) has caused a 0.11 V decrease in cell voltage. If a car battery consists of six such cells in series, a change of about 0.6 volts in battery voltage is observed. Voltage regulators, which control battery charging, operate by sensing the battery voltage.

<u>19.65</u> The overall equation we wish to attain is:

$$AgSCN(s) \rightleftarrows Ag^+(aq) + SCN^-(aq)$$

We have $AgSCN(s) + e^- \rightarrow Ag(s) + SCN^-(aq)$ $E° = +0.0895$ V. If we add to this:

$$Ag(s) \rightarrow Ag^+(aq) + e^- \qquad E° = -0.799 \text{ V}$$

$$AgSCN(s) \rightarrow Ag^+(aq) + SCN^-(aq) \quad E° = -0.710 \text{ V}$$

$$E° = \frac{0.0591}{1} \log K; \quad \log K = \frac{-0.710}{0.0591} = -12.0; \quad K = 1.0 \times 10^{-12}$$

<u>19.66</u> The cell shown in Figure 19.2 is rechargeable. Application of an external voltage of appropriate sign and magnitude could cause the cell reaction to run "backward," because all the components of the half-cell reactions remain in the system, in contact with each other as in the forward reaction. Thus, we could call the $Zn-Cu^{2+}$ cell a storage cell.

<u>19.67</u> The half-cell potential of interest is:

$$H_2O(\ell) + 2e^- \rightarrow H_2(g) + 2OH^-(aq) \qquad E° = -0.83 \text{ V}$$

$$E = -0.83 - \frac{0.0591}{2} \log\left(P_{H_2}[OH^-]^2\right).$$ At pH = 7, $[OH^-] = 10^{-7}$ M

$$E = -0.83 - 0.0295 \log (10^{-14}) = -0.42 \text{ V}$$

We can also approach this problem by using the half-reaction:

$$2H^+(aq) + 2e^- \rightarrow H_2(g) \qquad E° = 0$$

$$E = 0 - \frac{0.0591}{2} \log \left(\frac{P_{H_2}}{(H^+)^2}\right) = \frac{-0.0591}{2} \log(10^{14}) = -0.42 \text{ V.}$$

We obtain the same answer in either case, because the problem is equivalent whether we think of reducing $H^+(aq)$ or $H_2O(\ell)$.

19.68 During cell discharge, the PbO_2 electrode is positive; as the cathode it is the electrode to which electrons from the external circuit flow. During charging, it is the positively charged electrode, from which electrons are "pumped" by the external potential, to cause the cell reaction to reverse.

19.69 (a) The area of the coin is twice the area of the surface, plus the area of the rim: $2\pi(1.2)^2 + \pi(2.4)(0.10) = 9.80 \text{ cm}^2$.

$$(9.80 \text{ cm}^2)(1.0 \times 10^{-3} \text{ cm}) = (9.80 \times 10^{-3} \text{ cm}^3 \text{ Ag})\left(\frac{10.5 \text{ g Ag}}{1 \text{ cm}^3}\right)$$
$$= 0.103 \text{ g Ag}$$

(b)
$$\left(\frac{0.103 \text{ g Ag}}{\text{coin}}\right)(150 \text{ coins})\left(\frac{1 \text{ mol Ag}}{108 \text{ g Ag}}\right)\left(\frac{1 \text{ mol e}^-}{\text{mol Ag}}\right)\left(\frac{96{,}500 \text{ amp-sec}}{1 \text{ mol e}^-}\right)$$
$$\times \left(\frac{1}{0.200 \text{ amp}}\right)\left(\frac{1}{0.85}\right) = 8.10 \times 10^4 \text{ sec} = 22.5 \text{ hr}$$

19.70 Electrolysis of an aqueous Na_2SO_4 solution results in oxidation of water at the anode, and reduction of water at the cathode. This is so because neither ion of the electrolyte is readily oxidized or reduced. The electrode reactions are therefore:

anode: $2H_2O(\ell) \rightarrow O_2(g) + 4H^+(aq) + 4e^-$

cathode: $H_2O(\ell) + 2e^- \rightarrow H_2(g) + 2OH^-(aq)$

Note that $H^+(aq)$ is formed in the anode reaction, and that $OH^-(aq)$ is formed in the cathode reaction. These reactions account for the appearance of blue litmus (basic solution) at the cathode, and red litmus (acidic solution) at the anode.

19.71
$$\left(\frac{0.075 \text{ g Pb}}{1 \text{ min}}\right)\left(\frac{1 \text{ mol Pb}}{207 \text{ g Pb}}\right)\left(\frac{2 \text{ mol e}^-}{1 \text{ mol Pb}}\right)\left(\frac{96{,}500 \text{ amp-sec}}{1 \text{ mol e}^-}\right)\left(\frac{1 \text{ min}}{60 \text{ sec}}\right) = 1.2 \text{ amp}$$

19.72
$$(24.0 \text{ hr})(20.0 \text{ amp})\left(\frac{3600 \text{ sec}}{1 \text{ hr}}\right)\left(\frac{1 \text{ mol e}^-}{96{,}500 \text{ amp-sec}}\right)\left(\frac{4 \text{ mol NaBO}_3}{8 \text{ mol e}^-}\right)$$
$$\times \left(\frac{81.8 \text{ g NaBO}_3}{1 \text{ mol NaBO}_3}\right) = 732 \text{ g NaBO}_3$$

19.73 The half-cell reaction of interest is Equation [19.36].

$$(2.5 \text{ g NiO}_2)\left(\frac{1 \text{ mol NiO}_2}{91 \text{ g NiO}_2}\right)\left(\frac{2 \text{ mol e-}}{1 \text{ mol NiO}_2}\right)\left(\frac{96{,}500 \text{ amp-sec}}{1 \text{ mol e}^-}\right)\left(\frac{1}{2.0 \text{ amp}}\right)$$
$$= 2651 \text{ sec} = \text{about } 45 \text{ min}$$

19.74 The half-cell reaction of interest is:

$$PbSO_4(s) + 2H_2O(\ell) \rightarrow PbO_2(s) + HSO_4^-(aq) + 3H^+(aq) + 2e^-$$

$$(50.0\ g\ PbSO_4)\left(\frac{1\ mol\ PbSO_4}{303\ g\ PbSO_4}\right)\left(\frac{2\ mol\ e^-}{1\ mol\ PbSO_4}\right)\left(\frac{96,500\ amp\text{-}sec}{1\ mol\ e^-}\right)\left(\frac{1}{10.0\ amp}\right)$$
$$= 3.18 \times 10^3\ sec$$

Since the charging efficiency is specified as only 85 percent, it will require $3.18 \times 10^3/0.85 = 3.75 \times 10^3$ sec = 1.04 hr.

19.75 The maximum work obtainable is given by $W = n\mathcal{F}E$.

$$n\mathcal{F} = (100\ amp\text{-}hr)\left(\frac{3600\ sec}{hr}\right)\left(\frac{1\ coul}{1\ amp\text{-}sec}\right) = 3.6 \times 10^5\ coul$$

$$W = (3.6 \times 10^5\ coul)(12\ V) = 4.3 \times 10^6\ coul\text{-}V = 4.3 \times 10^6\ J$$

19.76 At the graphite electrode oxidation can occur:

$$4OH^-(aq) \rightarrow O_2(g) + 2H_2O(\ell) + 4e^-$$

At the copper coin, reduction of the copper compound on the surface is possible. For example:

$$H_2O(\ell) + Cu_2S(s) + 2e^- \rightarrow 2Cu(s) + OH^-(aq) + HS^-(aq)$$

or

$$H_2O(\ell) + CuO(s) + 2e^- \rightarrow Cu(s) + 2OH^-(aq)$$

19.77 Apparently an electrochemical cell is established. The likely anode reaction is oxidation of zinc, which forms the galvanizing layer in the pan: $Zn(s) \rightarrow Zn^{2+}(aq) + 2e^-$. The cathode reduction probably involves reduction of the Ag_2S on the metal surface:

$$H_2O(\ell) + Ag_2S(s) + 2e^- \rightarrow 2Ag(s) + HS^-(aq) + OH^-(aq)$$

Whether this reaction will in fact proceed to the right depends on the concentration of HS^- (or S^{2-}) in the solution. K_{sp} for Ag_2S is very low, only about 10^{-50}. Thus, the half-cell reaction is really not favored. The fact that ZnS(s) may form as $[S^{2-}]$ increases could help reaction to proceed to the right.

19.78 (a)

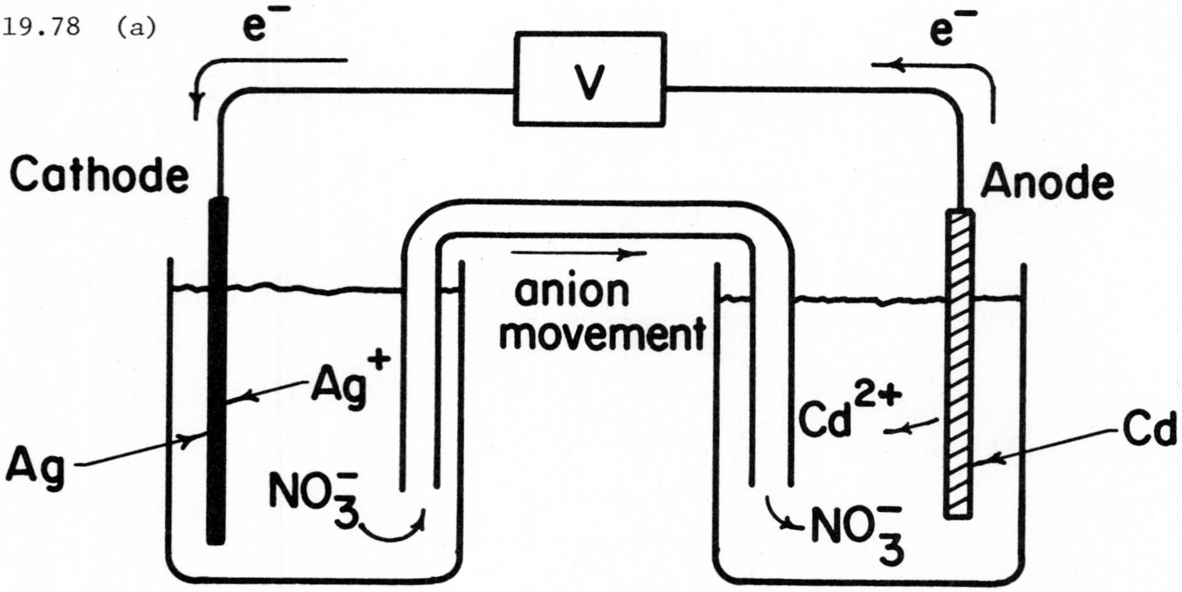

(b) The silver electrode gains weight, the Cd electrode loses.
(c) E° = 0.799 + 0.403 = 1.202 V. (d) This is a form of limiting reagent problem. The total charge the cell can deliver is determined by <u>either</u> the quantity of $Ag^+(aq)$ in solution that can be reduced: $Ag^+(aq) + e^- \rightarrow Ag(s)$, <u>or</u> the mass of Cd electrode that can be oxidized: $Cd(s) \rightarrow Cd^{2+}(aq) + 2e^-$. The mass of the silver electrode and the quantity of $Cd^{2+}(aq)$ present are of no consequence, since these substances are not consumed in the cell reaction.

$$(1.00\text{ g AgNO}_3)\left(\frac{1\text{ mol AgNO}_3}{170\text{ g AgNO}_3}\right)\left(\frac{1\text{ mol e}^-}{1\text{ mol AgNO}_3}\right) = 5.88 \times 10^{-3}\text{ mol e}^-$$

$$(1.00\text{ g Cd})\left(\frac{1\text{ mol Cd}}{112\text{ g Cd}}\right)\left(\frac{2\text{ mol e}^-}{1\text{ mol Cd}}\right) = 0.0178\text{ mol e}^-$$

We see that the total electrical charge the cell can deliver is limited by the quantity of Ag^+ present in solution: $(0.588 \times 10^{-3}$ mole $e^-)$ (96,500 coul/mol e^-) = 5.67×10^2 coul.

<u>19.79</u> (a) $(7 \times 10^8\text{ mol H}_2)\left(\frac{2F}{1\text{ mol H}_2}\right)\left(\frac{96{,}500\text{ coul}}{1F}\right) = 1.4 \times 10^{14}\text{ coul}$

(b)
$$\begin{array}{ll} 2H_2O(\ell) \rightarrow O_2(g) + 4H^+(aq) + 4e^- & E° = -1.23\text{ V} \\ 2[2H^+(aq) + 2e^- \rightarrow H_2(g)] & E° = 0\text{ V} \\ \hline 2H_2O(\ell) \rightarrow O_2(g) + 2H_2(g) & E° = -1.23\text{ V} \end{array}$$

$$E = E° - \frac{0.0591}{4}\log\,[O_2][H_2]^2 = -1.23\text{ V} - \frac{0.0591}{4}\log\,(300)^3$$

$$= -1.23\text{ V} - 0.11\text{ V} = -1.34\text{ V}$$

(c) $\text{Energy} = nEF = 2(7 \times 10^8\text{ mol})(1.34\text{ V})(96{,}500\ \frac{1}{\text{V-mol}})$
$= 2 \times 10^{14}\text{ J}$

(d) $(2 \times 10^{14}\text{ J})\left(\frac{1\text{ kwh}}{3.6 \times 10^6\text{ J}}\right)\left(\frac{\$0.23}{\text{kwh}}\right) = \$1.3 \times 10^7$

We see that it will cost more than ten million dollars for the electricity alone.

<u>19.80</u> $(1.62\text{ g Zn})\left(\frac{1\text{ mol Zn}}{65.4\text{ g Zn}}\right)\left(\frac{2\text{ mol e}^-}{1\text{ mol Zn}}\right)\left(\frac{96{,}500\text{ coul}}{1\text{ mol e}^-}\right)\left(\frac{1}{3.5 \times 10^{-3}}\right)$

$= 1.37 \times 10^6$ coul

20

Nuclear chemistry

20.1 (a) A nuclear is a proton or neutron. (b) The mass number is the total number of nucleons in a nucleus. (c) A nuclear transmutation is a nuclear reaction induced by collision of a nucleus with any other nucleus or a neutron. (d) an alpha (α) ray is a stream of helium nuclei. (e) A positron is a subatomic particle with the mass of an electron, but with opposite charge. (f) A gamma ray is a form of radiation emitted as part of a nuclear process; gamma rays are comparable in wavelength and energy to X-rays.

20.2 (a) 8 protons, 9 neutrons; (b) 42 protons, 57 neutrons; (c) 55 protons, 81 neutrons; (d) 47 protons, 68 neutrons.

20.3 (a) 58 protons, 79 neutrons, 137 nucleons; (b) 94 protons, 140 neutrons, 234 nucleons; (c) 48 protons, 65 neutrons, 113 nucleons; (d) 17 protons, 20 neutrons, 37 nucleons.

20.4 (a) ${}^{181}_{72}\text{Hf} \rightarrow {}^{0}_{-1}\text{e} + {}^{181}_{73}\text{Ta}$; (b) ${}^{226}_{88}\text{Ra} \rightarrow {}^{222}_{86}\text{Rn} + {}^{4}_{2}\text{He}$;
(c) ${}^{205}_{82}\text{Pb} \rightarrow {}^{0}_{1}\text{e} + {}^{205}_{81}\text{Tl}$; (d) ${}^{179}_{74}\text{W} + {}^{0}_{-1}\text{e} \rightarrow {}^{179}_{73}\text{Ta}$

20.5 (a) ${}^{93}_{40}\text{Zr} \rightarrow {}^{0}_{-1}\text{e} + {}^{93}_{41}\text{Nb}$; (b) ${}^{233}_{93}\text{Np} \rightarrow {}^{229}_{91}\text{Pa} + {}^{4}_{2}\text{He}$
(c) ${}^{222}_{89}\text{Ac} \rightarrow {}^{218}_{87}\text{Fr} + {}^{4}_{2}\text{He}$; (d) ${}^{246}_{99}\text{Es} + {}^{0}_{-1}\text{e} \rightarrow {}^{246}_{98}\text{Cf}$

20.6 (a) ${}^{32}_{16}\text{S} + {}^{1}_{0}\text{n} \rightarrow {}^{1}_{1}\text{H} + {}^{32}_{15}\text{P}$; (b) ${}^{7}_{4}\text{Be} + {}^{0}_{-1}\text{e}$ (orbital electron) $\rightarrow {}^{7}_{3}\text{Li}$;
(c) ${}^{187}_{75}\text{Re} \rightarrow {}^{187}_{76}\text{Os} + {}^{0}_{-1}\text{e}$; (d) ${}^{98}_{42}\text{Mo} + {}^{2}_{1}\text{H} \rightarrow {}^{1}_{0}\text{n} + {}^{99}_{43}\text{Tc}$;
(e) ${}^{235}_{92}\text{U} + {}^{1}_{0}\text{n} \rightarrow {}^{135}_{54}\text{Xe} + {}^{99}_{38}\text{Sr} + 2{}^{1}_{0}\text{n}$

20.7 (a) ${}^{252}_{98}\text{Cf} + {}^{10}_{5}\text{B} \rightarrow 3{}^{1}_{0}\text{n} + {}^{259}_{103}\text{Lw}$; (b) ${}^{2}_{1}\text{H} + {}^{3}_{2}\text{He} \rightarrow {}^{4}_{2}\text{He} + {}^{1}_{1}\text{H}$;
(c) ${}^{1}_{1}\text{H} + {}^{11}_{5}\text{B} \rightarrow 3{}^{4}_{2}\text{He}$; (d) ${}^{122}_{53}\text{I} \rightarrow {}^{122}_{54}\text{Xe} + {}^{0}_{-1}\text{e}$; (e) ${}^{59}_{26}\text{Fe} \rightarrow {}^{0}_{-1}\text{e} + {}^{59}_{27}\text{Co}$

20.8 (i) $^{250}_{100}Fm \rightarrow ^{246}_{98}Cf + ^{4}_{2}He$; (ii) $^{246}_{98}Cf \rightarrow ^{242}_{96}Cm + ^{4}_{2}He$; (iii) $^{242}_{96}Cm \rightarrow ^{238}_{94}Pu + ^{4}_{2}He$; (iv) $^{238}_{94}Pu \rightarrow ^{234}_{92}U + ^{4}_{2}He$

20.9 (a) $^{238}_{92}U + ^{1}_{0}n \rightarrow ^{239}_{92}U + ^{0}_{0}\gamma$; (b) $^{14}_{7}N + ^{1}_{1}H \rightarrow ^{11}_{6}C + ^{4}_{2}He$; (c) $^{18}_{8}O + ^{1}_{0}n \rightarrow ^{19}_{9}F + ^{0}_{-1}e$; (d) $^{59}_{26}Fe + ^{4}_{2}He \rightarrow ^{63}_{29}Cu + ^{0}_{-1}e$

20.10 The total mass number change is (235 - 207) = 28. Since each α particle accompanies a change of -4 in mass number, whereas emission of a β particle does not correspond to a mass change, there are 7 α particle emissions. The change in atomic number in the series is 10. Each α particle results in an atomic number lower by two. The seven α particle emissions would thus of themselves cause a decrease of 14 in atomic number. Each β particle emission raises the atomic number by one. To obtain the observed lowering of 10 in the series, there must have been four β particle emissions.

20.11 The criterion we employ in judging whether the nucleus is likely to be radioactive is the position of the nucleus on the plot shown in Figure 20.1. If the neutron/proton ratio is too high or low, or if the atomic number exceeds 83, the nucleus will be radioactive. (b) low neutron/proton ratio; (c) low neutron/proton ratio; (e) high atomic number.

20.12 We use the criteria listed in Table 20.1. (a) $^{19}_{9}F$; odd proton, even neutron more abundant; (b) $^{80}_{34}Se$ even-even more abundant; (c) $^{56}_{26}Fe$ even-even more abundant; (d) $^{118}_{50}Sn$ even-even more abundant than odd-odd.

20.13 Positron emission is most likely to occur from nuclei with low neutron/proton ratios. Positron emission, in effect, converts a proton into a neutron. $^{51}_{25}Mn$ has the lowest neutron/proton ratio among the three nuclei of comparable atomic numbers. It is, in fact, radioactive via positron emission.

20.14 (a) no - low neutron/proton ratio; should be a positron emitter. (b) no - low neutron/proton ratio. Should be a positron emitter, or possibly undergoes orbital electron capture. (c) no - high neutron/proton ratio; should be a beta emitter. (d) no - high atomic number. It should be an alpha emitter.

20.15 (a) Rather low neutron/proton ratio - should be a positron emitter. (b) high neutron/proton ratio - beta emitter. (c) high neutron/proton ratio - beta emitter. (d) low neutron/proton ratio - positron emitter.

20.16 The electrons most tightly bound to the nucleus are most likely to have a finite probability density (Section 5.5) within the volume of the nucleus itself. Only s orbitals do not have a node at the nucleus. One therefore expects that one of the 1s electrons will be captured. The 1s electrons will be ever more tightly pulled in toward the nucleus as atomic number increases, because they are not effectively shielded by the other electrons. The probability of electron capture should thus be greater for atoms of high atomic number.

20.17 (a) Thallium-210 has a rather higher neutron/proton ratio than the stable nuclei. Beta emission in effect converts a neutron into a proton, thus lowering the neutron/proton ratio. (b) The neutron/proton ratio for $^{204}_{89}Ac$ is only 1.29. Orbital electron capture in effect converts a proton into a neutron, thus increasing the neutron/proton ratio. (c) The neutron/proton ratio for $^{197}_{83}Bi$ is only 1.37. Alpha particle emission lowers both the neutron and proton count, and lowers the atomic number toward a value for which a lower neutron/proton ratio is stable.

20.18 The fact that a nucleus is radioactive has no chemical significance for the atom containing that nucleus, up to the time radioactive decay occurs. A ^{35}S atom has the same number of electrons as a ^{32}S atom, and behaves exactly the same chemically. However, once radioactive decay has occurred, the ^{35}S becomes ^{35}Cl. It now behaves chemically like a chlorine atom that has been placed in a chemical environment not typical of chlorine.

20.19 $^{66}_{32}Ge \rightarrow ^{66}_{31}Ga + ^{0}_{1}e$. Ten hours represents exactly four half lives. The amount of ^{66}Ge remaining is then $(25.0)(1/2)^4 = 1.56$ mg.

20.20 After 12.3 yr, one half-life, there are (1/2)48.0 = 24.0 mg. 49.2 yr are exactly four half-lives. There are then $(48.0)(1/2)^4 = 3.0$ mg tritium remaining.

20.21 The number of disintegrations per second is proportional to the number of radioactive nuclei that remain. Using Equation [20.24], solve for $t_{1/2}$:

$$t_{1/2} = \frac{t}{3.32\ \log(N_o/N_t)} = \frac{1\ \text{yr}}{3.32\ \log(3012/2921)} = 22.6\ \text{yr}$$

20.22 $\log(N_o/N_t) = \dfrac{t}{3.32\ t_{1/2}} = \dfrac{(16\ \text{hr})(60\ \text{min/hr})}{(3.32)(85\ \text{min})} = 3.40$

$N_o = 2.5 \times 10^3\ N_t$. The number of radionuclides is proportional to mass. Therefore, $m_o = 2.5 \times 10^3\ m_t$; $m_t = 24.0 \times 10^{-6}\text{g}/2.5 \times 10^3 = 9.6 \times 10^{-9}$ g $= 9.6$ ng ^{139}Ba.

20.23 $\log\left(\dfrac{N_o}{N_t}\right) = \dfrac{t}{3.32\ t_{1/2}} = \dfrac{1019}{(3.32)(2.4 \times 10^4)} = 1.28 \times 10^{-2}$

$\dfrac{N_o}{N_t} = 1.030$. Thus, $N_t/N_o = 1/1.030 = 0.971$.

About 97% of the ^{239}Pu will remain in 3000 A.D.

20.24 Let us assume that no depletion of iodide from the water due to plant uptake has occurred. The activity after 32 days would then be

$$\log\left(\frac{89}{N_t}\right) = \frac{32\ \text{days}}{(3.32)(8.1\ \text{days})} = 1.190;\quad 89/N_t = 15.5;$$

$N_t = 5.7$ counts per minute. Note that this is precisely the activity measured in the water. If the plants had absorbed some of the iodide, the activity would be lower than this level.

20.25 $t_{1/2} = \dfrac{t}{(3.32)\ \log(N_o/N_t)} = \dfrac{30.0\ \text{days}}{3.32\ \log(4600/3130)} = 54.0$ days

20.26 We follow the procedure outlined in Sample Exercise 20.7. The original quantity of ^{238}U is 50.0 mg plus the amount that gave rise to 14.0 mg ^{206}Pb. This amount is 14.0(238/206) = 16.2 mg. Thus,

$$t = 3.32(4.5 \times 10^9\ \text{yr})\ \log\left(\frac{66.2}{50.0}\right) = 1.8 \times 10^9\ \text{yr}$$

20.27 $t = 3.32(5.7 \times 10^3 \text{ yr}) \log\left(\frac{31.7}{25.8}\right) = 1.7 \times 10^3$ yr

20.28 If the mass of ^{40}Ar is 3.6 times that of ^{40}K, then the original mass of ^{40}K must have been 3.6 + 1 = 4.6 times that now present: $t = 3.32(1.27 \times 10^9 \text{ yr}) \log(4.6) = 2.8 \times 10^9$ yr.

20.29 Figure a shows a graph of the activity (disintegrations per minute) vs. time. We pick off $t_{1/2}$ as the time at which the activity is half the initial value.

Rearranging Equation 20.24, we can obtain: $t = 3.32\, t_{1/2} \log N_o - 3.32\, t_{1/2} \log N_t$; $\log N_t = 3.32\, t_{1/2} \log N_o - t/3.32\, t_{1/2}$. Thus, a graph of log (activity) vs. t should yield a straight line with slope = $-1/3.32\, t_{1/2}$. The slope of such a graph, Figure b, is -4.93×10^{-2}. Thus,

$$t_{1/2} = \frac{-1}{3.32(\text{slope})} = \frac{1}{3.32(4.93 \times 10^{-2})} = 6.1 \text{ hr}$$

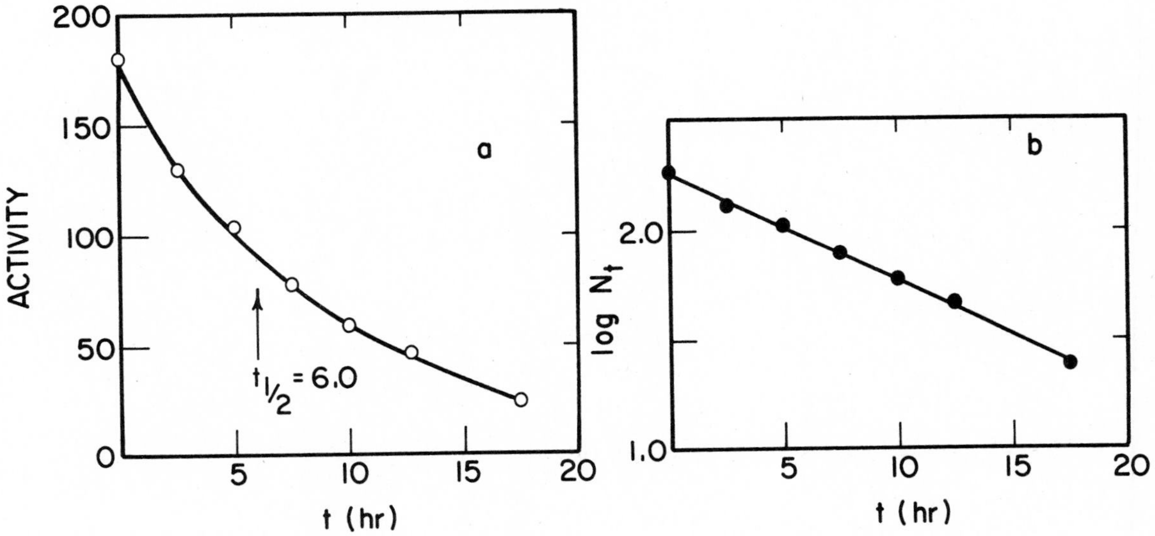

20.30 First, we calculate k:

$$k = \frac{0.693}{12.26 \text{ yr}}\left(\frac{1 \text{ yr}}{365 \text{ days}}\right)\left(\frac{1 \text{ day}}{24 \text{ hr}}\right)\left(\frac{1 \text{ hr}}{3600 \text{ sec}}\right) = 1.79 \times 10^{-9}/\text{sec}$$

From Equation [20.20], $1.50 \times 10^3/\text{sec} = (1.79 \times 10^{-9}/\text{sec})(N)$; $N = 8.38 \times 10^{11}$. In 26.0 g of water, there are

$$(26.0 \text{ g } H_2O)\left(\frac{1 \text{ mol } H_2O}{18 \text{ g } H_2O}\right)\left(\frac{6.02 \times 10^{23}\ H_2O}{1 \text{ mol } H_2O}\right)\left(\frac{2\text{ H}}{1\ H_2O}\right) = 1.74 \times 10^{24} \text{ H atoms}$$

The mole fraction of ^3_1H atoms in the sample is thus $8.38 \times 10^{11}/1.74 \times 10^{24} = 4.8 \times 10^{-13}$.

20.31 Calculate in each case the mass of nucleons plus atomic electrons: (a) 6(1.00728) + 6(1.00867) + 6(0.000548) = 12.09899 amu; binding energy per nucleon is (12.09899 − 12.00000)/12 = 8.25×10^{-3} amu.

$$\Delta E = (8.25 \times 10^{-3}\ \text{amu})\left(\frac{1\ \text{g}}{6.02 \times 10^{23}\ \text{amu}}\right)\left(\frac{1\ \text{kg}}{10^3\ \text{g}}\right)\left(\frac{9.00 \times 10^{16}\ \text{m}^2}{\text{sec}^2}\right)$$
$$= 1.23 \times 10^{-12}\ \text{J}$$

(b) $(61.50529 - 60.93106)/61 = 9.41 \times 10^{-3}$ amu

$$\Delta E = (9.41 \times 10^{-3}\ \text{amu})\left(\frac{1\ \text{g}}{6.02 \times 10^{23}\ \text{amu}}\right)\left(\frac{1\ \text{kg}}{10^3\ \text{g}}\right)\left(\frac{9.00 \times 10^{16}\ \text{m}^2}{\text{sec}^2}\right)$$
$$= 1.41 \times 10^{-12}\ \text{J}$$

(c) $(207.71698 - 205.97447)/206 = 8.46 \times 10^{-3}$ amu

$$\Delta E = (8.46 \times 10^{-3}\ \text{amu})\left(\frac{1\ \text{g}}{6.02 \times 10^{23}\ \text{amu}}\right)\left(\frac{1\ \text{kg}}{10^3\ \text{g}}\right)\left(\frac{9.00 \times 10^{16}\ \text{m}^2}{\text{sec}^2}\right)$$
$$= 1.26 \times 10^{-12}\ \text{J}$$

20.32 (a) 30(1.00728) + 34(1.00867) + 30(0.000548) = 64.52962 amu
Total binding energy = 64.52962 - 63.92914 = 0.60048 amu

$$\Delta E = (0.60048\ \text{amu})\left(\frac{1\ \text{g}}{6.02 \times 10^{23}\ \text{amu}}\right)\left(\frac{1\ \text{kg}}{10^3\ \text{g}}\right)\left(\frac{9.00 \times 10^{16}\ \text{m}^2}{\text{sec}^2}\right)$$
$$= 8.98 \times 10^{-11}\ \text{J}$$

Binding energy per nucleon = 8.98×10^{-11} J/64 = 1.40×10^{-12} J

(b) 37.30648 - 36.96590 = 0.34058 amu; total binding energy $\Delta E = 5.09 \times 10^{-11}$ J
Binding energy per nucleon = 5.09×10^{-11} J/37 = 1.38×10^{-12} J.
(c) 4.03300 - 4.00260 = 0.0304 amu; total binding energy = 4.54×10^{-12} J
Binding energy per nucleon = 1.14×10^{-12} J.

20.33 We calculate the binding energy (note that nuclear mass is provided here): 3(1.00728) + 3(1.00867) = 6.04785 amu; binding energy = (6.04785 - 6.01347) = 0.03438 amu. The energy equivalent of this mass we can evaluate using Einstein's equation, $\Delta E = c^2(\Delta m)$.

$$= (3.00 \times 10^8\ \text{m/sec})^2(0.03438\ \text{amu})\left(\frac{1\ \text{g}}{6.023 \times 10^{23}\ \text{amu}}\right)\left(\frac{1\ \text{kg}}{10^3\ \text{g}}\right)$$
$$= 5.14 \times 10^{-12}\ \text{kg-m}^2/\text{sec}^2 = 5.14 \times 10^{-12}\ \text{J}$$

For 1 mole of ^{6}Li nuclei, $\Delta E = (5.14 \times 10^{-12}\ \text{J})(6.02 \times 10^{23})$
$= 3.09 \times 10^{12}$ J/mol

20.34 If ΔH is -910.8 kJ/mol, we have

$$-910.8 \times 10^3\ \text{J} = c^2(\Delta m);\quad \Delta m = \frac{-910.8 \times 10^3 (\text{kg-m}^2/\text{sec}^2)}{(3.00 \times 10^8\ \text{m/sec})^2}$$
$$= -1.01 \times 10^{-11}\ \text{kg}$$

20.35
$$\left(\frac{1.07 \times 10^{16}\ \text{kJ}}{\text{min}}\right)\left(\frac{60\ \text{min}}{1\ \text{hr}}\right)\left(\frac{24\ \text{hr}}{1\ \text{day}}\right) = 1.54 \times 10^{19}\ \frac{\text{kJ}}{\text{day}} = 1.54 \times 10^{22}\ \text{J/day}$$

$$\Delta m = \frac{1.54 \times 10^{22} (\text{kg-m}^2/\text{sec}^2)}{(3.00 \times 10^8\ \text{m/sec})^2} = 1.71 \times 10^5\ \text{kg/day}$$

We now calculate the mass change in the given nuclear reaction: Δm = 140.9140 + 91.9218 + 2(1.00867) - 235.0439 = -0.19076 amu. Thus, it requires 235 g ^{235}U to produce energy equivalent to 0.191 g. Now 0.1% of 1.7×10^5 kg is 1.7×10^2 kg = 1.7×10^5 g. Then,

$$(1.7 \times 10^5 \text{ g})\left(\frac{235 \text{ g } ^{235}\text{U}}{0.191 \text{ g}}\right) = 2.09 \times 10^8 \text{ g } ^{235}\text{U}$$

This is about 230 tons of ^{235}U. Since a typical reactor fuel contains about 3 percent ^{235}U, this energy is the equivalent of about 7600 tons of reactor fuel per day.

20.36 (a) $\Delta m = 4.00260 + 1.00867 - 3.01605 - 2.01410 = -0.01888$ amu.

$$\Delta E = (0.01888 \text{ amu})\left(\frac{1 \text{ g}}{1 \text{ amu}}\right)\left(\frac{1 \text{ kg}}{10^3 \text{ g}}\right)(3.00 \times 10^8 \text{ m/sec})^2$$
$$= 1.70 \times 10^{12} \text{ J/mol}$$

(b) $\Delta m = 3.01603 + 1.00867 - 2(2.01410) = -3.5 \times 10^{-3}$ amu;
$\Delta E = 3.15 \times 10^{11}$ J/mol
(c) $\Delta m = 4.00260 + 1.00782 - 3.01603 - 2.01410 = -1.97 \times 10^{-2}$ amu
$\Delta E = 1.77 \times 10^{12}$ J/mol

20.37 Examination of Figure 20.8 shows that the binding energy per nucleon is largest for nuclei with mass numbers in the vicinity of 50. Thus, ^{48}Ca should be among the most stable of all nuclei. There is no possibility therefore of gaining energy by breaking up the ^{48}Ca nucleus into smaller nuclei, or of merging it to form nuclei of still higher mass.

20.38 We can use Figure 20.8 to see that the binding energy per nucleon (which gives rise to the mass defect) is greatest for nuclei of mass numbers around 50. Thus $^{59}_{27}Co$ should possess the greatest mass defect per nucleon.

20.39 (a) The rem is a unit of radiation dosage which measures the total biological damage caused by the radiation. It is the product of the amount of energy deposited by the radiation in the body and an index of its relative biological effectiveness. (b) The curie is a measure of the number of nuclear disintegrations occurring per second. 1 curie = 3.7×10^{10} disintegrations/sec. (c) A moderator is an element in a nuclear reactor used to control the rate of the nuclear reaction by slowing neutrons so that they are more readily captured by the fuel. (d) A breeder reactor is one in which the neutrons produced are used in part to create new nuclear fuel which can be later separated from the fuel rods. (e) The critical mass is that mass of fissionable material that is sufficient to sustain a chain nuclear reaction.

20.40 Both ^{90}Sr and ^{133}Ba are likely to be incorporated in the food chain in the environment, where they would serve as replacement for calcium or perhaps zinc. Neither H_2 nor Kr are likely to undergo incorporation into living systems. The H_2 could, of course, eventually react to form H_2O, but it would not be in a chemical form that would lead to any special concentration effects.

20.41 Among the three types of radiation, alpha rays are least penetrating, beta rays are next and gamma rays are the most penetrating. Thus, biological damage from external sources is least for alpha rays. However, when alpha emitting substances are ingested or inhaled, the damage which the alpha emissions can cause for a given level of energy deposition is ten times greater than for beta or gamma rays.

20.42 The danger from inhalation of plutonium-containing particles arises from the extensive damage that alpha particles can cause in biological tissues. Alpha particles have only a short range; that is, they cause damage only in a few mm of tissue as they move away from the radiation source. Only when alpha-emitting substances are ingested or inhaled does the radiation actually penetrate sensitive tissues and organs. Unfortunately, once plutonium gets inside the body it is not readily excreted, so this long-lived source of radioactivity continues to cause damage over a long period of time.

20.43 For a nuclear explosion to occur, the critical mass of material must be held together long enough for the branching chain reaction to proceed for awhile, so that the energy released generates enormous temperature increase. This will not occur in a nuclear reactor because there is not enough pressure to hold materials together, and also because neutron-capturing control rods are present as well. In a nuclear weapon, the critical mass is brought together with explosive force to develop the required high neutron flux.

20.44 The ^{59}Fe would be incorporated into the diet component, which in turn is then fed to the rabbits. After a time blood samples could be removed from the animals, the red blood cells separated, and the radioactivity of the sample measured. If the iron in the dietary compound has been incorporated into blood hemoglobin, the blood cell sample should show beta emission. Samples could be taken at various times to determine the rate of iron uptake, rate of loss of the iron from the blood, and so forth.

20.45 In principle, one could apply some fertilizer enriched with a radiotracer, then collect samples from water supplies into which water from the fertilized land might drain. However, there are problems with such a procedure. The dilution of the radiotracer might be so great that a useful level of radioactivity might not be expected in the water. Secondly, the time scale for the experiment might be too long for the half-life of the process giving rise to the radioactivity. It turns out that there is no radioactive nuclide of nitrogen with sufficiently long half-life. Nitrogen-13 is the longest-lived radioisotope, with a half-life of only 10 minutes.

20.46 (a) Add ^{36}Cl to water as a chloride salt. Then dissolve ordinary CCl_3COOH. After a time, distill the volatile materials away from the salt; CCl_3OOH is volatile, and will distill with water. Count radioactivity in the volatile material. If chlorine exchange has occurred, there will be radioactivity. (b) Prepare a saturated solution of $BaCl_2$ containing a small amount of solid $BaCl_2$. Add to this solution solid $BaCl_2$ containing ^{36}Cl. If the solid-solution equilibrium is dynamic, some of the ^{36}Cl in the solid will find itself in solution as chloride ion. After allowing some time for equilibrium to become established, filter the solution, count radioactivity in the solution that is separated from the solid. If there were no dynamic equilibrium, the $^{36}Cl^-$ would remain in the added solid, since the solution is already saturated before the addition of more solid. (c) Utilize ^{36}Cl in the soils of various pH, grow plants for a given period of time. Remove plants, and directly measure radioactivity in samples from stems, leaves and so forth, or reduce the volume of plant sample by some form of digestion and evaporation of solution to give a dry residue that can be counted.

20.47 2×10^{-12} curies$\left(\frac{3.7 \times 10^{10} \text{ disint/sec}}{1 \text{ curie}}\right) = 7.4 \times 10^{-2}$ disint/sec

$$\left(\frac{7.4 \times 10^{-2}\ \text{disint/sec}}{75\ \text{kg}}\right)\left(\frac{8 \times 10^{-13}\ \text{J}}{\text{disint}}\right)\left(\frac{1\ \text{rad}}{1 \times 10^{-2}\ \text{J/kg}}\right)\left(\frac{3600\ \text{sec}}{\text{hr}}\right)\left(\frac{24\ \text{hr}}{1\ \text{day}}\right)$$

$$\times \left(\frac{365\ \text{day}}{1\ \text{yr}}\right) = 2.5 \times 10^{-6}\ \text{rad/yr}$$

$$\left(\frac{2.5 \times 10^{-6}\ \text{rad}}{\text{yr}}\right)\left(\frac{10\ \text{rem}}{\text{rad}}\right) = 2.5 \times 10^{-5}\ \text{rem/yr}$$

Recall that there are 10 rem/rad for alpha particles.

20.48 The major problem has to do with the very high neutron fluxes. These neutrons cause nuclear reactions to occur which alter the material in a fundamental way. For example, an (n,p) type nuclear reaction converts an atom into an atom of the element with one lower atomic number. In addition, radiation can lead to other kinds of damage in the solid, causing structural defects.

20.49 In fusion the idea is to jam two nuclei together. Because nuclei have positive charge, bringing two nuclei the very short distances necessary to initiate nuclear fusion requires enormous energy input, to overcome the huge coulombic repulsive forces. A similar energy is not required for fission; here a large nucleus that already exists is bombarded with neutral neutrons to initiate a falling apart of the unstable nucleus. There is no huge coulombic repulsion to overcome in this process.

20.50 (a) The electron and positron possess the same mass, 0.000548 amu, but opposite electrical charge; the electron has a charge of -1.60×10^{-19} coul, the positron has a charge of $+1.60 \times 10^{-19}$ coul. (b) The binding energy of a nucleus is the energy required to decompose a nucleus into individual protons and neutrons from which it is formed. The mass defect is this binding energy expressed as the difference in mass of the nucleus and the sum of masses of the nucleons from which it may be considered to have formed. (c) A curie is a measure of the number of nuclear disintegrations that occur in a sample per second. A rem is a measure of the effective dose of radiation received by a human. (d) Gamma rays are short wavelength, high energy photons, comparable to x-rays, but emitted in radioactive decay processes. A beta ray, also emitted in radioactive decay processes, consists of high energy electrons.

20.51
1) ${}^{238}_{92}U \rightarrow {}^{234}_{90}Th + {}^{4}_{2}He$
2) ${}^{234}_{90}Th \rightarrow {}^{234}_{91}Pa + {}^{0}_{-1}e$
3) ${}^{234}_{91}Pa \rightarrow {}^{234}_{92}U + {}^{0}_{-1}e$
4) ${}^{234}_{92}U \rightarrow {}^{230}_{90}Th + {}^{4}_{2}He$
5) ${}^{230}_{90}Th \rightarrow {}^{226}_{88}Ra + {}^{4}_{2}He$
6) ${}^{226}_{88}Ra \rightarrow {}^{222}_{86}Rn + {}^{4}_{2}He$
7) ${}^{222}_{86}Ra \rightarrow {}^{218}_{84}Po + {}^{4}_{2}He$
8) ${}^{218}_{84}Po \rightarrow {}^{214}_{82}Pb + {}^{4}_{2}He$
9) ${}^{214}_{82}Pb \rightarrow {}^{214}_{83}Bi + {}^{0}_{-1}e$
10) ${}^{214}_{83}Bi \rightarrow {}^{214}_{84}Po + {}^{0}_{-1}e$
11) ${}^{214}_{84}Po \rightarrow {}^{210}_{82}Pb + {}^{4}_{2}He$
12) ${}^{210}_{82}Pb \rightarrow {}^{210}_{83}Bi + {}^{0}_{-1}e$
13) ${}^{210}_{83}Bi \rightarrow {}^{210}_{84}Po + {}^{0}_{-1}e$
14) ${}^{210}_{84}Po \rightarrow {}^{206}_{82}Pb + {}^{4}_{2}He$

20.52 Chemical reactions do not affect the character of atomic nuclei. The energy changes involved in chemical reactions are much too small to allow us to alter nuclear properties via chemical processes. Therefore, the nuclei that are formed in a nuclear reaction will continue to emit radioactivity regardless of any chemical changes we bring to bear. However, we can hope to use chemical means to separate radioactive substances, or remove them from foods or a portion of the environment.

20.53 Using Equation [20.24]:

$$\log\left(\frac{N_o}{N_t}\right) = \frac{1}{3.32\ t_{1/2}} = \frac{13}{(3.32)(5.25)} = 0.746$$

$$\frac{N_o}{N_t} = 5.57. \quad \text{Thus,}\ \frac{N_t}{N_o} = 0.180$$

20.54 $$(1 \times 10^{-6}\ \text{curie})\left(\frac{3.7 \times 10^{10}\ \text{disint/sec}}{\text{curie}}\right) = 3.7 \times 10^{4}\ \text{disint/sec}$$

$$\text{rate} = 3.7 \times 10^{4}\ \text{nuclei/sec} = kN$$

$$k = \frac{0.693}{t_{1/2}} = \left(\frac{0.693}{27.6\ \text{yr}}\right)\left(\frac{1\ \text{yr}}{365 \times 24 \times 3600\ \text{sec}}\right) = 7.96 \times 10^{-10}/\text{sec}$$

$$3.7 \times 10^{4}\ \text{nuclei/sec} = (7.96 \times 10^{-10}/\text{sec})\ N; \quad N = 4.6 \times 10^{13}\ \text{nuclei}$$

$$\text{mass}\ {}^{90}\text{Sr} = 4.6 \times 10^{13}\ \text{nuclei}\left(\frac{90\ \text{g Sr}}{6.02 \times 10^{23}\ \text{nuclei}}\right) = 6.9 \times 10^{-9}\ \text{g Sr}$$

20.55 (a) ${}^{6}_{3}\text{Li} + {}^{63}_{28}\text{Ni} \rightarrow {}^{69}_{31}\text{Ga}$; (b) ${}^{48}_{20}\text{Ca} + {}^{248}_{96}\text{Cm} \rightarrow {}^{296}_{116}\text{X}$

(c) ${}^{88}_{38}\text{Sr} + {}^{84}_{36}\text{Kr} \rightarrow {}^{116}_{46}\text{Pd} + {}^{56}_{28}\text{Ni}$; (d) ${}^{48}_{20}\text{Ca} + {}^{238}_{92}\text{U} \rightarrow {}^{70}_{20}\text{Ca} + 4{}^{1}_{0}\text{n} + 2{}^{106}_{46}\text{Pd}$

20.56 $$\log\left(\frac{N_o}{N_i}\right) = \frac{t}{3.32\ t_{1/2}}\ ; \quad \log\left(\frac{7000}{1000}\right) = \frac{t}{3.32(28.8\ \text{yr})};\ t = 80.8\ \text{yr}$$

20.57 $4{}^{1}_{1}\text{H} \rightarrow {}^{4}_{2}\text{He} + 2{}^{0}_{1}\text{e}$. See Sample Exercise 20.10 for an analogous case. The products of the nuclear reaction are an alpha particle, He^{2+}, and two electrons. The mass of these is just the mass of the helium atom.

$$\Delta m = 4(1.00782) - 4.00260 = 0.02868\ \text{amu}$$

$$\Delta E = (0.02868\ \text{amu})\left(\frac{3.00 \times 10^{8}\ \text{m}}{\text{sec}}\right)^{2}\left(\frac{1\ \text{kg}}{1 \times 10^{3}\ \text{amu}}\right) = 2.58 \times 10^{12}\ \text{J/mol}$$

The extremely high temperature is required to overcome the electrostatic charge repulsions between the nuclei in forcing them together to react.

20.58 The most massive radionuclides will have the highest neutron/proton ratios. Thus, they are most likely to decay by a process that lowers this ratio; i.e., beta emission. The least massive nuclides, on the other hand, will decay by a process that increases the neutron/proton ratio; i.e., positron emission or orbital electron capture.

20.59 We have that $\Delta m = \Delta E/c^2$: $$\Delta m = \frac{3.9 \times 10^{26}\ \text{kg-m/sec}^2}{(3.0 \times 10^{8}\ \text{m/sec})^2} = 4.3 \times 10^{9}\ \text{kg}$$

This is the mass lost in one second.

20.60 Using Equation [20.24], $t = 3.32(5.7 \times 10^3 \text{ yr}) \log\left(\frac{18.4}{9.6}\right)$
$= 5.3 \times 10^3$ yr

In this calculation, we have assumed that the relative numbers of ^{14}C nuclei, originally and at present, are proportional to the number of observed disintegrations per second.

20.61 The suggestion is not reasonable. The energies of nuclear states are very large relative to ordinary temperatures. Thus, merely changing the temperature by less than 100 K would not be expected to significantly affect the behavior of nuclei with regard to nuclear decay rates.

20.62 This decay series represents a change of 28 mass units. Since only alpha emissions change the nuclear mass, and each changes the mass by four, there must be a total of seven alpha emissions. Each alpha emission causes a decrease of two in atomic number. Therefore, the seven alpha emissions, by themselves, should have caused a decrease in atomic number of 14. The series as a whole involves a decrease of 10 in atomic number. Thus, there must be a total of four beta emissions, each of which increase atomic number by one.

20.63 The half-life for a long-lived radio nuclide can be determined by utilizing Equations [20.20] and [20.22]. Experimentally, one must know accurately N, the number of radionuclides in the sample, and be able to accurately count all the disintegration events that occur in a unit time. When the decay rate is slow (long-lived radionuclides), it is a demanding task to obtain good data. From the rate of decay (disintegrations per unit time) and N, one can calculate k, and then $t_{1/2}$.

20.64 A major difference is that the charge on the nitrogen nucleus, +7, is much smaller than on the gold nucleus, +79. Thus, the alpha particle could more easily penetrate the coulomb barrier (that is, the repulsive energy barrier due to like charges) to make contact with the nitrogen nucleus than in the case of gold. Rutherford used alpha particles that were being emitted from some radioactive source. He did not have access to machines that can accelerate particles to very high energy. It would be necessary to do just that to observe reaction of an alpha particle with a gold nucleus.

20.65 In each case, the product nucleus is underlined:

(a) $^{75}_{33}As + ^{4}_{2}He \rightarrow \underline{^{78}_{35}Br} + ^{1}_{0}n$; (b) $^{7}_{3}Li + ^{1}_{1}H \rightarrow \underline{^{7}_{4}Be} + ^{1}_{0}n$

(c) $^{31}_{15}P + ^{2}_{1}H \rightarrow \underline{^{32}_{15}P} + ^{1}_{1}H$

20.66 The fusion reaction will generate very high temperatures. Although there is not direct contact with a material container in the hottest part of the system, any materials used in extracting energy from the reaction zone must be able to withstand high temperatures. For the most promising fusion reactions, there is a very high flux of energetic neutrons. These can cause severe damage to structural materials; further, nuclear transformations in these materials, caused by the incident neutrons and other radiation from the fusion zone, will generate radioactive nuclides.

20.67 First we calculate k from a knowledge of $t_{1/2}$:

$$k = \left(\frac{0.693}{2.4 \times 10^4 \text{ yr}}\right)\left(\frac{1 \text{ yr}}{365 \times 24 \times 3600 \text{ sec}}\right) = 9.2 \times 10^{-13}/\text{sec}$$

We must now calculate N: $N = (0.500 \text{ g Pu})\left(\frac{1 \text{ mol Pn}}{239 \text{ g Pu}}\right)\left(\frac{6.02 \times 10^{23} \text{ Pu atoms}}{1 \text{ mol Pu}}\right)$

$= 1.3 \times 10^{21}$ Pu atoms

rate = $(9.2 \times 10^{-13}/\text{sec})(1.3 \times 10^{21} \text{ Pu atoms}) = 1.2 \times 10^9$ disint/sec

20.68 We determine the wavelengths of the photons by first calculating the energy equivalent of the mass of an electron or positron. (Since two photons are formed by annihilation of two particles of equal mass, we need calculate the energy equivalence of just one particle.)

$$\Delta E = (5.49 \times 10^{-4} \text{ amu})(3.0 \times 10^8 \text{ cm/sec})^2\left(\frac{1 \text{ g}}{6.02 \times 10^{23} \text{ amu}}\right)$$

$$\times \left(\frac{1 \text{ kg}}{10^3 \text{ g}}\right) = 8.2 \times 10^{-14} \text{ J}$$

We have also that $\Delta E = h\nu$ (Equation [5.2]):

$$\Delta E = \frac{hc}{\lambda}. \quad \text{Thus, } \lambda = \frac{hc}{\Delta E}; \; \lambda = \frac{(6.63 \times 10^{-34} \text{ J-sec})(3.0 \times 10^8 \text{ m/sec})}{8.2 \times 10^{-14} \text{ J}}$$

$$= 2.4 \times 10^{-12} \text{ m}$$

$$= 2.4 \times 10^{-3} \text{ nm}$$

This is a very short wavelength indeed; it lies at the short wavelength end of the range of observed gamma ray wavelengths (see Figure 5.3).

21

Chemistry of nonmetallic elements

21.1 We can best answer this question by referring to Table 21.1. Note that the first three characteristics in each list are physical properties, the other two are chemical. To the characteristics listed there, we might add that metals tend to lose electrons in reactions with nonmetals, whereas nonmetals tend to gain electrons in such reactions.

21.2

	Zinc	Oxygen
melting point	419°C	-219°C
boiling point	911°C	-183°C
density	7.1 g/cm^3	1.33 x 10^{-3} g/cm^3
structure	close-packed	O_2 molecules
ionization energy	906 kJ/mol	1314 kJ/mol
reduction potential	-0.76 V[a]	+1.23 V[b]

a. for $Zn^{2+} + 2e^- \rightarrow Zn$ b. for $O_2 + 4H^+ + 4e^- \rightarrow 2H_2O$

Note that oxygen forms a diatomic molecule, and in this form it is low-boiling and low-melting. Zinc, on the other hand, exists in a solid with many nearest neighbors about each zinc. Furthermore, the solid is a good conductor of electricity (Chapter 19). Note finally that whereas Zn is easily oxidized (reverse of reaction a), oxygen is very easily reduced.

21.3 Metal oxides are generally ionic solids with relatively high melting points, whereas nonmetal oxides are often gases, liquids or perhaps low-melting solids. The oxides of metals, when they are soluble in water, generally form basic solutions; metal oxides tend to be more soluble in acidic solutions. Nonmetal oxides are generally soluble in water; they dissolve to form acidic solutions.

21.4 Calcium oxide is a hard, high-melting solid (m.p. = 2927°C), whereas SO_3 melts at 62°C or 33°C, depending on its form. CaO is not very soluble in water (0.13 g/100 g H_2O at 25°C), whereas SO_3 is miscible with water. Solutions of CaO are basic, whereas solutions of SO_3 are acidic:

$$CaO(s) + H_2O(\ell) \rightarrow Ca^{2+} + 2OH^-(aq)$$
$$SO_3(s) + H_2O(\ell) \rightarrow HSO_4^-(aq) + H^+(aq)$$

21.5 Bismuth should be the most metallic in character. Metallic character would be manifested in electrical conductivity of the solid element; in the structure of the solid (large number of near-neighbor atoms bonded to one another); in the tendency of the oxide to form a basic solution on dissolving in water; in a tendency to form ionic compounds with the more electronegative nonmetallic elements.

21.6 In each case the more polar bonds are associated with the larger electronegativity difference between the two elements. (a) IF_3; (b) Bi_2O_3; (c) BF_3; (d) HF; (e) SnO_2.

21.7 Two factors contribute; the nitrogen atom is too small to permit as many as five fluorine atoms to be bonded to it without their experiencing strong repulsive interactions. Secondly, nitrogen does not have available more than four valence level orbitals that can be utilized to form bonds. In NF_3 it uses these to form 3 N-F bonds and provide for an unshared pair. Both P and As can make use of valence level d orbitals (3d for P, 4d for As) to form the five bonds to fluorine in PF_5 or AsF_5, respectively. Further, P and As are larger than N, so there is room for the five fluorine atoms.

21.8 Only N_2 among the group 5A elements exists as a diatomic molecule. The N_2 molecule is very stable. The heavier elements of the group do not form stable diatomic molecules because they cannot form strong π bonds. The sideways overlap of p orbitals to form a π bond (Figure 21.5) is extensive only for the lightest element in each nonmetal family. For this reason, the heavier elements adopt other bonding modes, to form σ bonds to three other atoms rather than utilize π bonds.

21.9 (a) The electronegativity difference between As and H is not as large as between N and H. Thus, the AsH_3 molecule more readily suffers As-H bond rupture, so AsH_3 acts as a good source of hydrogen for reduction. In addition, since As is not so electronegative as N, it is more easily oxidized to a positive oxidation state. (b) Since nitrogen is more electronegative than phosphorus, the element is more stable in compounds in which it has a lower oxidation state. In addition, nitrogen readily forms N_2 because it is such a stable molecule. (c) Chlorine is more electronegative than sulfur, thus is more stable in combination with hydrogen. (The electronegativity difference is larger for HCl than for H_2S.)

21.10 Recall that electronegativity is a measure of the ability of an atom to attract electrons to itself in a bond to another atom. We expect that when the charge on an atom increases because electrons are lost, any remaining electrons shared with other elements will be more strongly attracted to the positively charged atom. Thus, electronegativity should increase with increase in oxidation number.

Both $HClO_4$ and $HMnO_4$ are very strong acids (completely ionized) in aqueous solution. The MnO_4^- and ClO_4^- ions are both tetrahedral; both perchlorate and permanganate salts tend to be water-soluble; both ions have the central atom in the +7 oxidation state; both ions are strong oxidizing agents:

$$MnO_4^-(aq) + 8H^+(aq) + 5e^- \rightarrow Mn^{2+}(aq) + 4H_2O(\ell) \qquad E° = +1.51\ V$$

$$ClO_4^-(aq) + 8H^+(aq) + 7e^- \rightarrow (1/2)Cl_2(aq) + 4H_2O(\ell) \qquad E° = +1.34\ V$$

We expect that in the high (+7) oxidation state both Mn and Cl will have rather high effective electronegativities. They thus behave rather similarly. In lower oxidation states, they are much different, because Mn is a metal, and Cl is a nonmetal.

21.11 (a) $HClO_3$(+5); (b) BrF_3 (Br is +3, F is -1); (c) $XeOF_4$ (Xe is +6, F is -1); (d) H_5IO_6 (+7); (e) IO_3^- (+5); (f) $KClO_2$ (+3); (g) HBrO (+1); (h) KI_3 (2I are 0, one I is -1); the average oxidation state is thus -1/3; (i) PI_3 (I is -1).

21.12 Fluorine: $2KHF_2(\ell) \rightarrow H_2(g) + F_2(g) + 2KF(\ell)$ (electrolysis)

chlorine: $2NaCl(\ell) \rightarrow 2Na(\ell) + Cl_2(g)$ (electrolysis)

bromine: $2Br^-(aq) + Cl_2(g) \rightarrow Br_2(aq) + 2Cl^-(aq)$

iodine: $2I^-(aq) + Cl_2(g) \rightarrow I_2(s) + 2Cl^-(aq)$

21.13 $$\left(\frac{1.155\ \text{g soln}}{\text{cm}^3}\right)\left(\frac{0.50\ \text{g HF}}{1.00\ \text{g Soln}}\right)\left(\frac{1\ \text{mol HF}}{20.0\ \text{g HF}}\right)\left(\frac{1000\ \text{cm}^3}{1\ \text{L}}\right) = 28.9\ \text{M}$$

21.14 The halogens possess a common outer electron configuration, ns^2np^5. The energy required to remove an electron from the valence p orbital set should grow smaller as the average distance of the p electrons from the nucleus increases, because the electrostatic attraction is inversely proportional to the average distance between oppositely charged particles. As we move downward in the halogen family, the atomic size increases; thus, first ionization energy should decrease.

21.15 Fluorine reacts vigorously with water:

$$2F_2(g) + 2H_2O(\ell) \rightarrow 4HF(aq) + O_2(g)$$

Thus, if F_2 were to be formed at an electrode, it would immediately react with the water. However, fluorine may never form in the first place because water is oxidized at the electrode before fluorine is.

Electrolysis reactions producing F_2 or Cl_2 commercially are alike in that the oxidation process is analogous in each case, oxidation of halide to free halogen. However, they differ in the character of the cathode reaction. In F_2 formation, the reaction occurring at the cathode leads to H_2 rather than halogen gas. In the Downs cell utilized for formation of Cl_2, the cathode reaction is reduction of $Na^+(\ell)$ to $Na(\ell)$.

21.16 (a) $CaF_2(s) + H_2SO_4(\ell) \rightarrow 2HF(g) + CaSO_4(s)$

(b) $2I^-(aq) + Cl_2(g) \rightarrow I_2(s) + 2Cl^-(aq)$

(c) $Xe(g) + 2F_2(g) \rightarrow XeF_4(s)$

(d) $Ca^{2+}(aq) + 2OH^-(aq) + Cl_2(aq) \rightarrow ClO^-(aq) + Cl^-(aq) + H_2O(\ell) + Ca^{2+}(aq)$

(e) $S_8(s) + 24F_2(g) \rightarrow 8SF_6(g)$

$$\text{or } S_8(s) + 24BrF_5(\ell) \rightarrow 8SF_6(g) + 24BrF_3(\ell)$$

(f) $\quad 8H^+(aq) + I_2(s) + 10NO_3^-(aq) \rightarrow 2IO_3^-(aq) + 10NO_2(g) + 4H_2O(\ell)$

<u>21.17</u> The acid strength of the oxyacid (for a given oxidation state) may be related to the electronegativity of the central atom. When the electronegativity is highest, the O-H bond is most polar, and hydrogen is most readily lost as H^+. Thus, we expect the oxyacid to exhibit a varying acid dissociation constant in the order chlorine > bromine > iodine, as observed.

<u>21.18</u> (a) $PBr_5(\ell) + 4H_2O(\ell) \rightarrow H_3PO_4(aq) + 5H^+(aq) + 5Br^-(aq)$

(b) $IF_5(\ell) + 3H_2O(\ell) \rightarrow H^+(aq) + IO_3^-(aq) + 5HF(aq)$

(c) $BrCl(g) + H_2O(\ell) \rightarrow HBrO(aq) + H^+(aq) + Cl^-(aq)$

(d) $Br_2(\ell) + H_2O(\ell) \rightarrow HBrO(aq) + H^+(aq) + Br^-(aq)$

(e) $2ClO_2(g) + H_2O(\ell) \rightarrow H^+(aq) + ClO_3^-(aq) + HClO_2(aq)$

(f) $HI(g) \rightarrow H^+(aq) + I^-(aq)$

<u>21.19</u>

$$\left[:\overset{..}{\underset{..}{F}} - \overset{..}{\underset{..}{Br}} - \overset{..}{\underset{..}{F}}: \right]^+ \qquad \left[F - \underset{\underset{F}{|}}{\overset{\overset{F}{|}}{Sb}} \overset{F}{\underset{F}{\diagup}} F \right]^-$$

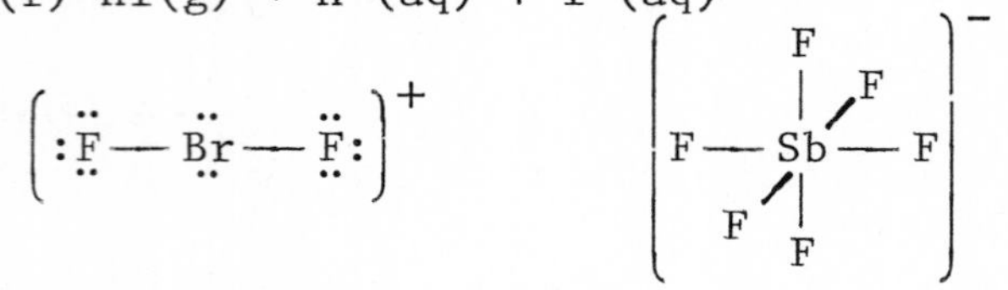

(For clarity, we omit 3 unshared electron pairs on each F in the anion.) Using VSEPR model, we predict a bent structure for the cation, and an octahedral geometry about Sb in the anion.

<u>21.20</u> The most important reason is steric. There is not room about chlorine to pack seven fluorines at the correct Cl-F distance; even in ClF_5 there is crowding. Iodine is larger and can accomodate a larger number of fluorines. A second reason is that the oxidation state of the central atom grows increasingly positive as the number of fluorines increase. Because Cl is more electronegative than I, it less readily gives up electrons to fluorine.

<u>21.21</u> (a) $Br_2(\ell) + 2OH^-(aq) \rightarrow BrO^-(aq) + Br^-(aq) + H_2O(\ell)$

(b) $Br_2(aq) + H_2O_2(aq) \rightarrow O_2(g) + 2H^+(aq) + 2Br^-(aq)$

(c) $3CaBr_2(s) + 2H_3PO_4(\ell) \rightarrow Ca_3(PO_4)_2(s) + 6HBr(g)$

(d) $AlBr_3(s) + H_2O(\ell) \rightarrow 3HBr(g) + Al(OH)_3(s)$

(e) $2HF(aq) + CaCO_3(s) \rightarrow CaF_2(s) + H_2O(\ell) + CO_2(g)$

<u>21.22</u>
$$2ClO_3^-(aq) + 12H^+ + 10e^- \rightarrow Cl_2(g) + 6H_2O(\ell)$$
$$10[Fe^{2+}(aq) \rightarrow Fe^{3+}(aq) + e^-]$$
$$\overline{2ClO_3^-(aq) + 10Fe^{2+}(aq) + 12H^+(aq) \rightarrow Cl_2(g) + 10Fe^{3+}(aq) + 6H_2O(\ell)}$$
$$E° = +1.47 + (-0.771) = +0.70 \text{ V}$$

<u>21.23</u>
$$2ClO_3^-(aq) + 12H^+(aq) + 10e^- \rightarrow Cl_2(g) + 6H_2O(\ell)$$
$$5[2Cl^-(aq) \rightarrow Cl_2(g) + 2e^-]$$
$$\overline{ClO_3^-(aq) + 5Cl^-(aq) + 6H^+(aq) \rightarrow 3Cl_2(g) + 3H_2O(\ell)}$$
$$E° = 1.47 + (-1.36) = +0.11 \text{ V}$$

21.24 (a) $2HClO_2(aq) + 6H^+(aq) + 6e^- \rightarrow Cl_2(g) + 4H_2O(\ell)$.

We know E° for each of the following half-cell reactions (from Figure 21.10):

$$HClO_2(aq) + 2H^+(aq) + 2e^- \rightarrow HClO(aq) + H_2O(\ell) \quad E° = +1.64 \text{ V}$$

$$2HClO(aq) + 2H^+(aq) + 2e^- \rightarrow Cl_2(g) + 2H_2O(\ell) \quad E° = +1.63 \text{ V}$$

If we double the first of these and add, we obtain the desired half-reaction. Using the procedure described in Section 21.3, we then have:

$$-6\mathcal{F}E° = -4\mathcal{F}(1.64) - 2\mathcal{F}(1.63); \quad E° = +1.64 \text{ V}$$

(b)

$$2Cl^-(aq) \rightarrow Cl_2(g) + 2e^- \quad E° = -1.36 \text{ V}$$

$$Cl_2(g) + 4H_2O(\ell) \rightarrow 6H^+(aq) + 2HClO_2(aq) + 6e^- \quad E° = -1.64 \text{ V}$$

$$2Cl^- + 4H_2O(\ell) \rightarrow 6H^+(aq) + 2HClO_2(aq) + 8e^-$$

$$-8\mathcal{F}E° = -2\mathcal{F}(-1.36 \text{ V}) - 6\mathcal{F}(-1.64 \text{ V}); \quad E° = 1.57 \text{ V}$$

21.25 (a) H_2S (−2); (b) H_2SeO_3 (+4); (c) TeO_3 (+6); (d) $KHSO_3$ (+4); (e) CaO_2 (−1); (f) Al_2S_3 (−2); (g) H_6TeO_6 (+6); (h) SeO_4^{2-} (+6); (i) $S_2O_3^{2-}$ (The average oxidation state of the two sulfurs is +2. However, they are not equivalent; one is effectively in the +4 oxidation state, the other in the zero oxidation state.)

21.26 (a) potassium selenate; (b) hydrogen telluride; (c) hydrogen sulfate (or bisulfate) ion; (d) ozone; (e) iron pyrite (or iron(II) persulfide); (f) selenium hexafluoride; (g) sodium hydrogen sulfate or sodium bisulfate.

21.27 (a) Se can be recovered from electrolysis residues by oxidizing it to $H_2SeO_3(aq)$ (Equations [21.40] and [21.41]). The element is then recovered by reduction with $SO_2(g)$:

$$H_2O(\ell) + H_2SeO_3(aq) + 2SO_2(g) \rightarrow Se(s) + 2H_2SO_4(aq)$$

(b) Use a telluride salt such as Na_2Te, in aqueous medium, with acid:

$$Te^{2-}(aq) + 2H^+(aq) \rightarrow H_2Te(g)$$

Alternatively, one could treat an insoluble telluride with an acid such as H_2SO_4.

(c) $SeO_2(s) + H_2O(\ell) \rightarrow H_2SeO_3(aq)$

(d) To produce sulfur from its compounds, we need an appropriate oxidizing or reducing agent. Another sulfur compound often serves best, e.g.,

$$16H_2S(g) + 8SO_2(g) \rightarrow 3S_8(s) + 16H_2O(g)$$

or $8H_2S(g) + 8Cl_2(g) \rightarrow 16HCl + S_8(s)$

or $8S_2O_3^{2-}(aq) + 16H^+(aq) \rightarrow 8H_2SO_3(aq) + S_8(s)$

(e) $H_2SO_4(\ell) + SO_3(s) \rightarrow H_2S_2O_7(\ell)$; (f) $8SO_3^{2-}(aq) + S_8(s) \rightarrow 8S_2O_3^{2-}(aq)$

21.28 Recall that electronegativity is related to the average of the ionization energy and electron affinity (Section 7.9). The ionization energy varies much more throughout the group 6A elements than does the electron affinity.

21.29 The electron affinities of the halogens are much larger because the effective nuclear charge seen by a valence level p electron is much greater for the halogen than for its group 6A neighbor. This greater effective nuclear charge, you will recall, arises because the p electrons do not completely shield one another from the nuclear charge.

21.30 $2H_2S(g) + 3O_2(g) \rightarrow 2SO_2(g) + 2H_2O(g)$

$$(600 \text{ g } H_2S)\left(\frac{1 \text{ mol } H_2S}{34.0 \text{ g } H_2S}\right)\left(\frac{1 \text{ mol } SO_2}{1 \text{ mol } H_2S}\right) = 17.6 \text{ mol } SO_2$$

$$V = \frac{(17.6 \text{ mol})\left(\frac{0.0821 \text{ L-atm}}{\text{mol-K}}\right)(473 \text{ K})}{(10/760) \text{ atm}} = 5.2 \times 10^4 \text{ L } SO_2$$

21.31 In forming a solid salt such as Na_2S, the energetics of forming the gas phase ions, which is what we measure in evaluating the electron affinity or ionization energy, is only part of the overall process. The lattice energy measures the energy released when a mol each of the ions making up the ionic lattice is brought together from infinite separation. This is a very large stabilizing term. The larger lattice energy for the higher charge lattice (e.g., Na_2S rather than NaS) more than compensates for the energy required to add the second electron to the sulfur.

21.32 (a) $SeO_2(s) + H_2O(\ell) \rightarrow H_2SeO_3(aq)$

(b) $2NaO_2(s) + 2H_2O(\ell) \rightarrow O_2(g) + H_2O_2(aq) + 2Na^+(aq) + 2OH^-(aq)$

(c) $ZnS(s) + 2H^+(aq) \rightarrow Zn^{2+}(aq) + H_2S(g)$

(d) $8KI(s) + 9H_2SO_4(\ell) \rightarrow 8KHSO_4(s) + H_2S(g) + 4I_2(g) + 4H_2O(g)$

(e) $2H_2O_2(aq) \rightarrow 2H_2O(\ell) + O_2(g)$

21.33 $8Fe(s) + S_8(s) \rightarrow 8FeS(s)$

$S_8(s) + 16F_2(g) \rightarrow 8SF_4(g)$ or $S_8(s) + 24F_2(g) \rightarrow 8SF_6(g)$

$S_8(s) + 8O_2(g) \rightarrow 8SO_2(g)$

$S_8(s) + 8H_2(g) \rightarrow 8H_2S(g)$

Sulfur acts as an oxidizing agent in reactions with Fe or H_2, as a reducing agent in reactions with O_2 or F_2. Incidentally, these reactions are often written using the symbol S rather than S_8 for sulfur.

21.34 $SO_2(g) + 2H_2S(g) \rightarrow 3S(s) + 2H_2O(g)$. Or, if we assume S_8 is the product, $8SO_2(g) + 16H_2S(g) \rightarrow 3S_8(s) + 16H_2O(g)$.

$$(2000 \text{ lb coal})\left(\frac{0.035 \text{ lb S}}{1 \text{ lb coal}}\right)\left(\frac{454 \text{ g S}}{1 \text{ lb S}}\right)\left(\frac{1 \text{ mol S}}{32 \text{ g S}}\right)\left(\frac{1 \text{ mol } SO_2}{1 \text{ mol S}}\right)\left(\frac{2 \text{ mol } H_2S}{1 \text{ mol } SO_2}\right)$$

$$= 2.0 \times 10^3 \text{ mol } H_2S; \quad V = \frac{2.0 \times 10^3 \text{ mol}(0.0821 \text{ L-atm/mol-K})(300 \text{ K})}{(740/760) \text{ atm}}$$

$$= 5.1 \times 10^4 \text{ L } H_2S$$

$$(2.0 \times 10^3 \text{ mol } H_2S)\left(\frac{3 \text{ mol S}}{2 \text{ mol } H_2S}\right)\left(\frac{32 \text{ g S}}{1 \text{ mol S}}\right) = 9.6 \times 10^4 \text{ g S}$$

This is about 200 lb S per ton of coal combusted.

21.35 (a)  bent

(b) 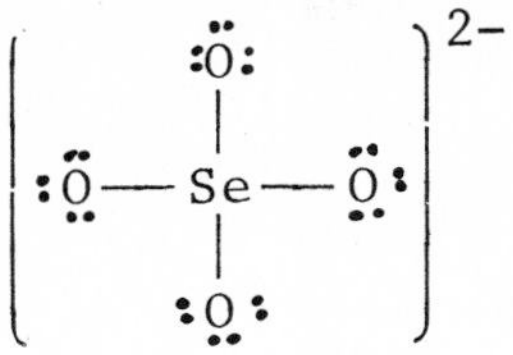 tetrahedral

(c) 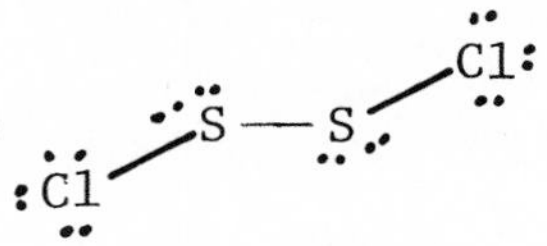 non-linear

(d) 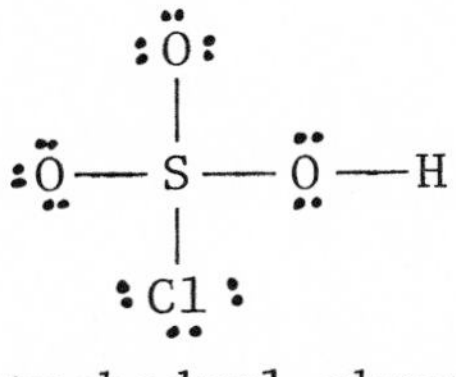 tetrahedral about S

(e) 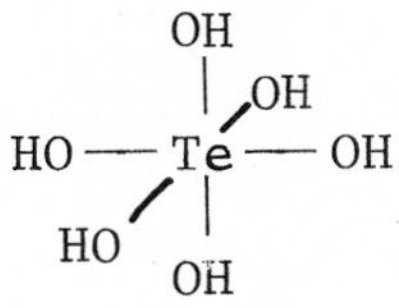 Each oxygen carries 2 unshared pairs, not shown. Geometry about Te is octahedral.

21.36 (a)

$$2[MnO_4^-(aq) + 8H^+(aq) + 5e^- \rightarrow Mn^{2+}(aq) + 4H_2O(\ell)]$$
$$5[H_2SO_3(aq) + H_2O(\ell) \rightarrow SO_4^{2-}(aq) + 4H^+(aq) + 2e^-]$$
$$2MnO_4^-(aq) + 5H_2SO_3(aq) \rightarrow 2MnSO_4(s) + 5SO_4^{2-}(aq) + 3H_2O(\ell) + 4H^+(aq)$$

(b)

$$Cr_2O_7^{2-}(aq) + 14H^+(aq) + 6e^- \rightarrow 2Cr^{3+}(aq) + 7H_2O(\ell)$$
$$3[H_2SO_3(aq) + H_2O(\ell) \rightarrow SO_4^{2-}(aq) + 4H^+(aq) + 2e^-]$$
$$Cr_2O_7^{2-}(aq) + 3H_2SO_3(aq) + 2H^+(aq) \rightarrow 2Cr^{3+}(aq) + 3SO_4^{2-}(aq) + 4H_2O(\ell)$$

(c)

$$Hg_2^{2+}(aq) + 2e^- \rightarrow 2Hg(\ell)$$
$$H_2SO_3(aq) + H_2O(\ell) \rightarrow SO_4^{2-}(aq) + 4H^+(aq) + 2e^-$$
$$Hg_2^{2+}(aq) + H_2SO_3(aq) + H_2O(\ell) \rightarrow 2Hg(\ell) + SO_4^{2-}(aq) + 4H^+(aq)$$

21.37

$$2[Fe(CN)_6^{4-}(aq) \rightarrow Fe(CN)_6^{3-}(aq) + e^-]$$
$$2H^+(aq) + H_2O_2(aq) + 2e^- \rightarrow 2H_2O(\ell)$$
$$2Fe(CN)_6^{4-}(aq) + H_2O_2(aq) + 2H^+(aq) \rightarrow 2Fe(CN)_6^{3-} + 2H_2O\ (\ell)$$

$$2[Fe(CN)_6^{3-}(aq) + e^- \rightarrow Fe(CN)_6^{4-}(aq)]$$
$$H_2O_2(aq) + 2OH^-(aq) \rightarrow O_2(g) + 2H_2O(\ell) + 2e^-$$
$$2Fe(CN)_6^{3-}(aq) + H_2O_2(aq) + 2OH^-(aq) \rightarrow 2Fe(CN)_6^{4-}(aq) + O_2(g) + 2H_2O(\ell)$$

21.38 (a) HNO_2 (+3); (b) K_2HPO_3(+3) (Note that there is one hydrogen on the phosphorus - see Figure 21.28); (c) P_2O_3 or P_4O_6 (+3); (d) N_2H_4 (-2); (e) $Ca(H_2PO_4)_2$ (+5); (f) KN_3 (average oxidation state of nitrogen atoms is -1/3); (g) H_3AsO_4 (+5); (h) Sb_2S_3 (+3).

21.39 (a) $2Ca_3(PO_4)_2(s) + 6SiO_2(s) + 10C(s) \xrightarrow{\Delta} P_4(g) + 6CaSiO_3(\ell) + 10CO(g)$

(b) $N_2(g) + 3H_2(g) \rightarrow 2NH_3(g)$

(c) $4NH_3(g) + 5O_2(g) \xrightarrow[1000°C]{catalyst} 4NO(g) + 6H_2O(g)$

$2NO(g) + O_2(g) \rightarrow 2NO_2(g)$

$3NO_2(g) + H_2O(\ell) \rightarrow 2H^+(aq) + 2NO_3^-(aq) + NO(g)$

(d) $Ca_3(PO_4)_2(s) + 4H^+(aq) + 3SO_4^{2-}(aq) \rightarrow 3CaSO_4(s) + 2H_2PO_4^-(aq)$

21.40 (a) 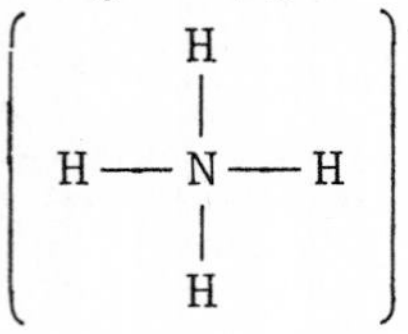 (b) 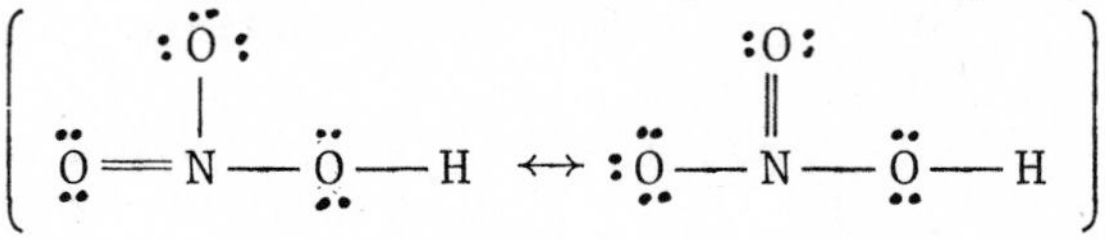

(c) 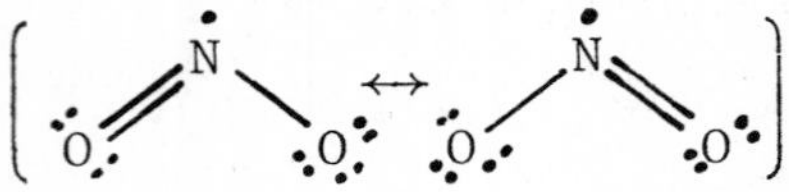

(d) The structure is the same as that shown in Figure 21.27 for P_4O_{10}, with each oxygen replaced by sulfur. Recall also that there are lone pairs on each sulfur to complete the octet.

(e) 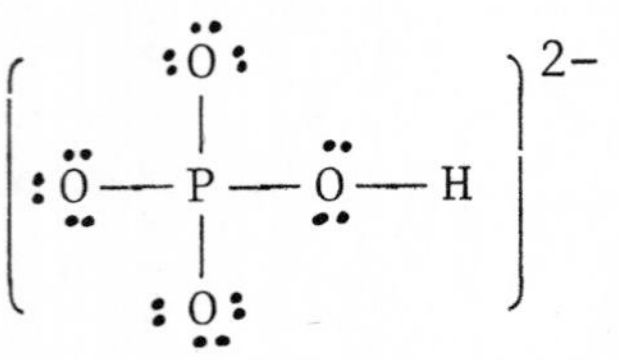 (f) 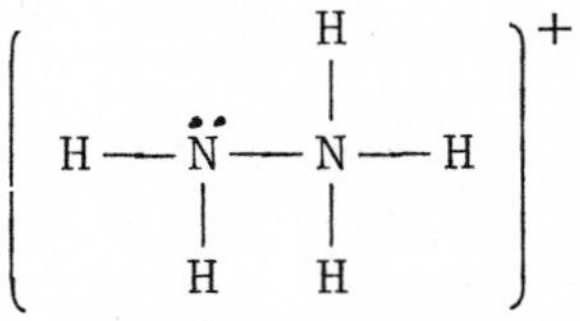

21.41 (a) acidic (NH_4^+ is the acid); (b) acidic (forms $H_3PO_3(aq)$; (c) acidic (see Equation [21.69]); (d) basic ($N_2H_4(aq) + H_2O(\ell) \rightarrow N_2H_5^+(aq) + OH^-(aq)$); (e) basic ($AsO_4^{3-}(aq) + H_2O(\ell) \rightarrow HAsO_4^{2-} + OH^-(aq)$).

21.42 The bond energies decrease in the order $N_2 > O_2 > F_2$. This variation is easily accounted for in terms of the bond orders in these bonds, as spelled out in detail in Table 8.7.

21.43 In N_2 the triple bond between nitrogen atoms consists of one σ and two π bonds. The π bonds are quite stable and contribute substantially to the N-N bond energy. In P_2 the π bonds are very weak, because the sideways overlap of the 3p orbitals is not very large in P_2. Thus, the P_2 molecule is not as stable as an alternative arrangement in which each phosphorus forms bonds to three other P atoms, and in which the bonds are σ in character.

21.44 $P_4(s) \rightarrow P_4(g)$ $\Delta H = 112$ kJ/mol; $\Delta H = \Delta H_f^\circ(P_4(g)) - \Delta H_f^\circ(P_4(s))$

$112 = \Delta H_f^\circ(P_4(g)) - 17$; $\Delta H_f^\circ(P_4(g)) = 129$ kJ/mol; $P_4(g) \rightleftharpoons 2P_2(g)$;

$\Delta H = 2\Delta H_f^\circ(P_2(g)) - \Delta H_f^\circ(P_4(g)) = 2(146) - 129 = 163$ kJ/mol

21.45 Rather surprisingly at first glance, phosphorus actually has a higher ionization energy than sulfur. This arises because the electron to be removed from phosphorus is taken from the half-filled 3p subshell. These three electrons experience an effective nuclear charge that is determined by the incomplete shielding of the nuclei by the 3p electrons. At the same time, the inter-electronic repulsions are not so large since each electron is in a different orbital. In sulfur, there is one additional electron, that has had to go into an orbital already occupied by another electron. The inter-electronic repulsions that result offset the expected greater effective nuclear charge that this electron should experience. Thus, the first ionization energies of phosphorus and sulfur are about the same.

21.46 (a) $N_2H_4(g) + 5F_2(g) \rightarrow 2NF_3(g) + 4HF(g)$

(b) $SeO_3^{2-}(aq) + 3H_2O(\ell) + 4e^- \rightarrow Se(s) + 6OH^-(aq)$

$N_2H_4(aq) + 4OH^-(aq) \rightarrow N_2(g) + 4H_2O(\ell) + 4e^-$

$SeO_3^{2-}(aq) + N_2H_4(aq) \rightarrow Se(s) + N_2(g) + H_2O(\ell) + 2OH^-(aq)$

(c) $Cu^{2+}(aq) + 2e^- \rightarrow Cu(s)$

$2NH_2OH(aq) \rightarrow N_2(g) + 2H_2O(\ell) + 2H^+(aq) + 2e^-$

$Cu^{2+}(aq) + 2NH_2OH(aq) \rightarrow Cu(s) + N_2(g) + 2H_2O(\ell) + 2H^+(aq)$

(d) $2N_3^-(aq) + Cl_2(g) \rightarrow 3N_2(g) + 2Cl^-(aq)$

21.47 (a) $2NO_3^-(aq) + 10H^+(aq) + 8e^- \rightarrow N_2O(g) + 5H_2O(\ell)$

$4[Zn(s) \rightarrow Zn^{2+}(aq) + 2e^-]$

$2NO_3^-(aq) + 4Zn(s) + 10H^+(aq) \rightarrow N_2O(g) + 4Zn^{2+}(aq) + 5H_2O(\ell)$

(b) $32[NO_3^-(aq) + 2H^+(aq) + e^- \rightarrow NO_2(g) + H_2O(\ell)]$

$S_8(s) + 16H_2O(\ell) \rightarrow 8SO_2(g) + 32H^+(aq) + 32e^-$

$S_8(s) + 32NO_3^-(aq) + 32H^+(aq) \rightarrow 8SO_2(g) + 32NO_2(g) + 16H_2O(\ell)$

(c) $2[NO_3^-(aq) + 4H^+(aq) + 3e^- \rightarrow NO(g) + 2H_2O(\ell)]$

$3[SO_2(g) + H_2O(\ell) \rightarrow SO_3(g) + 2H^+(aq) + 2e^-]$

$2NO_3^-(aq) + 3SO_2(g) + 2H^+(aq) \rightarrow 2NO(g) + 3SO_3(g) + H_2O(\ell)$

(d) $CON_2H_4(aq) + H_2O(\ell) \rightarrow 2NH_3(aq) + CO_2(g)$

21.48 $$\left(\frac{18 \text{ g N}}{100 \text{ g fert}}\right)\left(\frac{1 \text{ mol N}}{14 \text{ g N}}\right)\left(\frac{1 \text{ mol } CON_2H_4}{2 \text{ mol N}}\right)\left(\frac{60 \text{ g } CON_2H_4}{1 \text{ mol } CON_2H_4}\right)$$

$= 39 \text{ g } CON_2H_4/100 \text{ g fertilizer} = 39\%$ wt percentage

21.49 The electronegativity of nitrogen is greater than that of phosphorus. Thus, the electron pair in the O-H bond is attracted more strongly toward oxygen in nitric acid, because the oxygen in nitric acid is bound to a more electron-withdrawing atom.

There is a second reason why nitric acid is strong. Ionization of the proton from HNO_3 leaves the very stable nitrate ion, stabilized by resonance as illustrated in Section 7.6. The stability of the NO_3^- ion acts as a kind of driving force for loss of the proton from HNO_3.

21.50 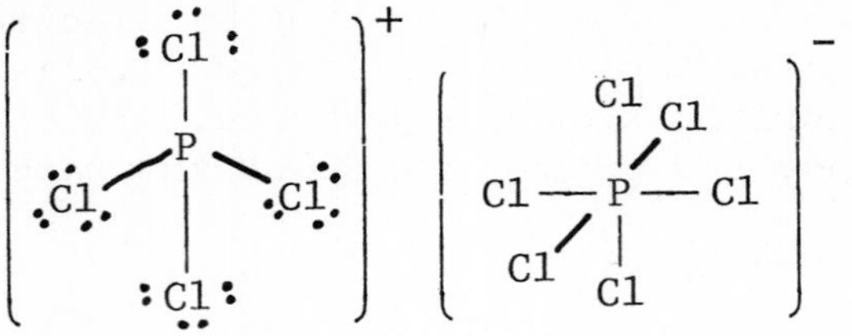

(Each chlorine in PCl_6^- also has three unshared pairs, not shown for clarity.)

In PCl_4^+ the phosphorus employs an sp^3 hybrid set. In PCl_6^- the phosphorus employs an sp^3d^2 hybrid set. The ionic form is stabilized in the solid state by the lattice energy that is gained by forming the ions. This is a case of spontaneous ion formation between two identical species, driven by the gain in stability of the ionic lattice compared with the covalent lattice.

21.51 (a) $NO_3^-(aq) + 4H^+(aq) + 3e^- \rightarrow NO(g) + 2H_2O(\ell)$ $\quad E° = 0.96$ V

(b) $N_2O(aq) + H_2O(\ell) \rightarrow 2NO(aq) + 2H^+(aq) + 2e^-$ $\quad E° = -1.59$ V

21.52 (a) We first write the balanced half-reaction from Figure 21.23 that together will add to give the balanced half-reaction we want:

$$NO_2(aq) + H^+(aq) + e^- \rightarrow HNO_2(aq) \qquad E° = 1.12 \text{ V}$$

$$HNO_2(aq) + H^+(aq) + e^- \rightarrow NO(g) + H_2O(\ell) \qquad E° = 1.00 \text{ V}$$

$$NO_2(aq) + 2H^+(aq) + 2e^- \rightarrow NO(g) + H_2O(\ell)$$

$$-2FE° = -F(1.12) - F(1.00); \quad E° = 2.12/2 = 1.06 \text{ V}$$

(b) We again use the half-reactions given in Figure 21.23.

$$2NH_4^+(aq) \rightarrow N_2(g) + 8H^+(aq) + 6e^- \qquad E° = -0.27 \text{ V}$$

$$N_2(g) + 6H_2O(\ell) \rightarrow 2NO_3^-(aq) + 12H^+(aq) + 10e^- \qquad E° = -1.25 \text{ V}$$

$$2NH_4^+(aq) + 6H_2O(\ell) \rightarrow 2NO_3^-(aq) + 20H^+(aq) + 16e^-$$

$$-16FE° = -6F(-0.27) - 10F(1.25); \quad E° = \frac{-1.62 - 12.5}{16} = -0.88 \text{ V}$$

Note that we can divide our final balanced half-reaction through by 2 to obtain the equation with the lowest possible set of integer coefficients. However, this does not change the value of E°. We must use the equation as written above to calculate E° from the other two balanced half-reactions we used. The point to remember, as discussed in the text, is that we are adding free energies, which are given by $\Delta G° = -nFE°$.

21.53 The XeO_6^{4-} ion can be visualized as an octahedral array of oxygens about xenon. There is a total of 48 valence shell electrons. Each oxygen has an octet consisting of three unshared pairs and one pair shared with the xenon, thus accounting for all 48 electrons. This means that there are 6 electron pairs in the xenon valence orbitals. We expect an sp^3d^2 hybrid orbital set, consistent with an octahedral geometry.

21.54 (a) $MnO_2(s) + 2Cl^-(aq) + 4H^+(aq) \rightarrow Mn^{2+}(aq) + Cl_2(g) + 2H_2O(\ell)$

(b) $10Cl^-(aq) + 2MnO_4^-(aq) + 16H^+(aq) \rightarrow 2Mn^{2+}(aq) + 5Cl_2(g) + 8H_2O(\ell)$

(c) $6Cl^-(aq) + Cr_2O_7^{2-}(aq) + 14H^+(aq) \rightarrow 3Cl_2(g) + 2Cr^{3+}(aq) + 7H_2O(\ell)$

21.55 (a) White phosphorus consists of P_4 molecules. The relatively low molecular mass of this molecule, in addition to its lack of polarity, lead to high volatility. On the other hand, red phosphorus consists of chains of phosphorus atoms. The "molecules" in red phosphorus are therefore of high molecular weight. (b) The compound hydrolyzes as follows:

$$SF_4(aq) + 2H_2O(\ell) \rightarrow SO_2(g) + 4HF(aq)$$

Obviously, an aqueous solution of HF will exhibit acidic character. (c) The 4s and 4p electrons in Kr are more tightly bound to the nucleus than in Xe. Therefore, the utilization of these electrons in forming covalent bonds to fluorine "costs" more in energy in the case of Kr than in Xe. The result is that the Kr-F bond energy in the bonds that would result from forming KrF_6 would not compensate for the energy required to make the krypton electrons available for bonding.

A second factor is that Kr is a smaller atom than Xe; there may not be room about the Kr atom for six fluorine atoms. (d) The N_2 molecule has an extremely high bond energy. For this reason, N_2 represents an especially stable bonding form for nitrogen. For phosphorus the situation is not comparable; red phosphorus, the most stable allotrope, consists of phosphorus atoms singly bonded to one another. In nature, such bonds would be easily broken under high temperature conditions or by irradiation, leading to formation of stable phosphorus-oxygen bonds.

21.56 Xenon has a considerably lower ionization energy than argon. Thus, the electrons of the octet about xenon can be "promoted" to higher energy orbitals (e.g., the 5d) to allow formation of bonds. With argon, this promotion is much more costly in energy terms, and the bonds that can be formed are not sufficiently stable to make such a process feasible. In addition, there is the steric factor; argon is smaller, and cannot accomodate the number of surrounding atoms that xenon can.

21.57 The fact that the enthalpies of formation of the xenon oxides are positive tells us that the reverse reactions, decompositions of those oxides to yield the free elements, are exothermic processes. Compounds for which this is true must be handled with great care. They have a tendency to explode violently, because once the decomposition begins, the heat evolved increases the temperature within the sample, thus accelerating the reaction even further.

21.58 From Appendix D, we need only ΔH_f° for F(g), so that we can estimate ΔH for the process: $F_2(g) \rightarrow F(g) + F(g)$ $\Delta H^\circ = +160$ kJ. Then for XeF_2 we have

$$XeF_2(g) \rightarrow Xe(g) + F_2(g) \qquad -\Delta H_f^\circ = +109 \text{ kJ}$$
$$F_2(g) \rightarrow 2F(g) \qquad \Delta H^\circ = +160 \text{ kJ}$$
$$XeF_2(g) \rightarrow Xe(g) + 2F(g) \qquad \Delta H^\circ = 269 \text{ kJ}$$

The average Xe-F bond enthalpy is thus 269/2 = 134 kJ. Similarly,

$$XeF_4(g) \rightarrow Xe(g) + 2F_2(g) \qquad -\Delta H_f^\circ = +218 \text{ kJ}$$
$$2F_2(g) \rightarrow 4F(g) \qquad \Delta H^\circ = 320 \text{ kJ}$$
$$XeF_4(g) \rightarrow Xe(g) + 4F(g) \qquad \Delta H^\circ = 538 \text{ kJ}$$

Average Xe-F bond energy = 538/4 = 134 kJ

$XeF_6(g) \rightarrow Xe(g) + 3F_2$ $\quad -\Delta H_f^\circ = 298$ kJ

$3F_2(g) \rightarrow 6F(g)$ $\quad \Delta H^\circ = 480$ kJ

$XeF_6(g) \rightarrow Xe(g) + 6F(g)$ $\quad \Delta H^\circ = 778$ kJ

Average Xe-F bond energy = 778/6 = 130 kJ

The average Xe-F bond enthalpies are remarkable constant in the series.

21.59 (a) $2Na(\ell) + 2HCl(g) \rightarrow 2NaCl(s) + H_2(g)$

(b) $H_2SO_3(aq) + Br_2(\ell) + H_2O(\ell) \rightarrow HSO_4^-(aq) + 2Br^-(aq) + 3H^+(aq)$

Note that two strong acids are formed, $H_2SO_4(aq)$ and HBr(aq). H_2SO_4 is not volatile, and remains behind when the HBr is distilled.

(c) The half reactions are:

$12OH^-(aq) + Br_2(\ell) \rightarrow 2BrO_3^-(aq) + 6H_2O(\ell) + 10e^-$

$5[ClO^-(aq) + H_2O(\ell) + 2e^- \rightarrow Cl^-(aq) + 2OH^-(aq)]$

$2OH^-(aq) + Br_2(\ell) + 5ClO^-(aq) \rightarrow 5Cl^-(aq) + 2BrO_3^-(aq) + H_2O(\ell)$

Now we must take account of the formation of $KBrO_3(s)$:

$2K^+(aq) + 2BrO_3^-(aq) \rightarrow 2KBrO_3(s)$

$2K^+(aq) + 2OH^-(aq) + Br_2(\ell) + 5ClO_3^-(aq) \rightarrow 5Cl^-(aq) + 2KBrO_3(s) + H_2O(\ell)$

(d) The half-reactions are:

$2BrO_3^-(aq) + 12H^+(aq) + 10e^- \rightarrow Br_2(\ell) + 6H_2O(\ell)$

$5[H_2SO_3(aq) + H_2O(\ell) \rightarrow HSO_4^-(aq) + 3H^+(aq) + 2e^-]$

$2BrO_3^-(aq) + 5H_2SO_3(aq) \rightarrow Br_2(\ell) + 5HSO_4^-(aq) + 3H^+(aq) + H_2O(\ell)$

(e) $3UCl_4(s) + 4ClF_3(g) \rightarrow 3UF_4(g) + 8Cl_2(g)$

21.60 (a) P_4O_6; (b) Cl_2O_7; (c) As_4O_{10}; (d) B_2O_3; (e) SO_3; (f) CO_2.

21.61 There are 12 electrons in the valence orbitals of sulfur. This suggests an octahedral disposition of electron pairs, with an unshared pair at one position. We expect the angle θ to be less than 90° because of the larger repulsion arising from the unshared pair.

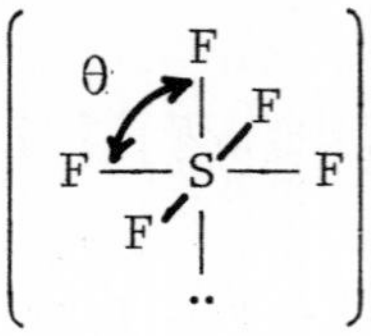

21.62 (a) $H_2SeO_3(aq) + N_2H_4(aq) \rightarrow Se(s) + N_2(g) + 3H_2O(\ell)$

(b) $H_6TeO_6(s) \xrightarrow{\Delta} TeO_3(s) + 3H_2O(\ell)$

(c) $Al_2Se_3(s) + 6H^+(aq) \rightarrow 2Al^{3+}(aq) + 3H_2Se(g)$

(d) The likely half-reactions are:

$$O_3(g) + H_2O(\ell) + 2e^- \rightarrow O_2(g) + 2OH^-(aq)$$
$$2I^-(aq) \rightarrow I_2(s) + 2e^-$$

$$O_3(g) + H_2O + 2I^-(aq) \rightarrow O_2(g) + I_2(s) + 2OH^-(aq)$$

(e) $2Pb(NO_3)_2(s) \xrightarrow{\Delta} 2PbO(s) + 4NO_2(g) + O_2(g)$

21.63 $3NaI(s) + H_3PO_4(\ell) \xrightarrow{\Delta} 3HI(g) + Na_3PO_4(s)$

$$\left(\frac{4.80 \text{ mol HI}}{1 \text{ L Soln}}\right)(2.50 \text{ L Soln})\left(\frac{1 \text{ mol NaI}}{1 \text{ mol HI}}\right)\left(\frac{150 \text{ g NaI}}{1 \text{ mol NaI}}\right) = 1.8 \text{ kg NaI}$$

21.64 $2XeO_3(s) \rightarrow 2Xe(g) + 3O_2(g)$

$$(0.654 \text{ g } XeO_3)\left(\frac{1 \text{ mol } XeO_3}{179 \text{ g } XeO_3}\right)\left(\frac{5 \text{ mol gas}}{2 \text{ mol } XeO_3}\right) = 9.13 \times 10^{-3} \text{ mol gas}$$

$$P = \frac{(9.13 \times 10^{-3} \text{ mol})(0.0821 \text{ L-atm/mol-K})(321 \text{ K})}{0.452 \text{ L}} = 0.533 \text{ atm}$$

21.65 The half reaction for oxidation in all these cases is: $H_2S(aq) \rightarrow S(s) + 2H^+ + 2e^-$. (We could write the product as $S_8(s)$, but this is not necessary. In fact, it is not necessarily the case that S_8 would be formed, rather than some other allotropic form of the element.)

(a) $2Fe^{3+}(aq) + H_2S(aq) \rightarrow 2Fe^{2+}(aq) + S(s) + 2H^+(aq)$

(b) $Br_2(\ell) + H_2S(aq) \rightarrow 2Br^-(aq) + S(s) + 2H^+(aq)$

(c) $2MnO_4^- + 6H^+(aq) + 5H_2S(aq) \rightarrow 2Mn^{2+}(aq) + 5S(s) + 8H_2O(\ell)$

(d) $2NO_3^-(aq) + H_2S(aq) + 2H^+(aq) \rightarrow 2NO_2(aq) + S(s) + 2H_2O(\ell)$

21.66 $$\left(\frac{0.28 \text{ mol } (NH_4)_2S_2O_3}{1 \text{ L Soln}}\right)(10.8 \text{ L Soln})\left(\frac{148 \text{ g } (NH_4)_2S_2O_3}{1 \text{ mol } (NH_4)_2S_2O_3}\right)$$

$= 448 \text{ g } (NH_4)_2S_2O_3$

21.67 Let us assume that the reactions occur in basic solution. The half-reaction for reduction of H_2O_2 is in all cases $H_2O_2(aq) + 2e^- \rightarrow 2OH^-(aq)$.

(a) $H_2O_2(aq) + S^{2-}(aq) \rightarrow 2OH^-(aq) + S(s)$

(b) $SO_2(g) + 2OH^-(aq) + H_2O_2(aq) \rightarrow SO_4^{2-}(aq) + 2H_2O(\ell)$

(c) $NO_2^-(aq) + H_2O_2(aq) \rightarrow NO_3^-(aq) + H_2O(\ell)$

(d) $As_2O_3(s) + 2H_2O_2(aq) + 6OH^-(aq) \rightarrow 2AsO_4^{3-}(aq) + 5H_2O(\ell)$

(e) We know that this reaction must be occurring in acidic solution, since $Fe(OH)_3$ would form if the solution were basic. The half-reactions are:

$$2H^+(aq) + H_2O_2(aq) + 2e^- \rightarrow 2H_2O(\ell)$$
$$2[Fe^{2+}(aq) \rightarrow Fe^{3+}(aq) + e^-]$$

$$2Fe^{2+}(aq) + H_2O_2(aq) + 2H^+(aq) \rightarrow 2Fe^{3+}(aq) + 2H_2O(\ell)$$

21.68 The half-reaction for oxidation of H_2O_2 in acidic solution is as follows: $H_2O_2(aq) \rightarrow O_2(g) + 2H^+(aq) + 2e^-$.

(a)
$$\begin{array}{l} 2[MnO_4^-(aq) + 8H^+(aq) + 5e^- \rightarrow Mn^{2+}(aq) + 4H_2O(\ell)] \\ 5[H_2O_2(aq) \rightarrow O_2(g) + 2H^+(aq) + 2e^-] \\ \hline 2MnO_4^-(aq) + 5H_2O_2(aq) + 6H^+(aq) \rightarrow 2Mn^{2+}(aq) + 5O_2(g) + 8H_2O(\ell) \end{array}$$

(b) $H_2O_2(aq) + Cl_2(aq) \rightarrow O_2(g) + 2Cl^-(aq) + 2H^+(aq)$

(c) $H_2O_2(aq) + 2Ce^{4+}(aq) \rightarrow O_2(g) + 2Ce^{3+}(aq) + 2H^+(aq)$

(d) $3H_2O_2(aq) + O_3(g) \rightarrow 3H_2O(\ell) + 3O_2(g)$

21.69 $NH_4NO_2(s) \rightarrow N_2(g) + 2H_2O(\ell)$

$$(3.26 \text{ g } NH_4NO_2)\left(\frac{1 \text{ mol } NH_4NO_2}{64.0 \text{ g } NH_4NO_2}\right)\left(\frac{1 \text{ mol } N_2}{1 \text{ mol } NH_4NO_2}\right) = 0.0509 \text{ mol } N_2$$

$$V = \frac{(0.0509 \text{ mol})(0.0821 \text{ L-atm/mol-K})(295 \text{ K})}{(760 - 20)/760 \text{ atm}} = 1.27 \text{ L}$$

Recall that we correct for the vapor pressure of water in the collection container, as described in Sample Exercise 9.12.

21.70 We will need to refer to Equations [21.69], [21.67] and [21.63] in working through this problem.

$$\left(\frac{7.3 \times 10^9 \text{ kg } HNO_3}{\text{yr}}\right)\left(\frac{10^3 \text{ g}}{1 \text{ kg}}\right)\left(\frac{1 \text{ mol } HNO_3}{63 \text{ g } HNO_3}\right)\left(\frac{3 \text{ mol } NO_2}{2 \text{ mol } HNO_3}\right)\left(\frac{1 \text{ mol NO}}{1 \text{ mol } NO_2}\right)$$

$$\times\left(\frac{4 \text{ mol } NH_3}{4 \text{ mol NO}}\right)\left(\frac{1}{0.92}\right) = 1.9 \times 10^{11} \text{ mol } NH_3/\text{yr}$$

In practice, the ratio (3 mol NO_3/2 mol HNO_3) can be reduced to 1 by recirculating the NO. Thus, only about 0.67 times this much NH_3, or 1.3×10^{11} mol NH_3, are required.

$$V = \frac{(1.3 \times 10^{11} \text{ mol})(0.0821 \text{ L-atm/mol-K})(298 \text{ K})}{45 \text{ atm}} = 7.0 \times 10^{10} \text{ L}$$

21.71 (a) $P_4O_{10}(s) + 6H_2O(\ell) \rightarrow 4H_3PO_4(aq)$

(b) $PBr_3(\ell) + H_3O(\ell) \rightarrow H_3PO_3(aq) + 3H^+(aq) + 3Br^-(aq)$

(c) $As_4O_6(s) + 6H_2O(\ell) \rightarrow 4H_3AsO_3(aq)$

(d) $Li_3N(s) + 3H_2O \rightarrow NH_3(g) + 3Li^+(aq) + 3OH^-(aq)$

(e) $3NO_2(g) + H_2O(\ell) \rightarrow 2H^+(aq) + 2NO_3^-(aq) + NO(g)$

21.72 $$(0.365 \text{ g Au})\left(\frac{1 \text{ mol Au}}{197 \text{ g Au}}\right)\left(\frac{3 \text{ mol } NO_2}{1 \text{ mol Au}}\right)\left(\frac{46.0 \text{ g } NO_2}{1 \text{ mol } NO_2}\right) = 0.256 \text{ g } NO_2$$

21.73 The nitrogen atom has a ground state electron configuration $1s^22s^22p^12p^12p^1$. The presence of a half-filled 2p subshell leads to a comparatively stable arrangement. When an electron is added to form N^-, it must go into an orbital that is already occupied. The inter-electronic repulsion that results is as large as the increase in total nuclear-electron attractive energy. In the other group 5A elements, the situation is similar. However, because the atoms are larger, the electrons can remain further apart on the average, and thus electron-electron repulsions are relatively less important. The electron affinities are therefore negative, though not very large. In the case of carbon, the added electron goes into a 2p orbital that was vacant. Thus, the electron-electron repulsions are not so large as for nitrogen. With oxygen, the added electron goes into an already occupied orbital, but by this point the incomplete shielding of the nucleus by the four 2p electrons leads to a relatively large effective nuclear charge. The electron-nuclear attraction for the added electron in oxygen thus more than compensates for the inter-electron repulsions.

21.74 $\Delta H° = (-207.3) - 0.5(-285.8) = -64.4$ kJ
$\Delta G° = (-111.3) - 0.5(-236.8) = 7.1$ kJ

Recall that $\Delta G° = -2.303\ RT \log K$;

$$\log K = \frac{7.1 \times 10^3 \text{ J}}{2.303(8.314 \text{ J/mol-K})\ 298 \text{ K}} = -1.24$$

$$K = 0.058 = \frac{[HNO_3]}{P_{N_2}^{1/2}[H_2O]P_{O_2}^{5/4}}\ ; \quad P_{N_2} = 0.8 \text{ atm}; \quad P_{O_2} = 0.2 \text{ atm};$$

$[H_2O] = 1$ (nearly pure liquid)

Thus, $[HNO_3] = (0.058)(0.89)(0.14) = 7.2 \times 10^{-3}$ M. We conclude that the process cannot lead to a substantial concentration of $HNO_3(aq)$.

21.75 We omit the unshared pairs on oxygen atoms for clarity.

(a) 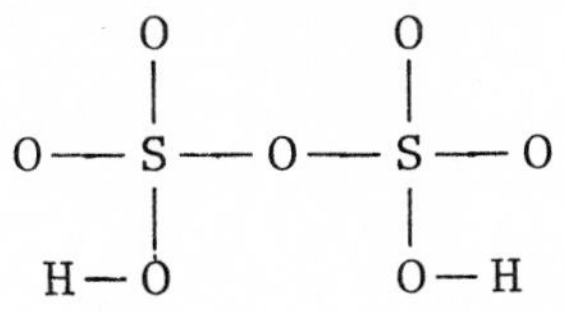

(b) 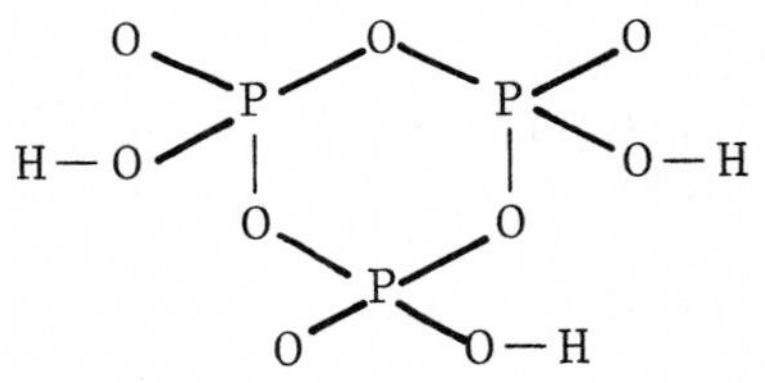

(c) 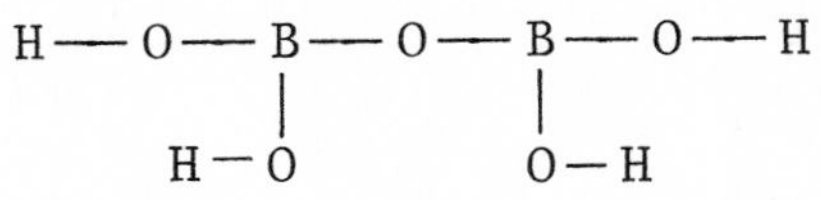

(d) 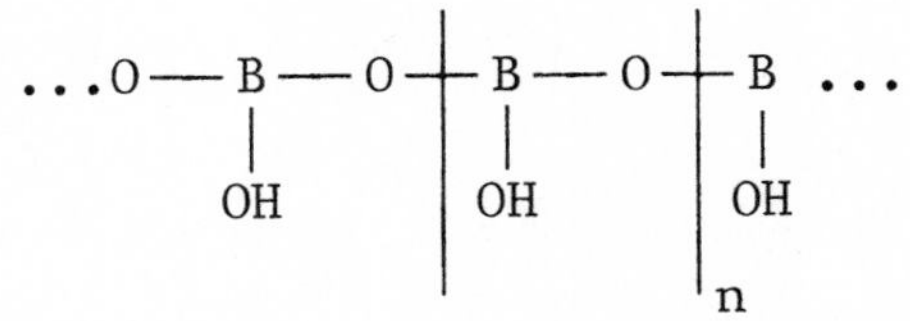

22

The lithosphere: geochemistry and metallurgy

22.1 For sodium, the most important mineral is NaCl. For iron, the most important minerals commercially are Fe_2O_3 and Fe_3O_4. Iron pyrite, FeS_2, is abundant, but not at present a commercially useful form. Deposits of $FeCO_3$ are also known.

22.2 The SiO_4^{4-} tetrahedron is the fundamental unit. These tetrahedra are joined by a sharing of oxygen or oxygens at a single corner, an edge or even a face. In aluminosilicates, some fraction of the Si^{4+} ions is replaced by Al^{3+} ions, while retaining the tetrahedral arrangement of surrounding oxygens.

22.3 (a) CaO; (b) $CaSO_4 \cdot 2H_2O$; (c) $CaCO_3$; (d) $Mg_3Si_4O_{10}(OH)_2$;

22.4 (a) A mineral is a more or less pure substance with definite composition; an ore is a mixture of several substances; it may contain one or more minerals of commercial interest. (b) Clays are hydrated aluminum silicates, whereas glass is an amorphous solid formed by melting together SiO_2 and other oxides, then allowing the melt to cool. (c) Calcite, $CaCO_3$, is a mineral; that is, a pure substance. Limestone is a naturally occurring rock, a mixture of several substances. However, calcite is the major component in limestone. (d) The term rock refers to any naturally occurring intimate mixture of several minerals; a common example is granite, which is largely SiO_2. Cement is a man-made mixture of $CaCO_3$, SiO_2 and clay, powdered and heated in a kiln to about 1500°C. A little $CaSO_4 \cdot 2H_2O$ is then added. When mixed with rocks and added water, the cement hardens to form concrete.

22.5 For a metal to occur in nature as the free element, it must have a substantial negative potential for oxidation. That is E° for the process $M(s) \rightarrow M^{n+} + ne^-$ must be rather negative. In addition, reducing conditions must have been present when the metal was deposited in the early history of the earth. The metal deposits must have since been protected from oxidizing conditions.

22.6 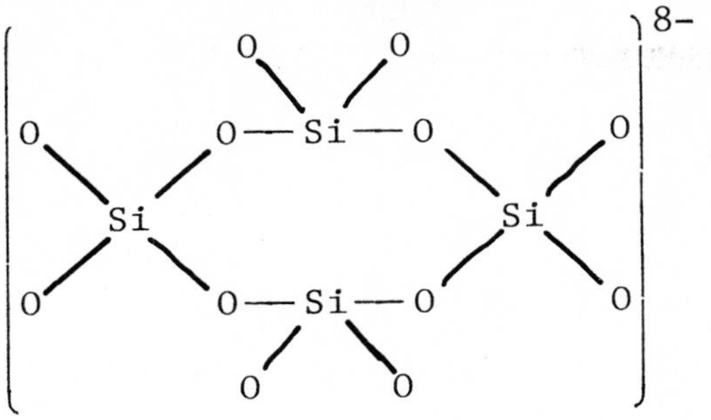

For clarity, we omit the unshared pairs, 2 on each bridging oxygen, 3 on each terminal oxygen.

22.7 (a) SiO_4^{4-}; (b) SiO_3^{2-}; (c) SiO_3^{2-}

22.8 Whereas the arrangement of four Si-O σ bonds represents the most stable bonding arrangement for silicon, carbon is more stable in CO_2, in which carbon forms two double bonds to 2 oxygens. Possibly also carbon is too electronegative to be stable in an ionic or highly polar environment of four surrounding oxygens, each carrying a net charge.

22.9 (a) yes; as Al^{3+} (size and charge state are appropriate). (b) no; Pd forms 2+ and 4+ oxidation states, but in these its stable geometrical environment is square-planar and octahedral, respectively. Palladium is too large to substitute for Si. (c) no; K^+ is too large and has too low a charge to effectively replace the small, highly charged Si^{4+}. (d) yes; as 4+. Ti^{4+} should be rather similar to Si^{4+}, and one expects that replacement should occur.

22.10 (a) Mica consists of sheets of $AlSi_3O_{10}^{5-}$, held together by relatively weaker electrostatic interactions with the cations that balance charge. The mineral can be readily cleaved along these sheets. (b) Metal ions are bound in soil to aluminosilicate structures. In acidic ground water, the proton may displace the metal ion (e.g., K^+) from the mineral, releasing it for use by plants. In a basic solution, this does not occur; the metal ion remains bound to the clay. (c) Mg^{2+} and Fe^{2+} have similar ionic radii (0.74 Å for Fe^{2+}, 0.65 Å for Mg^{2+}). Thus, they can "stand in" for one another in a mineral in which the metal ion is located in an interstitial hole in a close-packed oxide lattice. (d) When Al^{3+} substitutes for Si^{4+} to form aluminosilicates, the silicate structure acquires an even larger negative charge. More cations such as K^+ or Ca^{2+} are needed to maintain charge balance. Thus the electrostatic interactions between sheets are larger, and the mineral is harder.

22.11 The anorthite is acting as a base, or proton acceptor. Note that the overall effect of the reaction is that one water molecule and two protons are transferred to the $Al_2Si_2O_8^{2-}$ core, releasing Ca^{2+}. The OH^- groups that remain following proton transfer react with CO_2 to form HCO_3^-.

22.12 Assume we have 100 g of glass:

$$(16\ g\ B_2O_3)\left(\frac{1\ mol\ B_2O_3}{70\ g\ B_2O_3}\right)\left(\frac{2\ mol\ B}{1\ mol\ B_2O_3}\right) = 0.46\ mol\ B$$

$$(76\text{ g SiO}_2)\left(\frac{1\text{ mol SiO}_2}{60\text{ g SiO}_2}\right)\left(\frac{1\text{ mol Si}}{1\text{ mol SiO}_2}\right) = 1.27\text{ mol Si}$$

Thus, the ratio mol B/mol Si = 0.46/1.27 = 0.36.

<u>22.13</u> $$(21.3\text{ g Na}_2\text{O})\left(\frac{1\text{ mol Na}_2\text{O}}{62\text{ g Na}_2\text{O}}\right)\left(\frac{2\text{ mol Na}}{1\text{ mol Na}_2\text{O}}\right) = 0.69\text{ mol Na}$$

$$(5.2\text{ g CaO})\left(\frac{1\text{ mol CaO}}{56\text{ g CaO}}\right)\left(\frac{1\text{ mol Ca}}{1\text{ mol CaO}}\right) = 0.092\text{ mol Ca}$$

$$(73.5\text{ g SiO}_2)\left(\frac{1\text{ mol SiO}_2}{60.0\text{ g SiO}_2}\right)\left(\frac{1\text{ mol Si}}{1\text{ mol SiO}_2}\right) = 1.22\text{ mol Si}$$

Thus, the three elements are present in the ratio: Na: Ca: Si = 7.5:1:13

<u>22.14</u> (a) $Ba^{2+}(aq) + 2OH^-(aq) + CO_2(g) \rightarrow BaCO_3(s) + H_2O(\ell)$

(b) $CaCO_3(s) \xrightarrow{\Delta} CaO(s) + CO_2(g)$

$CaO(s) + H_2O(\ell) \rightarrow Ca(OH)_2(s)$

(c) $SrCO_3(s) \xrightarrow{\Delta} SrO(s) + CO_2(g)$

<u>22.15</u> $K_{sp} = [Fe^{2+}][CO_3^{2-}] = 3.2 \times 10^{-11}$. Assuming a simple solubility equilibrium, $x^2 = 3.2 \times 10^{-11}$; $x = 5.6 \times 10^{-6}$ M = $[Fe^{2+}] = [CO_3^{2-}]$

$$\left(\frac{5.6 \times 10^{-6}\text{ mol FeCO}_3}{1\text{ L}}\right)\left(\frac{116\text{ g FeCO}_3}{1\text{ mol FeCO}_3}\right) = 6.5 \times 10^{-4}\text{ g FeCO}_3\text{/L}$$

The solubility equilibrium involves $FeCO_3(s) \rightleftarrows Fe^{2+}(aq) + CO_3^{2-}(aq)$. However, CO_3^{2-} is a strong base, and will undergo hydrolysis: $CO_3^{2-}(aq) + H_2O(\ell) \rightleftarrows HCO_3^-(aq) + OH^-(aq)$. Water that has been exposed to air, as ground water has, contains dissolved CO_2, which further shifts the hydrolysis equilibrium to the right by acting as an acid: $CO_3^{2-}(aq) + H_2CO_3(aq) \rightleftarrows 2HCO_3^-(aq)$. Thus, the solubility of $FeCO_3(s)$ in water is enhanced by the existence of these equilibrium involving CO_3^{2-} ion.

<u>22.16</u> Mortar consists of a mixture of sand, water and CaO. On contact with air, the CaO reacts with CO_2 to form $CaCO_3$, which acts as the binder. Treating a sculpture containing $CaCO_3$ with a mixture of $Ba(OH)_2$ and urea, $CO(NH_2)_2$, leads to reactions [22.12] and [22.13]. $BaCO_3$ is less soluble than $CaCO_3$, thus does not so readily dissolve out on contact with acid. Furthermore, if the rain contains sulfate (as it usually does), $BaSO_4(s)$ may form (Equation [22.14]). $BaSO_4(s)$ has a small K_{sp}, and is not readily soluble in acidic solution.

<u>22.17</u> (a) octahedral; (b) tetrahedral; (c) octahedral; (d) octahedral

<u>22.18</u> (a) tetrahedral; (b) tetrahedral; (c) tetrahedral

<u>22.19</u> We expect Mn^{2+} to most readily substitute for cations of the same charge and radius. Zn^{2+} is closest; Ni^{2+} is next. The difference in radius from Ca^{2+} or Mg^{2+} is rather too large, but Mn^{2+} does often mimic Mg^{2+} in its behavior.

22.20 The various stages of ore processing in metallurgy include preliminary treatment, including various possible mechanical concentration stages; reduction of the ore to obtain the free metal; refining of the raw, impure metal.

21.21 (a) $2CoS_2(s) + 5O_2(g) \rightarrow 2CoO(s) + 4SO_2(g)$

(b) $2CoO(s) + C(s) \xrightarrow{\Delta} 2Co(\ell) + CO_2(g)$

(c) $Co(s) \rightarrow Co^{2+}(aq) + 2e^-$; $Co^{2+}(aq) + 2e^- \rightarrow Co(s)$

22.22 (a) Coke acts as a reducing agent for the iron, either directly or as CO [Equations [22.19] and [22.20]). Secondly, the combustion of coke produces the required high temperature in the furnace. (b) Air is needed to burn the coke and thus produce the high temperature required: $2C(s) + O_2(g) \rightarrow 2CO(g)$; $C(s) + O_2(g) \rightarrow CO_2(g)$. (c) Limestone is added to provides CaO which reacts with the gangue, or silicate waste material present along with the ore. The reaction, $CaO(s) + SiO_2(s) \rightarrow CaSiO_3(\ell)$, forms slag, a liquid that separates from molten iron, and that can be drawn off.

22.23 $(1 \times 10^3 \text{ g } CuFeS_2)\left(\frac{1 \text{ mol } CuFeS_2}{183 \text{ g } CuFeS_2}\right)\left(\frac{1 \text{ mol Cu}}{1 \text{ mol } CuFeS_2}\right)\left(\frac{63.5 \text{ g Cu}}{1 \text{ mol Cu}}\right)$

$= 347$ g Cu

22.24 $CaSiO_3$ has a great deal of ionic character, especially in the interaction of Ca^{2+} with the anionic silicate material. Thus, as a liquid it is likely to dissolve ionic substances. Molten iron is not ionic, nor is it even polar. Even in the liquid state it maintains a more or less close-packed arrangement of Fe atoms bound together through delocalized metal bonding. There is thus so little similarity in the two liquids that there is no driving force for them to mix to form a solution.

22.25 Any phosphorus or silicon would be present in the highest oxidation state, and as the oxide, i.e., as P_4O_{10} or SiO_2. We know that at high temperature the limestone decomposes with loss of CO_2:

$$CaCO_3(s) \xrightarrow{\Delta} CaO(s) + CO_2(g)$$

Then, the basic oxide, CaO, reacts with the relatively acidic oxides as follows: $CaO(s) + SiO_2(\ell) \rightarrow CaSiO_3(\ell)$; $6CaO(s) + P_4O_{10}(\ell) \rightarrow 2Ca_3(PO_4)_2(\ell)$.

22.26 (a) $Co_3O_4(s) + 4Mg(s) \rightarrow 3Co(s) + 4MgO(s)$

(b) $2NiS_2(s) + 5O_2(g) \rightarrow 2NiO(s) + 4SO_2(g)$

(c) $OsO_4(g) + 4H_2(g) \rightarrow Os(s) + 4H_2O(g)$

(Note: surprisingly, OsO_4 melts at 41°C, boils at 130°C.)

(d) $2Cs^+(\ell) + 2e^- \rightarrow 2Cs(\ell)$; $2Cl^-(\ell) \rightarrow Cl_2(g) + 2e^-$

22.27 (a) $HfCl_4(g) + Na(\ell) \xrightarrow{\Delta} Hf(s) + 4NaCl(s)$

(b) $HgCO_3(s) \rightarrow HgO(s) + CO_2(g) \rightarrow Hg(\ell) + (1/2)O_2(g) + CO_2(g)$

(c) $Re_2O_7(s) + 7H_2(g) \xrightarrow{\Delta} 2Re(s) + 7H_2O(g)$

(d) $MnO_2(s) + C(s) \xrightarrow{\Delta} Mn(\ell) + CO_2(g)$

22.28 An oxygen furnace could be used to rid copper of several possible impurities, such as sulfur or carbon. It would not result in a copper as pure as that obtained from electrolytic refining, but at least in principle the copper should not oxidize too readily. For magnesium, however, such an oxidizing atmosphere could never be used without oxidizing the metal completely to MgO.

22.29

Substance	Appearance	Flexibility	Melting Point (°C)	Thermal Conductivity
sulfur (S_8)	yellow	brittle	115	low
magnesium	white metallic	flexible	650	high
iron	metallic	flexible, no "spring"	1537	high
steel	metallic	"spring," harder than iron	1450-1500	good, but lower than iron

22.30 In silver metal each silver atom is surrounded by many nearest neighbors (it has the face-centered cubic structure, Figure 11.15). From the electronic structure of the Ag atom (Table 6.2), we can see that each Ag atom has but one valence electron to employ in bonding to all these nearest neighbors, The valence electrons are delocalized throughout the entire metal structure, but, on average, there is only a fraction of a bond order between any two Ag atoms. Now we know that bond distance and bond order are related; bond distances are longer for bonds of lower bond order. Thus, it is reasonable that the Ag-Ag distance should be longer for the metal (2.88 Å) than predicted for a full σ bond between a pair of Ag atoms (2 x 1.34 Å = 2.68 Å).

22.31 In subliming a mole of chromium metal, all of the metallic bonds between chromium atoms need to be broken. Each chromium atom has several valence shell electrons involved in this bonding. Thus we must expect that it will cost a good deal of energy to break all these bonds to form chromium vapor. (Actually, they may not all be broken; the vapor probably consists at least in part of Cr_2 molecules, possibly others.) On the other hand, to sublime iodine, which consists of I_2 molecules, we need only overcome the van der Waals type of intermolecular forces between the molecules. These forces are much weaker, so the enthalpy of vaporization is correspondingly lower.

22.32 In the series K, Ca, Sc, the number of valence electrons increases from 1 to 3. The average binding energy per metal atom increases with the number of valence electrons in the delocalized metal orbitals. Melting point and hardness parallel the number of metal-metal bonding electrons.

22.33 Each carbon atom in diamond has four nearest neighbors. There are just four valence shell electrons in each carbon available for bonding. The carbon atoms form highly localized, strong bonds between all pairs of carbon atoms, and there are just enough electron pairs to fill these orbitals. There is no tendency for the electrons to move throughout the solid because there are no accessible vacant orbitals into which the electrons can move. In silver, on the other hand, there is but one valence electron to bind each atom to several nearest neighbors. The valence electrons are distributed in molecular orbitals that extend over the entire structure. They move readily under the influence of an applied electric field.

22.34 The ammonia decomposes to leave nitrogen atoms in the metal lattice. These form an interstitial alloy. The metal becomes harder, more brittle and higher melting. The electrical conductivity probably decreases.

22.35 (a) $(67 \text{ g Cu})\left(\frac{1 \text{ mol Cu}}{63.5 \text{ g Cu}}\right) = 1.05 \text{ mol Cu};$

$(33 \text{ g Zn})\left(\frac{1 \text{ mol Zn}}{65.4 \text{ g Zn}}\right) = 0.50 \text{ mol Zn}$

$\text{mol \% Cu} = \left(\frac{1.05}{1.55}\right)(100) = 68\%; \text{ mol \% Zn} = \left(\frac{0.50}{1.55}\right)(100) = 32\%$

(b) $(67 \text{ g Pb})\left(\frac{1 \text{ mol Pb}}{207 \text{ g Pb}}\right) = 0.32 \text{ mol Pb};$

$(33 \text{ g Sn})\left(\frac{1 \text{ mol Sn}}{119 \text{ g Sn}}\right) = 0.28 \text{ mol Sn}$

$\text{mol \% Pb} = \left(\frac{0.32}{0.60}\right)(100) = 53\%; \text{ mol \% Sn} = \left(\frac{0.28}{0.62}\right)(100) = 47\% \text{ Sn}$

(c) $(64 \text{ g Sn})\left(\frac{1 \text{ mol Sn}}{119 \text{ g Sn}}\right) = 0.54 \text{ mol Sn};$

$(36 \text{ g Pb})\left(\frac{1 \text{ mol Pb}}{207 \text{ g Pb}}\right) = 0.17 \text{ mol Pb}$

$\text{mol \% Sn} = \frac{0.54}{0.71} = 76\%; \text{ mol \% Pb} = \frac{0.17}{0.71} = 24 \%$

22.36 (a) The statement is true. However, it should be recognized that the term "contact" here really means contacts that are sufficiently strong to push the anions out and away from one another, to reduce anion-anion repulsions. (b) False. Limestone deposits dissolve when in contact with acidic aqueous solution: $CuCO_3(s) + H^+(aq) \rightarrow Ca^{2+}(aq) + HCO_3^-(aq)$. (c) True; (d) False. In aluminosilicate minerals, the Al^{3+} ions replace Si^{4+} ions in the SiO_4 tetrahedral environments. (e) False. A few important metals are produced by electrolytic reduction methods. By and large, these are the more active metals (Al, Na, Mg). Electrolytic refining is also employed to produce relatively high purity forms of less active metals such as Zn, Cu, Ni.

22.37 Asbestos minerals have either double-stranded silicate chains or a sheet structure. In the first instance the general formula is $M_xM'_y\ Si_4O_{11}(OH)_2$. Where M and M' are both divalent, x + y must equal 8 for charge balance. In the second case, the general formula is $M_xM'_y\ (Si_2O_5)_2(OH)_2$. Where M and M' are both divalent (they can be the same, of course), x + y = 3 to achieve charge balance (see Figure 22.7). The fibrous asbestos minerals belong to the first group. The fibers are formed from the long silicate chains. The sheet structures are found in minerals such as talc, that consist of sheets of silicate structures that easily slide over one another.

22.38 (a) The formula suggests a three-dimensional SiO_2 lattice in which half the Si^{4+} ions are replaced by Al^{3+}. Na^+ ions are added in appropriate interstitial holes to maintain charge balance (a feldspar). (b) Here we have a three-dimensional SiO_2 lattice in which one third of the Si^{4+} ions are replaced by Al^{3+} ions. The K^+ ions are present in interstitial positions. (c) We can think of this as consisting of SiO_4^{4-} tetrahedra, but notice that Zr is an element of group 4B. We expect it to form Zr^{4+} ions, and thus it should substitute reasonably well for Si^{4+}. We can thus think of $ZrSiO_4$ as a three-dimensional SiO_2 lattice in which half of the Si^{4+} have been replaced by Zr^{4+} ions. Both cations should be in tetrahedral environments. (d) A derivative of $Si_2O_5^{2-}$, in which half the Si^{4+} are replaced by Ti^{4+}. Ca^{2+} is present for charge balance.

22.39 $FeCO_3(s) + H_2O(\ell) + CO_2(aq) \rightarrow Fe^{2+}(aq) + 2HCO_3^-(aq)$

$$4Fe^{2+}(aq) + O_2(aq) + 4H^+(aq) \rightarrow 4Fe^{3+}(aq) + 2H_2O(\ell)$$

$$4[Fe^{3+}(aq) + 3H_2O(\ell) \rightarrow Fe(OH)_3(s) + 3H^+(aq)]$$

$$4Fe^{2+}(aq) + O_2(g) + 10H_2O(\ell) \rightarrow 4Fe(OH)_3(s) + 8H^+(aq)$$

22.40 The carbonates of Ca^{2+}, Fe^{2+} and Mn^{2+} are insoluble in water. On contact with water that is saturated with CO_2 from the atmosphere the following reaction occurs: $MCO_3(s) + H_2O(\ell) + CO_2(aq) \rightarrow M^{2+}(aq) + 2HCO_3^-(aq)$.

22.41 $r+/r- = 0.98/1.33 = 0.74$ and $0.45/1.33 = 0.34$. We expect the Na^+ ions to be in octahedral holes, possibly cubic; the Al^{3+} ions should be in tetrahedral holes.

22.42 $$(1\text{ kg ore})\left(\frac{660\text{ g }Fe_3O_4}{1\text{ kg ore}}\right)\left(\frac{1\text{ mol }Fe_3O_4}{231\text{ g }Fe_3O_4}\right)\left(\frac{3\text{ mol Fe}}{1\text{ mol }Fe_3O_4}\right)\left(\frac{55.8\text{ g Fe}}{1\text{ mol Fe}}\right)$$
$$\times\left(\frac{1.03\text{ g pig iron}}{1\text{ g Fe}}\right) = 493\text{ g pig iron}$$

One can see that substantial shipping costs could be saved by shipping the pig iron rather than the raw ore. The northern refineries went out of business because the hardwood forests were decimated around the mills, and because reduction with coke became a much better procedure.

22.43 $\Delta G = \Delta G_f(CO) - \Delta G_f(NiO) = -520 - (-180) = -340$ kJ

or $\Delta G = \Delta G_f(CO_2) - 2\Delta G_f(NiO) = -400 - 2(-180) = -40$ kJ

ΔG is more negative for formation of CO as product.

22.44 The CO_3^{2-} ion is a discrete species with a Lewis structure best represented as three resonance structures:

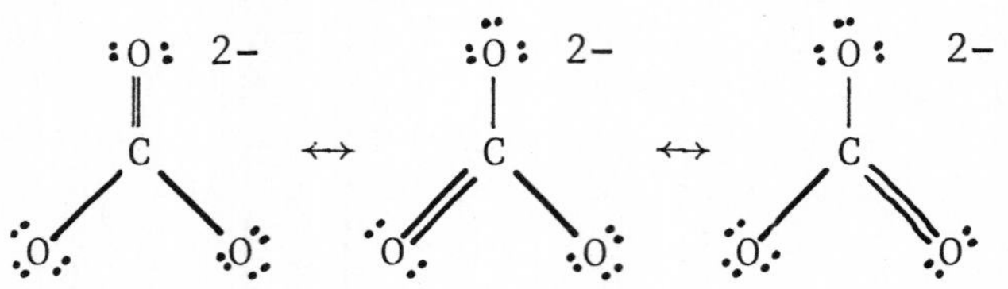

The SiO_3^{2-} ion has no substantial Si-O π bonding. This is in fact the empirical formula for a single chain silicate ion (Figure 22.7(a)), or a cyclic structure with formula such as $Si_3O_9^{6-}$ or $Si_4O_{12}^{8-}$.

22.45 The problem for this nation is to obtain some source of energy that can be used to carry out the reduction. We know that there are some metals for which reduction by coke or some reducing gas produced from coal (e.g., CO) is not feasible. These include Al and Mg. These metals are produced by electrolytic reduction. It would no doubt be possible to work out a suitable means for electrolytic reduction of Fe_2O_3, but the question then arises as to where the electrical capacity would come from. With no fossil fuel sources available, it would be necessary to use nuclear power or some as yet undeveloped source of electrical energy based on solar input. In terms of present technology, such a country would be critically dependent on other nations.

22.46 $$(1.0 \times 10^7 \text{ kg Zn})\left(\frac{10^3\text{g}}{1\text{ kg}}\right)\left(\frac{1\text{ mol Zn}}{65.4\text{ g Zn}}\right)\left(\frac{2F}{1\text{ mol Zn}}\right)\left(\frac{96{,}500\text{ coul}}{1F}\right)$$

$$= 3.0 \times 10^{13} \text{ coul}$$

$$\text{kilowatt-hrs} = (3.0 \times 10^{13}\text{ coul})(3.5\text{ V})\left(\frac{1\text{ J}}{\text{coul-V}}\right)\left(\frac{1\text{ kwh}}{3.6 \times 10^6\text{ J}}\right)\left(\frac{1}{0.80}\right)$$

$$= 3.6 \times 10^7 \text{ kwh}$$

22.47 The enthalpy of sublimation reflects the overall energy of metallic bonding in the metal lattice. (Actually, this is true only if the sublimation process leads to atomic gaseous products, M(g), rather than M_2(g) and so forth.) We thus expect that ΔH_s will reflect the number of valence electrons available for metal bonding. The graph shows that ΔH_s climbs steeply with an increase in the number of valence electrons, then dips in the middle, then again rises and finally decreases to a low value at Zn. Note the overall similarity to Figure 22.21. The dip in the middle is possibly due to the presence in the vapor phase of metal atom clusters in those cases where strong metal-metal bonds are formed. Thus, there are probably more Mn_2 molecules in the vapor of manganese than there are Zn_2 molecules in Zn vapor. This has the effect of lowering ΔH_s.

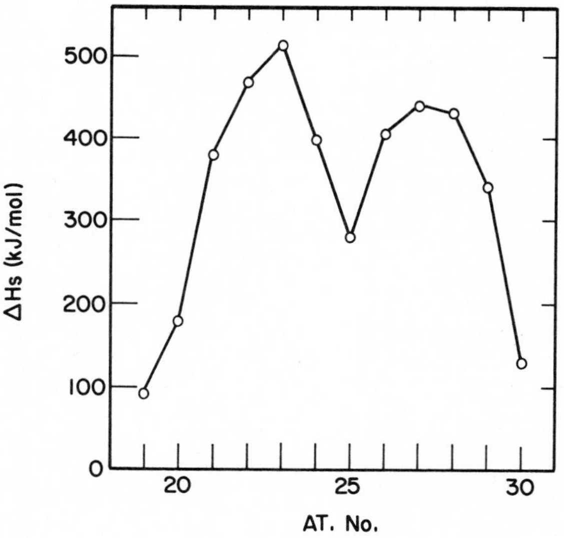

22.48 Think of all the observed physical properties of metals such as malleability, electrical and thermal conductivity. These can be explained only in terms of delocalized bonding. Consider also the observed structures, which show each atom with many nearest neighbors. Localized electron pair bonds could not produce such structures.

22.49 (18/24)(100) = 75%

22.50 Note the relative reduction potentials:

$Cu^{2+}(aq) \rightarrow Cu(s) + 2e^-$ $\quad E° = +0.337$ V

$Al^{3+}(aq) \rightarrow Al(s) + 3e^-$ $\quad E° = -1.66$ V

The values tell us that copper is reduced to the free metal relatively easily. Reducing conditions might have existed at some places in the earth's early history when the crust was cooling, so that copper was reduced to the metal. However, the potential for Al is so negative that no conditions could reasonably have existed under which the Al^{3+} could be reduced to metal.

22.51 (a) $2ZnS(s) + 3O_2(g) \rightarrow 2ZnO(s) + 2SO_2(g)$

(b) $ZnCO_3(s) \xrightarrow{\Delta} ZnO(s) + CO_2(g)$

Then, $ZnO(s) + C(s) \xrightarrow{\Delta} Zn(\ell) + CO(g)$. For most uses, zinc is further refined by electrolysis, after dissolving in aqueous sulfuric acid.

23

Chemistry of metals: coordination compounds

23.1 (a) In a metal complex the metal and the coordinating ligands that surround it are called the coordination sphere. (b) A molecule or ion that is coordinated (chemically bound) to the metal ion in a complex is called a ligand. (c) The coordination number is the number of ligands bound to the central metal ion in a complex. (d) A tridentate ligand is one that is coordinated, or bound, to the central metal at three different points on the ligand. It occupies three different coordination positions about the metal.

23.2 (a) four; (b) six; (c) six (en is bidentate); (d) four; (e) seven.

23.3

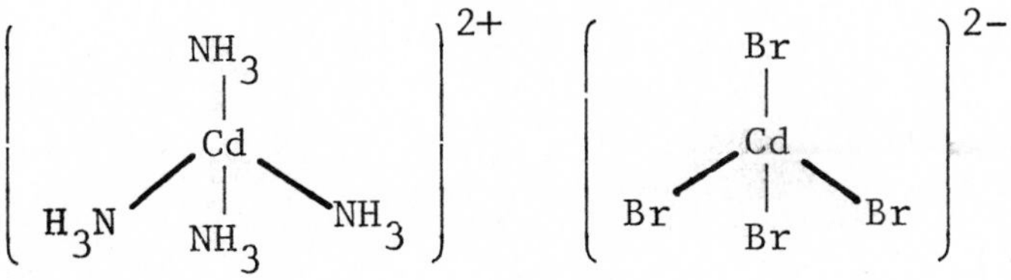

tetrahedral

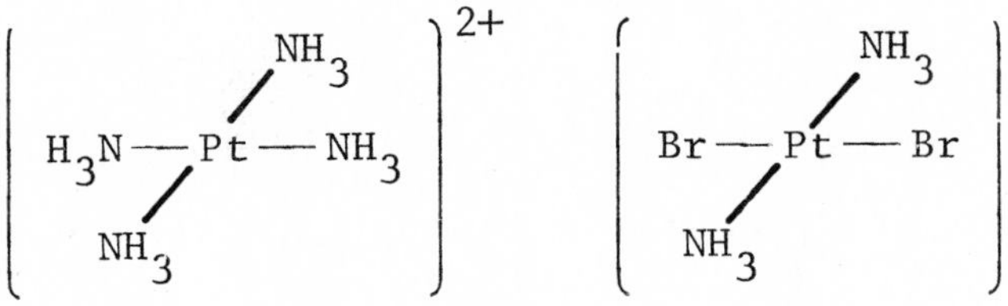

square-planar

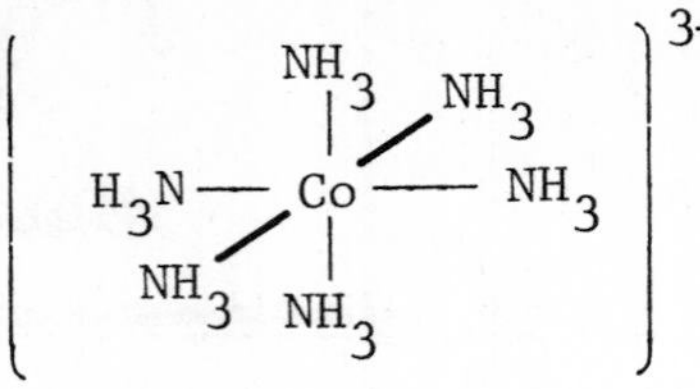 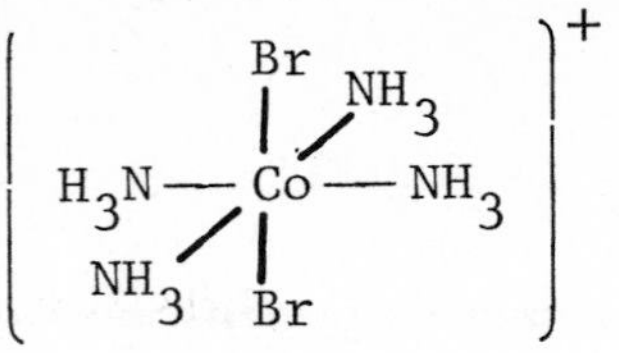 octahedral

23.4 (a) 3+; (b) 3+; (c) 4+; (d) 3+

23.5 $Ni(H_2O)_6{}^{2+}$, $Ni(H_2O)_4(en)^{2+}$, $Ni(H_2O)_2(en)_2{}^{2+}$, $Ni(en)_3{}^{2+}$

23.6

(a) 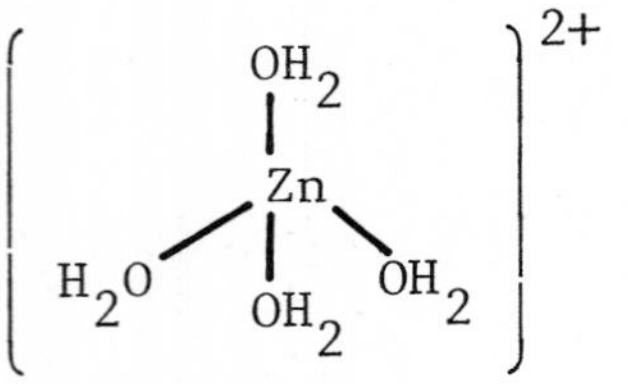

tetrahedral

(b) 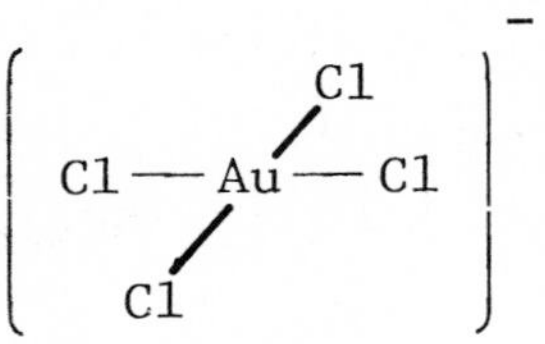

square-planar

(c) 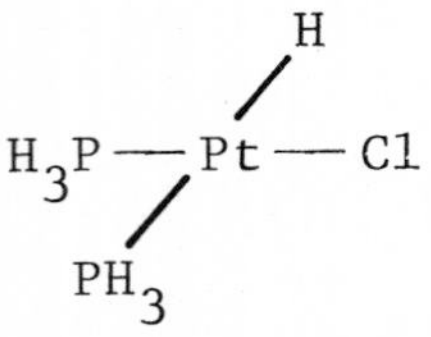

square-planar

(d) 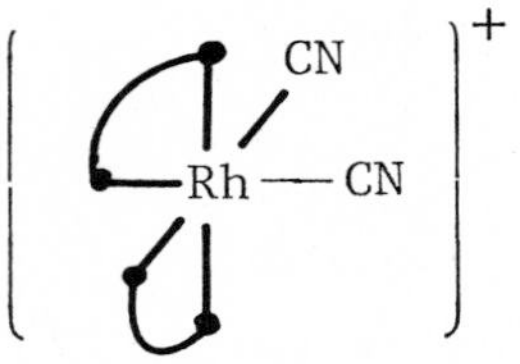

octahedral

(e) 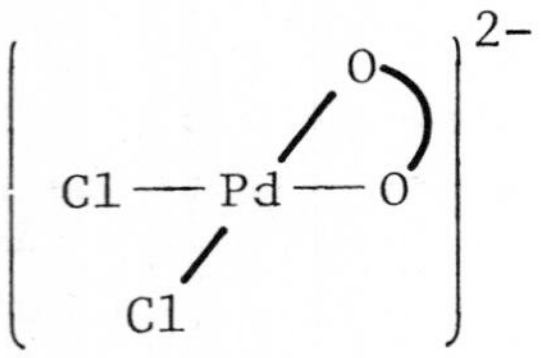

square-planar

23.7

(a) 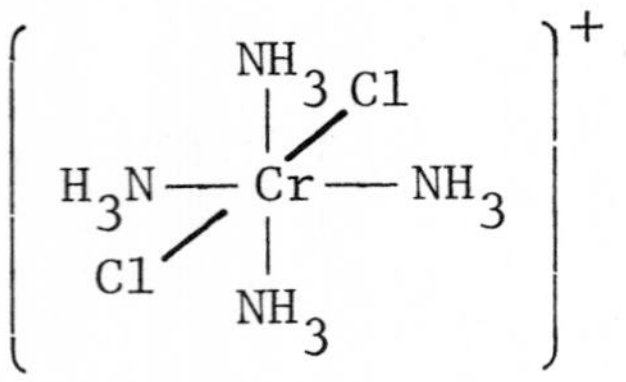

(b) 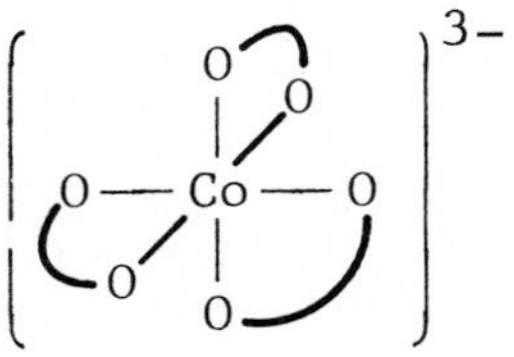

(c) 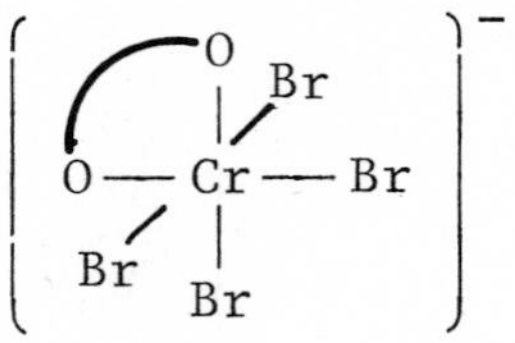

(d) 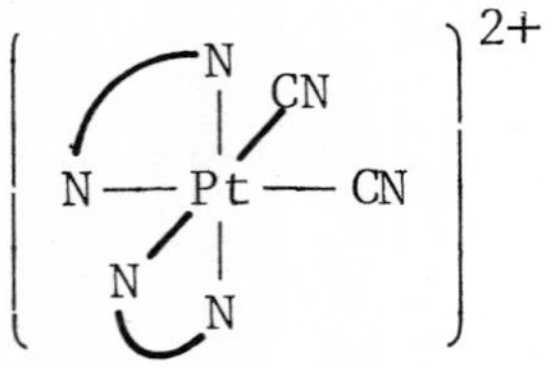

23.8 From 23.6: tetraaquazinc(II); tetrachloroaurate(III); cis-chlorohydridobis(phosphine)platinum(II); cis-dicyanobis(ethylenediamine)-rhodium(III); dichlorooxalatopalladate(II). From 23.7: trans-tetraammine-dichlorochromium(III); tris(oxalato)cobaltate(III); tetrabromooxalato-chromate(III); cis-dicyanobis(ethylenediammine)platinum(IV).

23.9 (a) cesium tetracyanonickelate(II); (b) tetraammineplatinum(II), diamminetetrachlorocobaltate(III); (c) sodium bis(glycinato)zincate(II); (d) tetraaquasulfatoiron(III) chloride; (e) potassium pentachloronitrito-osmate(IV); (f) ammineethylenediaminetri(hydroxo)cobalt(III).

23.10 (a) $[Ni(NH_3)_4](ClO_4)_2$; (b) $Na_3[Co(NO_2)_6]$; (c) $[Cr(en)_2(H_2O)_2](NO_3)_3$; (d) $[Co(en)_2Cl(SCN)]_2[CdCl_4]$; (e) $K_3[Ni(CN)_6]$; (f) $[Fe(en)_2SO_4]$; (g) $[Co(NH_3)_5N_3]SO_4$.

23.11 (a) $[Pd(NH_3)_2Br_4]$; (b) $[Cr(NH_3)_4Br_2]Br$; (c) $[V(NH_3)_4Br_2]Br$.

23.12 (a) SO_4^{2-} is bidentate. (b) EDTA is tetradentate. (c) en is bidentate (Zn^{2+} forms four-coordinate, tetrahedral complexes). (d) CO_3^{2-} is bidentate.

23.13 $[Pt(NH_3)_3Cl_3]Cl$ - KBr; $[Co(NH_3)_6]Cl_3$ - $ScBr_3$; $K_2[PtCl_6]$ - $Ca(NO_3)_2$

23.14 The observed isomerism could be due to linkage isomers, or to geometrical isomerism. It could not be due to coordination sphere isomerism, since there are no ions in the formulation outside the coordination sphere. It cannot be due to optical isomerism, because that form does not give rise to differences in solubility and color.

23.15 (a) geometrical isomerism:

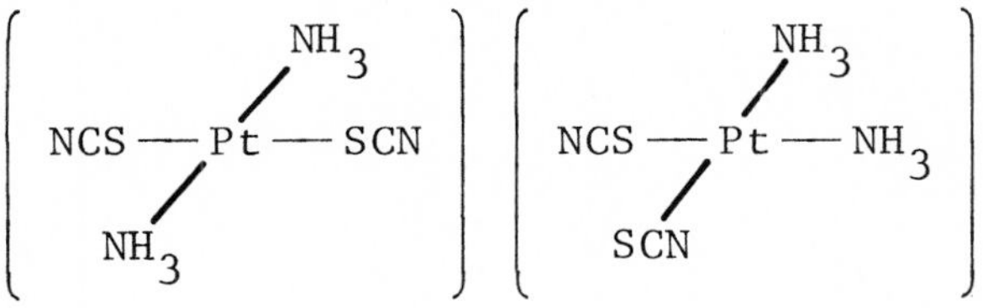

trans cis

(b) optical isomerism:

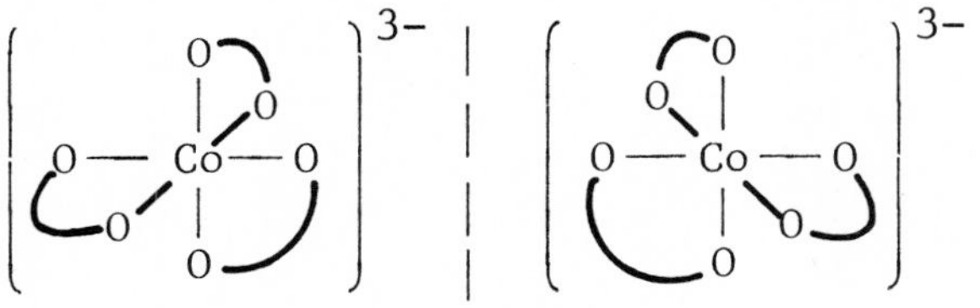

(O⌒O represents bidentate $C_2O_4^{2-}$ ion.)

(c) linkage isomerism:

$[Pt(NH_3)_2(SCN)_2]$

$[Pt(NH_3)_2(SCN)(NCS)]$

(d) coordination sphere isomerism:

$[Co(NH_3)_4CO_3]\ C_2O_4$

$[Co(NH_3)_4\ C_2O_4]\ CO_3$

23.16 (a) only one:

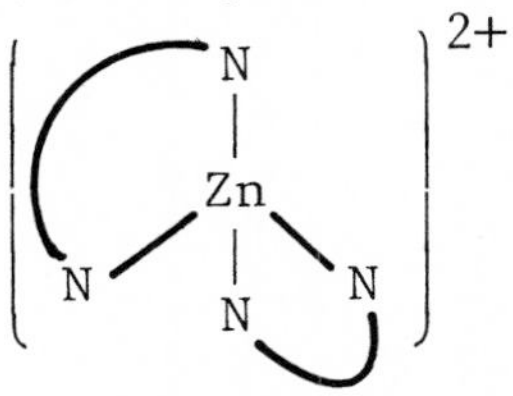

(b)

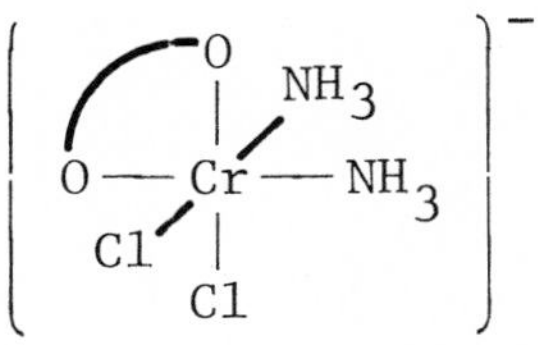

This isomer also has a non-superimposable mirror image

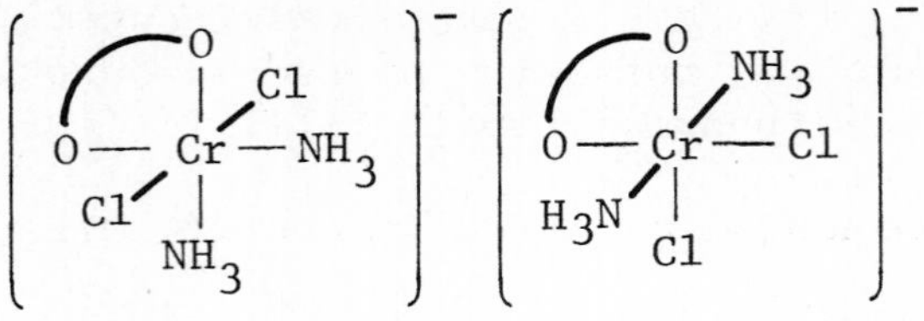

(c) 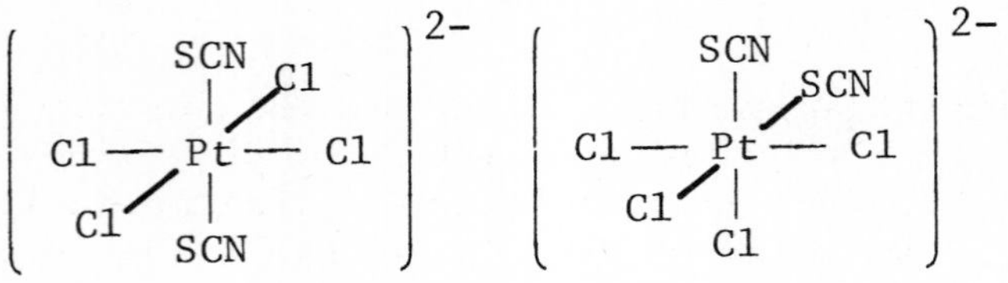 (There are also 4 other possible linkage isomers involving -NCS binding.)

(d) 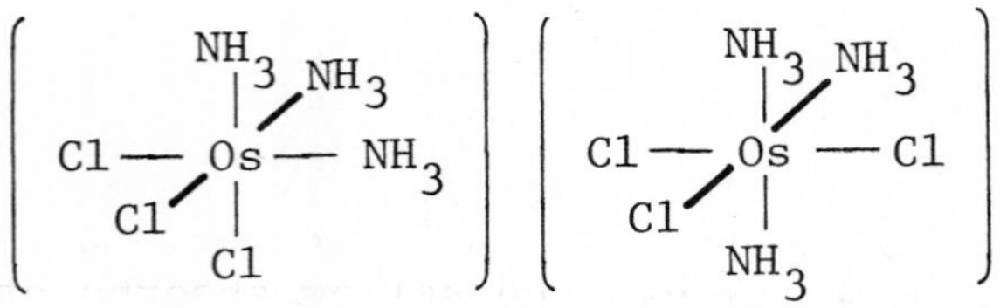

<u>23.17</u> (a) Only one isomer.

(b) 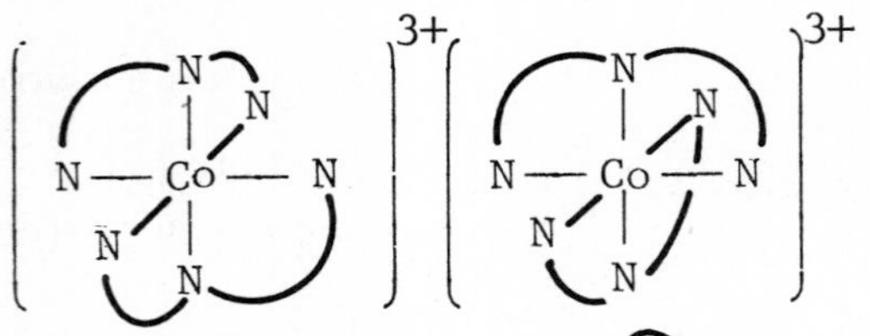 (c) 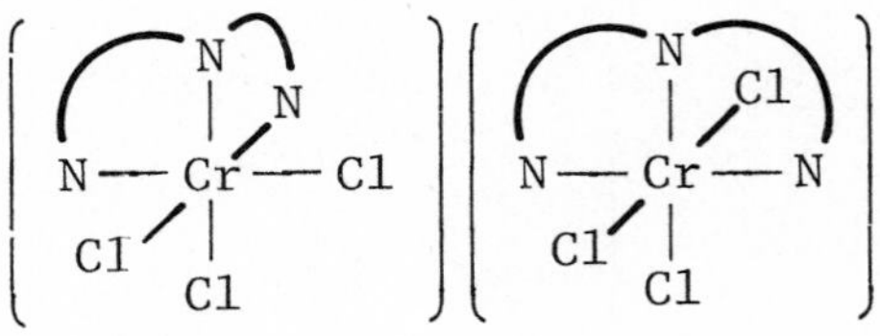

(d) 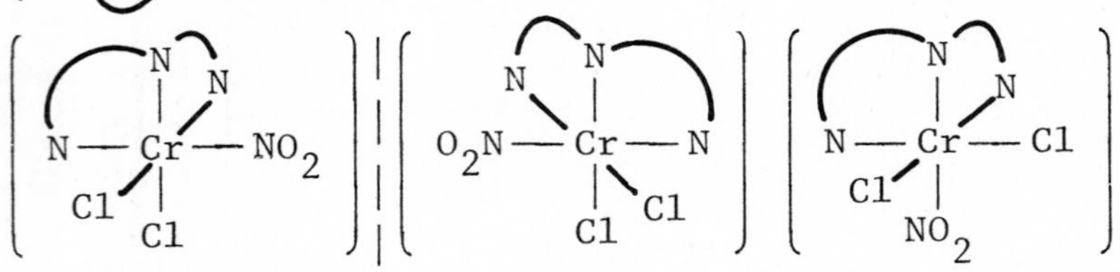

optical isomers

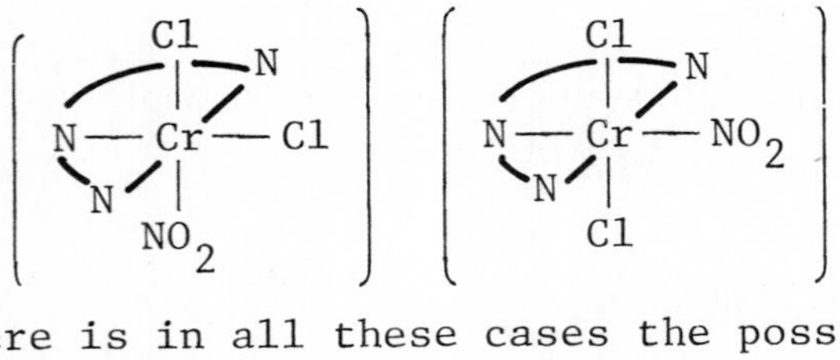

There is in all these cases the possibility also of linkage isomerism of the NO_2 group.

<u>23.18</u> 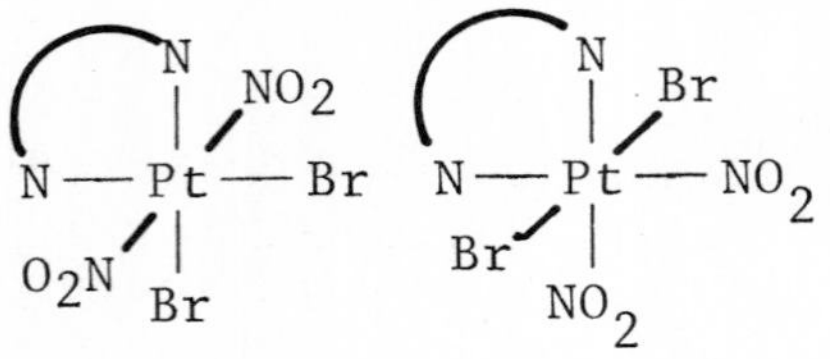

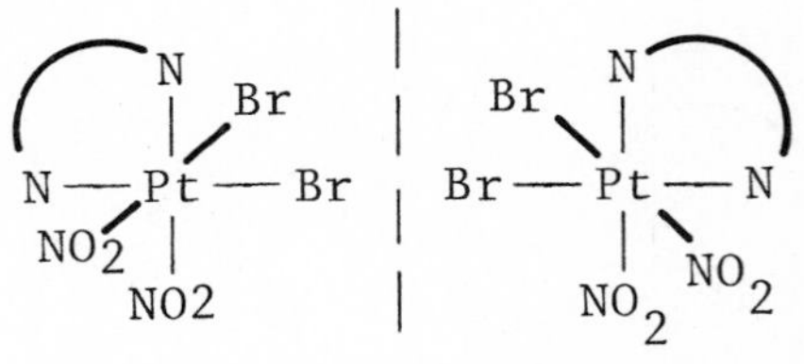

optical isomers

23.19 We recall that cobalt(III) complexes are generally inert; that is, they do not rapidly exchange ligands out of the coordination sphere. Therefore, the ions that form precipitates in these two cases are probably outside the coordination sphere. We can thus formulate the red complex as $[Co(NH_3)_5SO_4]Br$, pentaamminesulfatocobalt(III) bromide, and the violet compound as $[Co(NH_3)_5Br]SO_4$, pentaamminebromocobalt(III) sulfate.

23.20 $[Co(NH_3)_4Cl_2]OH$; tetraamminedichlorocobalt(III) hydroxide. $[Co(NH_3)_4Cl(OH)]Cl$; tetraamminechlorohydroxocobalt(III) chloride.

23.21 $[Pt(NH_3)_4][CuCl_4]$; tetraammineplatinum(II) tetrachlorocuprate(II). $[Cu(NH_3)_4][PtCl_4]$; tetraamminecopper(II) tetrachloroplatinate(II). $[PtCl(NH_3)_3][CuCl_3(NH_3)]$ and $[CuCl(NH_3)_3][PtCl_3(NH_3)]$ are also possible, but it is doubtful whether such compounds could be prepared.

23.22 purple, rose, magenta (something near a red)

23.23 (a) 8; (b) 8 (c) 3; (d) 1

23.24 The yellow compound probably absorbs at around 440-480 nm, the red at 490-500 nm and the green at 650-750 nm. Thus, the yellow absorbs at shortest wavelength, the green at longest (one must be careful about green, however; compounds absorbing very short wavelengths, around 380-430 nm, sometimes have a yellowish-green color).

23.25 The directional characteristics of the d orbitals account for the splitting. One subset points along the axes on which ligands are found in an octahedral complex, the other subset points between the ligands.

23.26

(a)	(b)	(c)
Cr: $[Ar]3d^54s^1$	Ru: $[Kr]4d^65s^2$	Ni: $[Ar]3d^84s^2$
Cr^{3+}: $[Ar]3d^3$	Ru^{3+}: $[Kr]4d^5$	Ni^{3+}: $[Ar]3d^7$
[ ][ ]	[ ][ ]	[↑][ ]
[↑][↑][↑]	[↑↓][↑↓][↑]	[↑↓][↑↓][↑↓]

23.27 (a) This complex contains Mn(III), which has a $3d^4$ electron configuration. It must be low spin, since there would be four unpaired electrons in the high spin case. (b) low spin; (c) high spin.

23.28 $$\left(\frac{120\ \text{kJ}}{\text{mol}}\right)\left(\frac{1\ \text{mol}}{6.02 \times 10^{23}\ \text{ions}}\right)\left(\frac{1\ \text{ion}}{1\ \text{photon}}\right) = 1.99 \times 10^{-19}\ \text{J/photon}$$

$$\nu = c/\lambda = \Delta E/h;\ \lambda = \frac{ch}{\Delta E}$$

$$= \frac{(3.0 \times 10^{8}\ \text{m/sec})(6.63 \times 10^{-34}\ \text{J-sec})}{1.99 \times 10^{-19}\ \text{J}} = 9.98 \times 10^{-7}\ \text{m} = 998\ \text{nm}$$

This wavelength falls just below the visible, in the infrared region of the spectrum (Figure 5.3).

23.29 The solid compound would be packed in a tube and its magnetic susceptibility measured as shown in Figure 23.15. If the apparent mass of the sample increases as the magnetic field is applied, it is paramagnetic. Nickel in $K_2Ni(SCN)_4$ is Ni(II). Thus, it has a $3d^8$ electron configuration. A tetrahedral geometry about Ni would produce the orbital energy diagram shown at the left. On the other hand, a square-planar arrangement (Figure 23.32) would be expected to lead to complete pairing of electrons.

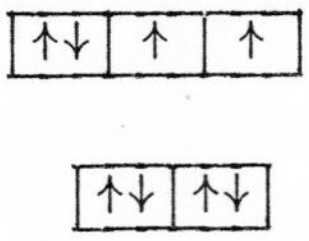

23.30 (a) $[Co(NH_3)_6]^{2+}$. NH_3 is a stronger field ligand; that is, it is higher in the spectrochemical series. It therefore produces a larger energy separation of the 3d orbitals. (b) $[FeCl_4]^-$. The energy splitting is larger for the more highly charged metal ion. (c) $[V(NO_2)_6]^{3-}$; nitrite is higher in the spectrochemical series.

23.31 (a) 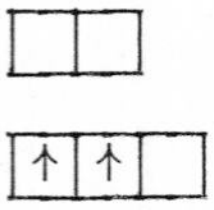 (b) 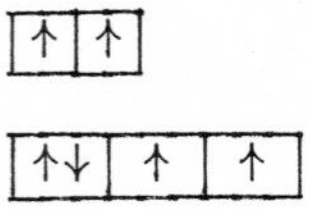 (c) 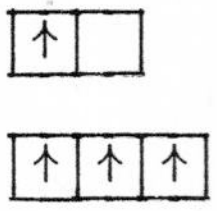

(d) 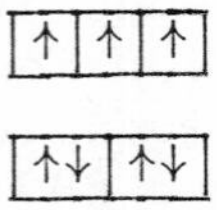 (e) 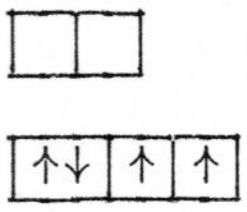 (f) 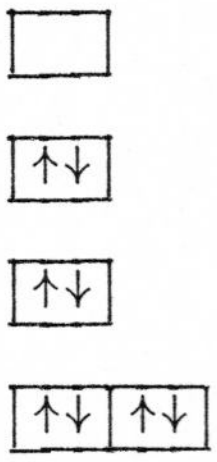

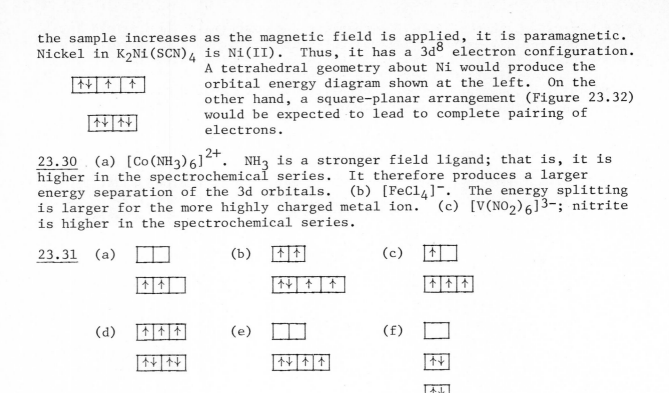

23.32 (a) The larger formation constant for $[Co(en)_3]^{3+}$ is related to the higher charge on the central metal ion. The ligands are bound more tightly. Of course, we must remember that water is being replaced by the en ligands. Since en is a stronger field ligand than OH_2, the tendency of en to replace OH_2 increases as the charge on the ion increases. (b) Here again we see the effect of higher charge. In this case even the Fe^{2+} complex, $[Fe(CN)_6]^{4-}$, is quite stable because it is a low spin complex of $3d^6$ electron configuration. (c) These two complexes have identical 3d electron configurations, $3d^6$. However, the Co^{3+} complex is low spin and very stable, whereas the Fe^{2+} complex is high spin and not very stable. The charge on the central ion makes the difference.

23.33 Let us draw the interactions with the d orbitals:

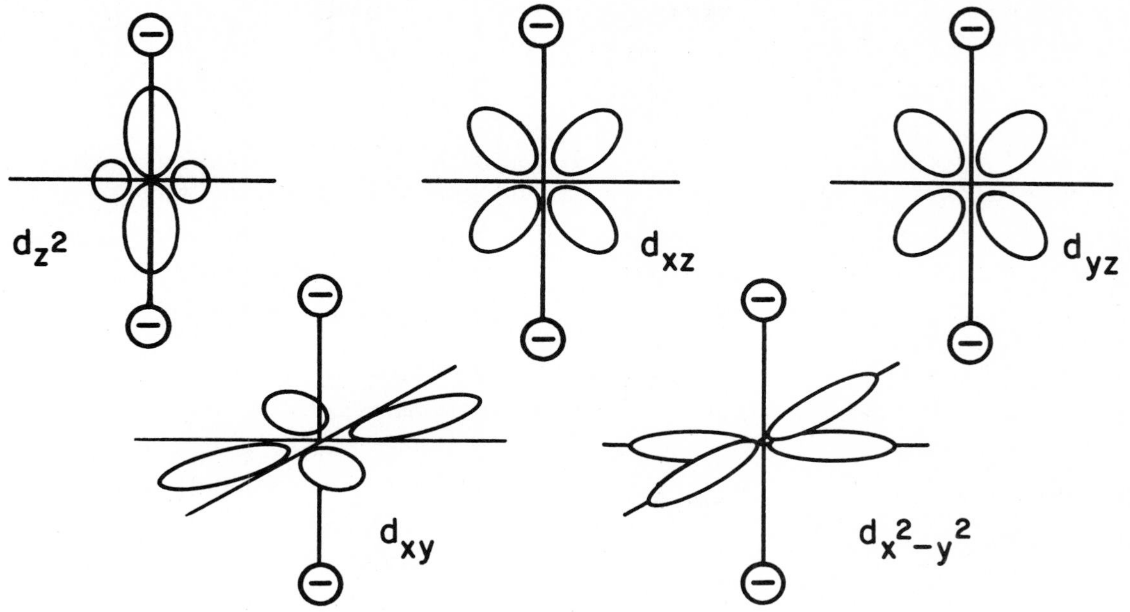

The relative ordering of the d orbitals is as follows:

[ ]	d_{z^2}	A metal complex with six d electrons would have 2 unpaired electrons if the complex were low spin.
[↑|↑]	d_{xz}, d_{yz}	
[↑↓|↑↓]	$d_{xy}, d_{x^2-y^2}$	

23.34 (a) Structural isomers have different bonding relationships between the parts of a molecule. Thus, structural isomers in general have different melting and boiling points, and so forth. Stereoisomers have the same relationships between the internal parts, but they lack symmetry, so that it is possible to have nonsuperimposable mirror images. (b) A labile complex undergoes rapid replacement of one or more of its ligands in aqueous solution. An inert complex undergoes such ligand replacement reactions only slowly. (c) In a high spin complex, the separation in energy of the two d orbital sets created by the ligand field is not as large as the spin-pairing energies. The electrons thus remain unpaired as much as possible. In a low spin complex, the reverse is true; the electrons pair up in lower energy orbitals. (d) A chiral complex is one that can have a nonsuperimposable mirror image. A cis-isomer is one of a pair of geometrical isomers that has two identical groups in adjacent coordination sites. (e) The spectrochemical series is a list of ligands in the order of their usual ability to cause a splitting of the metal d orbital energies. Crystal field stabilization energy is a measure of the extra stabilization that derives from putting electrons into the lower energy set of d orbitals under the influence of a ligand field.

23.35 $[Pt(NH_3)_6]Cl_4$; $[Pt(NH_3)_4Cl_2]Cl_2$; $[Pt(NH_3)_3Cl_3]Cl$; $[Pt(NH_3)_2Cl_4]$; $K[Pt(NH_3)Cl_5]$

23.36 The reaction that occurs produces an ion of higher charge. It would appear from the relative values that the reaction could be

$$[Co(NH_3)_4Br_2]^+(aq) + H_2O(\ell) \rightarrow [Co(NH_3)_4(H_2O)Br]^{2+} + Br^-$$

This reaction would convert the 1:1 electrolyte, $[Co(NH_3)_4Br_2]Br$, to a 2:1 electrolyte, $[Co(NH_3)_3(H_2O)Br]Br_2$. The reaction is not extremely rapid, so we would classify the starting complex as relatively inert, though it does undergo ligand replacement.

23.37

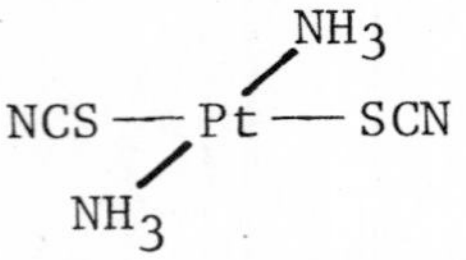

trans-diamminedithiocyanato-platinum(II)

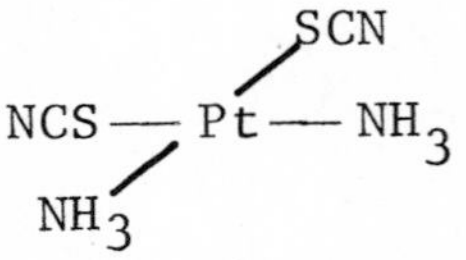

cis-diamminedithiocyanato-platinum(II)

There are two additional isomers that are named cis- or trans-diammine-diisothiocyanatoplatinum(II), and two that are named cis- or trans-diammineisothiocyanatothiocyanatoplatinum(II).

23.38 The $[Co(H_2O)_6]^{2+}$ complex should be rose-colored, or pink. The $CoCl_4^-$ solutions should be green or bluish green. The $CoCl_4^-$ ion is tetrahedral. We know that the d orbital splittings in tetrahedral complexes are only 4/9 as large as for the same ligand in an octahedral complex. Thus, the energy of the transition between the two sets of d orbitals is smaller in the tetrahedral complex, and the absorption is at longer wavelength.

23.39 (a) a square planar complex:

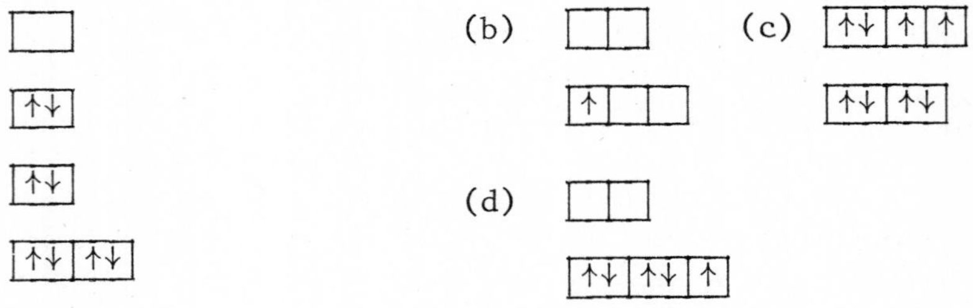

23.40 The crystal field splitting of the 3d orbital energies is larger in $Co(NH_3)_6^{3+}$ than in $Fe(NH_3)_6^{3+}$, because the metal charge is higher. Thus, the six 3d electrons are spin-paired in the cobalt case, but not in the iron case:

$Co(NH_3)_6^{3+}$ $Fe(NH_3)_6^{2+}$

23.41 We will represent the end of the bidentate ligand containing the CF_3 group by a closed circle, the other end by an open circle:

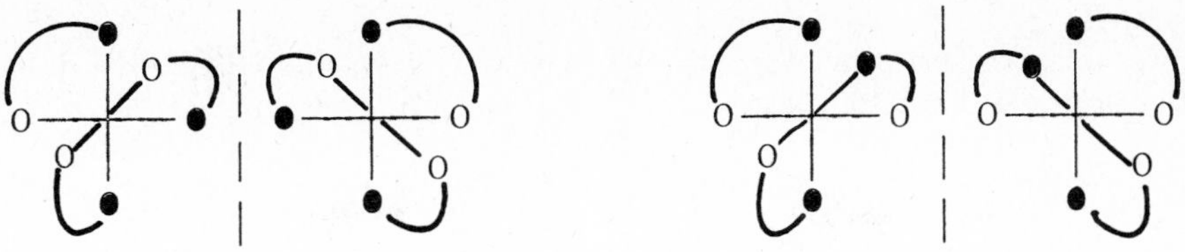

23.42 The V(III) complexes will all have a $3d^2$ electron configuration. The wavelength of the absorption due to a 3d electron transition should decrease in the order $V(H_2O)_6^{3+} > V(NH_3)_6^{3+} > V(CN)_6^{3-}$. This is so because the crystal field splitting is increasing in this order. An increased crystal field splitting results in a shorter wavelength transition.

23.43 (a) $AgBr(s) + 2S_2O_3^{2-}(aq) \rightleftarrows Ag(S_2O_3)_2^{3-}(aq) + Br^-(aq)$

(b) $[Cr(en)_2Cl_2]^+(aq) + H_2O(\ell) \rightarrow [Cr(en)_2(H_2O)Cl]^{2+}(aq) + Cl^-(aq)$
and $[Cr(en)_2(H_2O)Cl]^{2+}(aq) + H_2O(\ell) \rightarrow [Cr(en)_2(H_2O)_2]^{3+} + Cl^-(aq)$

The three moles of AgCl arise from the two chlorides shown in these two reactions, plus that which was outside the coordination sphere in the original reaction.

(c) $Zn(OH)_2(s) + 4NH_3(aq) \rightleftarrows Zn(NH_3)_4^{2+}(aq) + 2OH^-(aq)$

(d) $Co^{2+}(aq) + 4Cl^-(aq) \rightleftarrows CoCl_4^{2-}(aq)$

23.44 (a) orange-yellow; (b) yellow-orange; (c) orange-red; (d) red or red-violet

23.45 (a) only one; (b) two; (cis or trans arrangement of N and O ends); (c) four; two are geometrical, the other two are stereoisomers of each of these. See the figure for solution to exercise 23.41; the same figure applies to this problem as well.

23.46 We can calculate the CFSE for each ion assuming that the +2 ions are high spin, and using the idea that the CFSE is 4/9 as large in the tetrahedral case as in octahedral. (Also recall that the orbitals are reversed, Figure 23.31.) From Table 23.3 we see that CFSE is zero for Mn^{2+}, and 0.8Δ for Co^{2+}. For Cr^{3+}, a d^3 case, it will be 1.2Δ. Now remember also that Δ itself will be much larger for the +3 ion, so Cr^{3+} has by far the largest CFSE. In going to a tetrahedral hole, it loses much of this, because there is room for only two electrons in the lower energy set. Thus, the Cr^{3+} gains most in energy by placement in an octahedral hole.

23.47 (a) A d^3 complex; thus, CFSE is 1.2Δ. Thus, CFSE is 1.2(182 kJ) = 218 kJ. (b) A d^3 complex; CFSE = 1.2(258 kJ) = 310 kJ; (c) A d^3 complex; CFSE = 1.2Δ = 1.2(230 kJ) = 276 kJ. (d) A d^6 complex which will surely be <u>low spin</u>. Thus, CFSE = 2.4Δ = 2.4(545 kJ) = 1308 kJ.

23.48 □□ Application of pressure would result in shorter metal ion-oxide ion distances. This would have the effect of increasing the ligand-electron repulsions, and would result in a larger splitting in the d orbital energies.
↑↑↑
Thus, application of pressure should result in a shift in the absorption to higher energy (shorter wavelength).

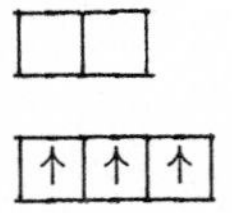

23.49 Let us first determine the empirical formula, assuming that the remaining mass of complex is all Pd.

		moles	mole ratios
$(37.6\ g\ Br)\left(\frac{1\ mol\ Br}{79.9\ g\ Br}\right)$	=	0.470 mol Br	1.98
$(28.3\ g\ C)\left(\frac{1\ mol\ C}{12.0\ g\ C}\right)$	=	2.36 mol C	9.95
$(6.60\ g\ N)\left(\frac{1\ mol\ N}{14.0\ g\ N}\right)$	=	0.471 mol N	1.99
$(2.37\ g\ H)\left(\frac{1\ mol\ H}{1.01\ g\ H}\right)$	=	2.35 mol H	9.91
$(25.13\ g\ Pd)\left(\frac{1\ mol\ Pd}{106\ g\ Pd}\right)$	=	0.237 mol Pd	1.00

We obtain $Pd(NC_5H_5)_2Br_2$. This should be a square-planar complex of Pd(II), a non-electrolyte. Because the dipole moment is zero, we can infer that it must be the trans- isomer.

23.50 Iron(II) has a $3d^6$ electron configuration. In an octahedral environment, the low spin state has all six electrons paired in the lower energy orbitals that point between the ligands. When the iron is in a high spin state, the higher energy orbitals that point directly along the metal ion-ligand axes each contain one electron. This added electron-ligand repulsion causes the ligands to be pushed out a bit further away from the metal ion. The result is that we observe an apparently larger ionic radius.

24

Organic compounds

24.1 alkane, C_6H_{14}; alkene, C_6H_{12}; alkyne, C_6H_{10}; aromatic, C_6H_6.

24.2 There are 5 isomers. Their carbon skeletons are as follows:

n-hexane

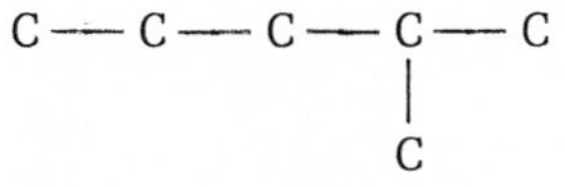
2-methylpentane

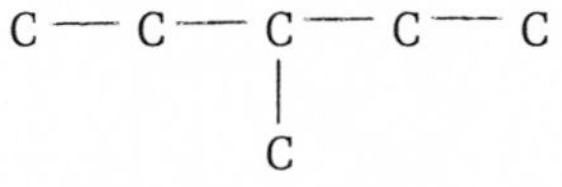
3-methylpentane

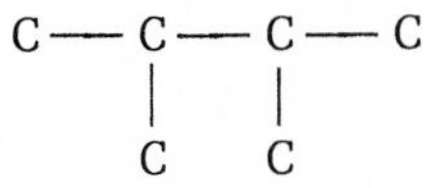
2,3-dimethylbutane

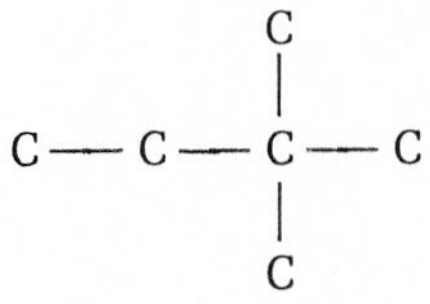
2,2-dimethylbutane

24.3

(a) 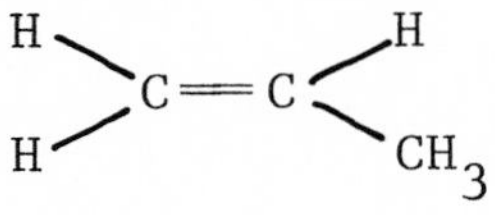

(b) 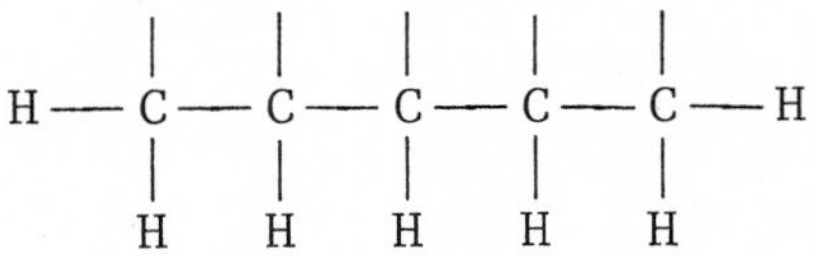

(c) 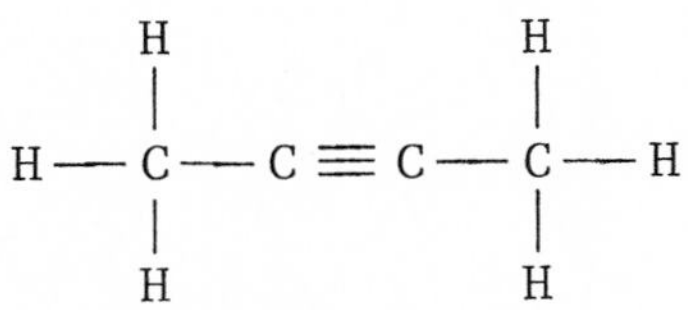

(d)

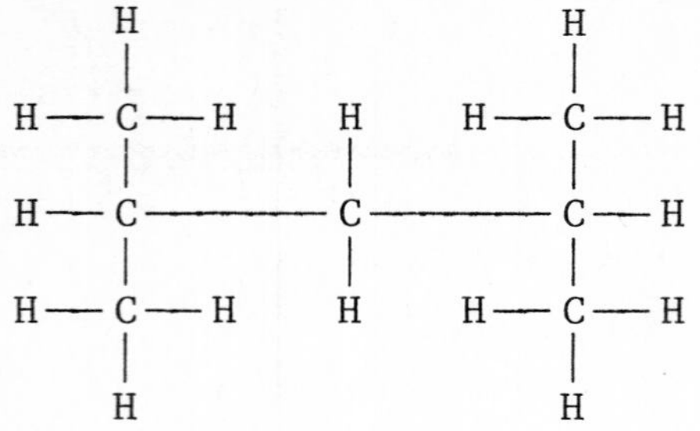

(e)

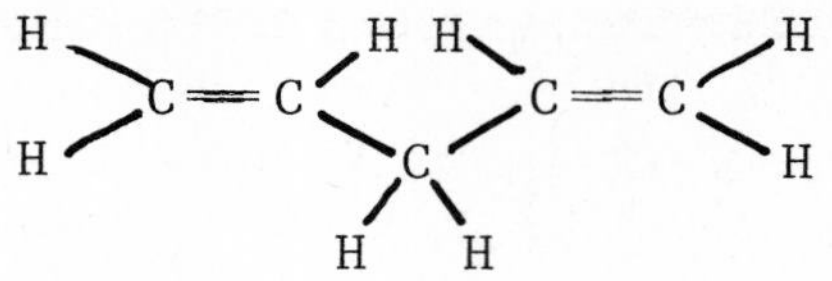

24.4 $CH_3CH_2CH_2C \equiv CH$ $CH_3CH_2C \equiv CCH_3$ $CH_3CH = C = CHCH_3$

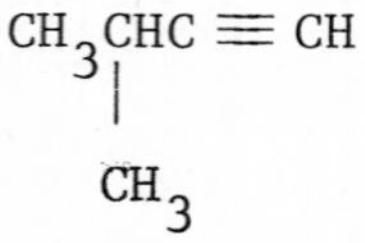

$CH_3\underset{\overset{|}{CH_3}}{CH}C \equiv CH$ $CH_2 = CHCH_2CH = CH_2$ $CH_2 = CH - \underset{\overset{|}{CH_3}}{C} = CH_2$

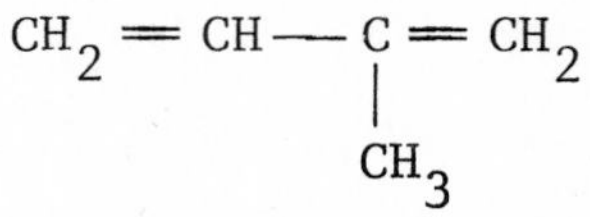

$CH_2 = CHCH = CHCH_2$

$CH_2 = C = CHCH_2CH_3$ $CH_2 = C = \underset{\overset{|}{CH_3}}{C} - CH_3$

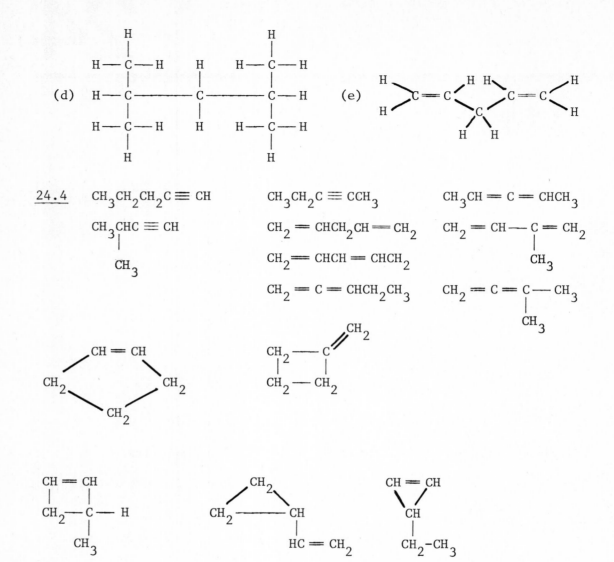

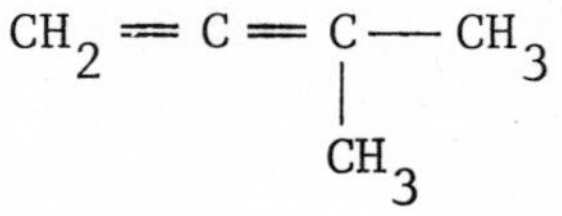

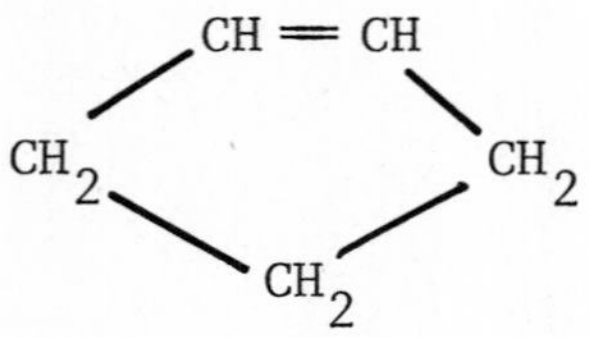

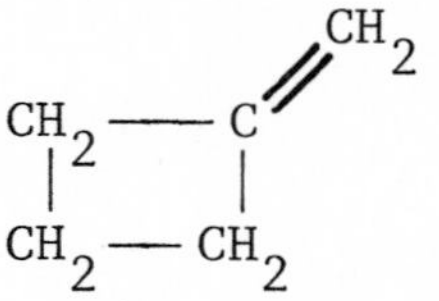

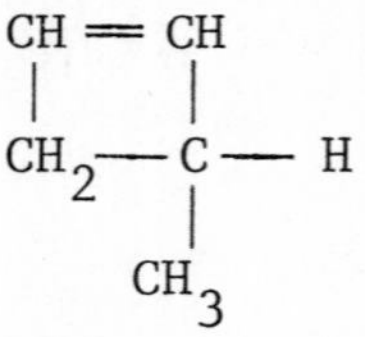

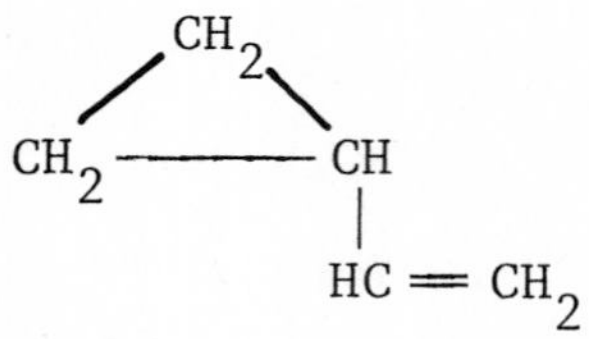

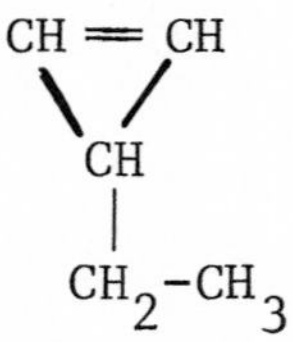

We probably don't have them all!

24.5   (a) 109°;   (b) 120°;   (c) 180°

24.6 (a)(b) $HC \equiv C - CH_2Cl$

(c)

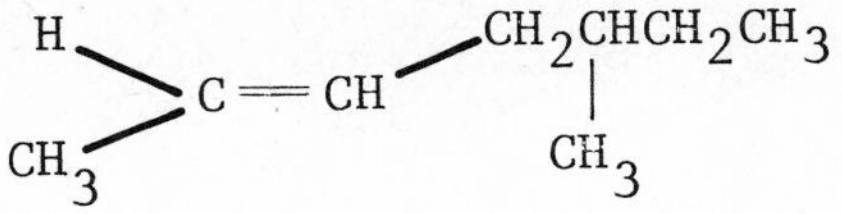

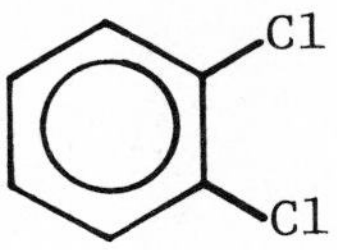

(d)

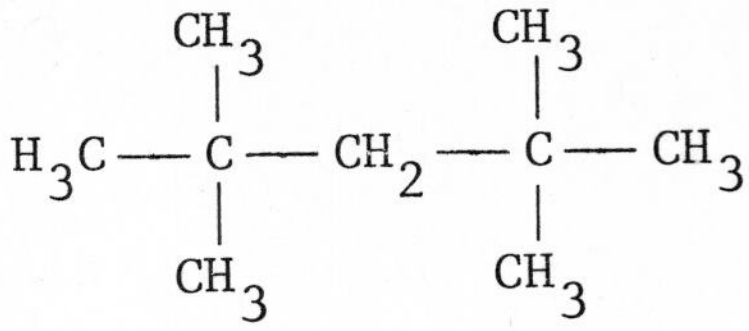

(e) $CH_3 - CH - CHCH_2CH_2CH_3$
    $\qquad\;\; | \qquad\;\; |$
    $\qquad\;\; CH_3 \quad CH_2CH_3$

(f)

(g)

(h) $CH_2 \!-\! CHCH_3$
    $\;\;| \qquad\quad |$
    $\;CH_2 \!-\! CH_2$

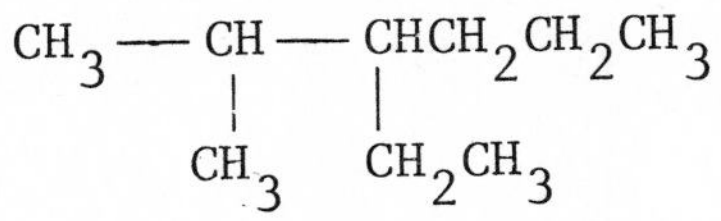

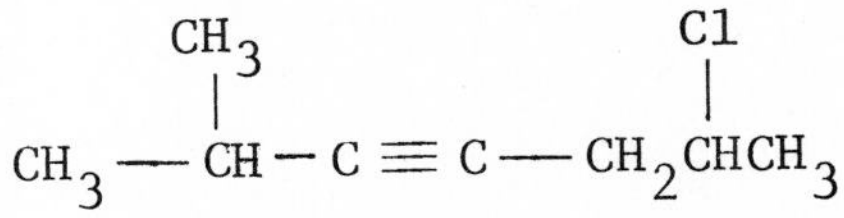

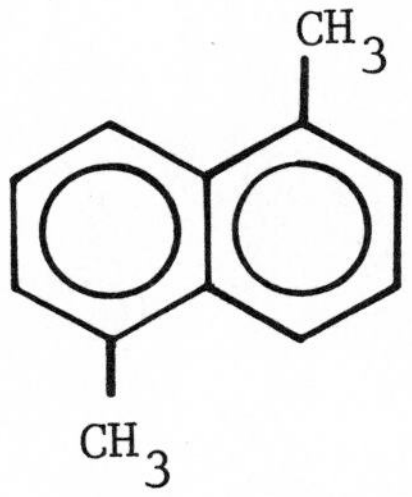

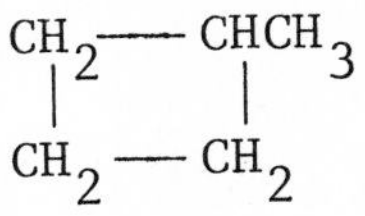

24.7 To save space here we show the condensed structural formulas. You can easily write the complete structural formula by showing each individual C-H and C-C bond.

$\qquad\qquad\quad CH_3$
$\qquad\qquad\quad |$
(a) $CH_3CCH_2CH_2CH_3$
$\qquad\qquad\quad |$
$\qquad\qquad\quad CH_3$

$\qquad\qquad\quad CH_3$
$\qquad\qquad\quad |$
(b) $CH_3CHCHCH_2CH_2CH_3$
$\qquad\qquad\qquad |$
$\qquad\qquad\qquad CH_3$

(c)(d)

(e) $CH_3CHCH_2CH_3$
$\qquad\quad\; |$
$\qquad\quad\; Cl$

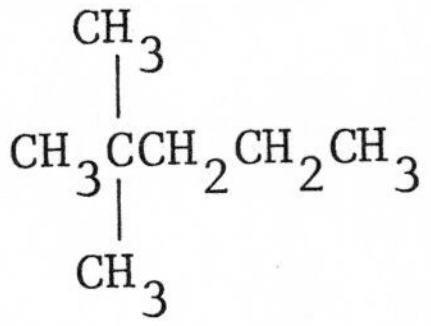

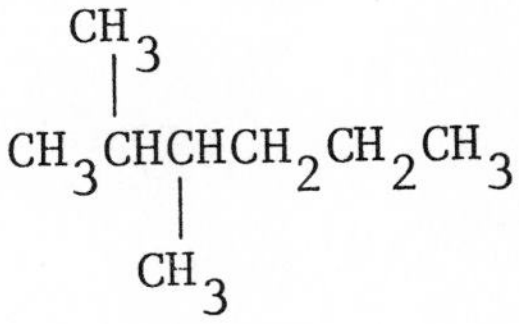

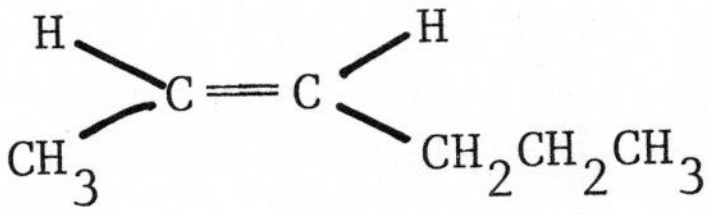

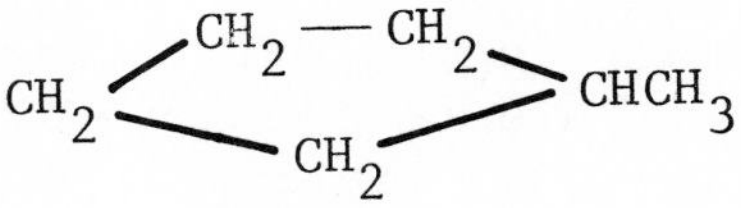

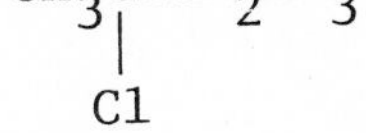

(f)

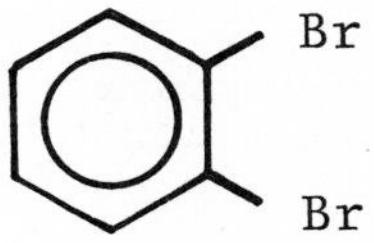

24.8 (a) 2,3-dimethylhexane; (b) 6-methyl-cis-3-heptene; (c) 2,3-dichloronapthalene; (d) 4,4-dimethyl-1-pentyne; (e) 2-bromobutane; (f) 2,5-heptadiene; (g) 2-phenyl-3-bromohexane

24.9 (a) no; (b) (c) no;
(d) no.

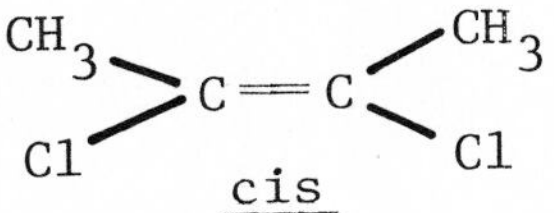

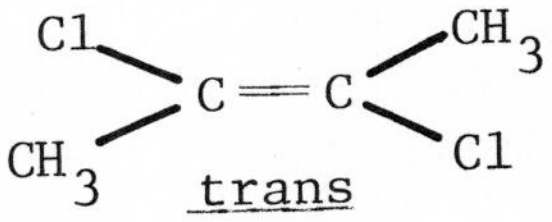

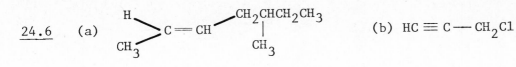

24.10

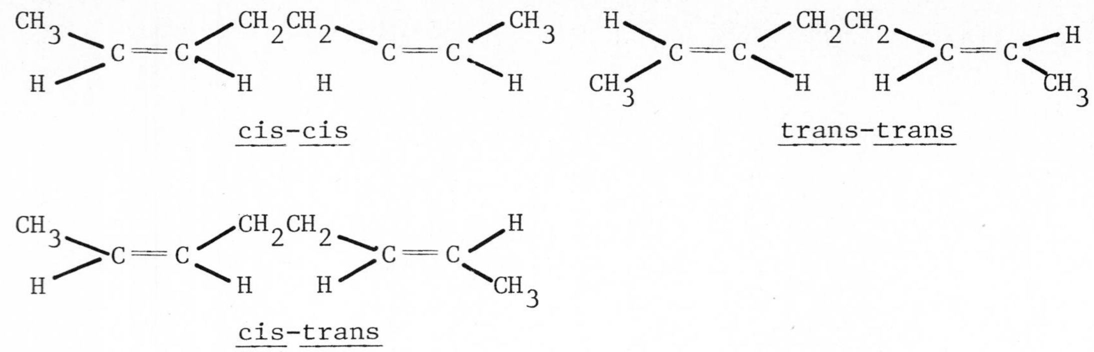

24.11

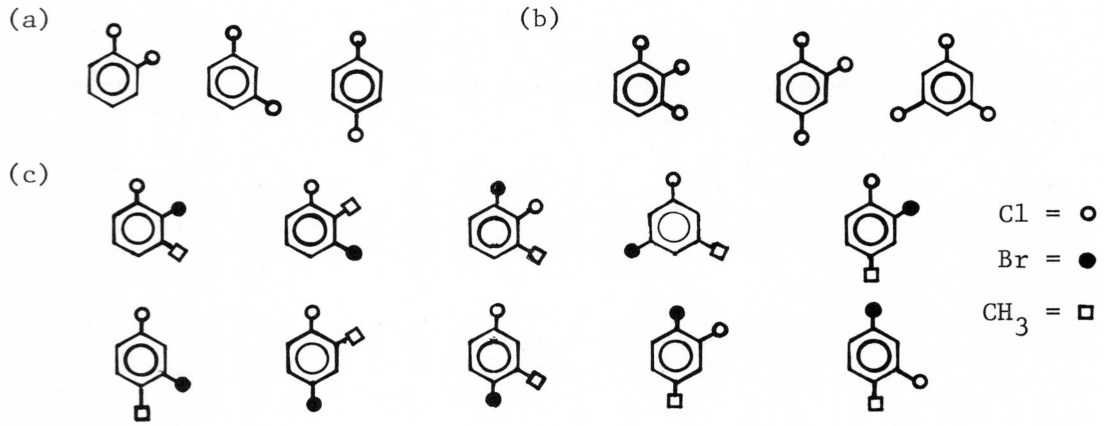

24.12 (a) Cracking is a high temperature, catalyzed process in which straight-chain hydrocarbons are converted to branched-chain molecules. (b) An isomerization reaction is any process in which a molecule is converted into another structural isomer. For example, a cis-alkene could be converted to a trans compound, or a straight-chain alkane could be converted to a branched-chain compound with the same number of carbon atoms. (c) An anti-knock agent is a chemical added to gasolines to improve their burning characteristics in internal combustion engines. (d) The octane number of a gasoline mixture is a measure of its burning characteristics. A high octane number is indicative of a slow-burning fuel, a desirable characteristic in modern high compression engines. (e) A branched-chain alkane is one in which the carbon skeleton contains one or more shorter carbon-carbon side chains attached at intermediate carbons along the longest chain in the molecule. (f) Refining is the process by which crude oil is separated into several fractions according to boiling point.

24.13 Assuming that each component has the same effective octane number in the mixture (and this isn't necessarily always the case), we obtain: octane number = 0.20(0) = 0.30(75) + 0.50(100) = 72.5.

24.14 Octane number can be increased by increasing the fraction of branched-chain alkanes or aromatics, since these have high octane numbers. This can be done by cracking. The octane number can also be increased by adding an anti-knock agent such as tetraethyl lead, $Pb(C_2H_5)_4$, or methycyclopentadienylmanganesetricarbonyl, $CH_3C_5H_4Mn(CO)_3$.

24.15 A major problem in starting a car in cold weather can be that the carburetor is cold, and the gas is not properly volatized as it is mixed with air. Use of a lower-boiling mixture helps to overcome this problem. The lower molecular weight components are, of course, lower boiling.

24.16 (a) $CH_3CH_2CH_2CH_2CH_3(g) + Cl_2(g) \xrightarrow{h\nu} CH_3CH_2CH_2CHClCH_3(g) + HCl(g)$

(b) $2CH_3CH{=}CH_2(g) + 9O_2(g) \rightarrow 6CO_2(g) + 6H_2O(g)$

(c) $CH_3CH_2C{\equiv}CH + Br_2 \rightarrow CH_3CH_2CBr{=}CHBr$

(d) $C_6H_6 + Br_2 \xrightarrow{FeBr_3} C_6H_5Br + HBr$

or $C_6H_6 + HNO_3 \xrightarrow{H_2SO_4} C_6H_5NO_2 + H_2O$

24.17 (a) cyclobutane. There is some strain in the four-membered ring because the C-C-C bond angles must be less than the desired 109°.
(b) cyclohexene. The C=C bond is capable of addition reactions, for example with Br_2. (c) 1-hexene. Whereas the alkene readily undergoes addition reactions, the aromatic hydrocarbon is extremely stable.
(d) 2-hexyne. Alkynes undergo addition reactions even more readily than alkenes.

24.18 (a) $CH_2{=}CHCH_2CH_3 + H_2 \rightarrow CH_3CH_2CH_2CH_3$

(b) $CH_3CH{=}CHCH_3 + H_2O \xrightarrow{H_2SO_4} CH_3CH(OH)CH_2CH_3$

(c) $\underbrace{CH_2CH_2CH_2CH_2}_{} + 6O_2 \rightarrow 4CO_2 + 4H_2O$

(d) $\underline{n}CH_2{=}CHCl \rightarrow -CH_2\left|\underset{Cl}{\underset{|}{CH}}-CH_2\right|_n\underset{Cl}{\underset{|}{CH}}-$

24.19 (a) $CH_3\underset{Cl}{\underset{|}{C}}HCH_3$ (b) $(CH_3)_2\underset{Cl}{\underset{|}{C}}CH_2CH_3$ (c) $CH_3\overset{Cl}{\overset{|}{\underset{Cl}{\underset{|}{C}}}}CH_3$ or $CH_3\underset{Cl}{\underset{|}{C}}{=}CH_2$

24.20 The energy gained by forming two σ bonds (or four σ bonds) more than compensates for the loss of one (or two) π bonds when addition occurs to an alkene (or alkyne). However, in benzene the aromatic ring is especially stabilized by the delocalization of π electrons around the ring. It therefore requires a substantial activation energy to cause the loss of this aromatic character. The most usual reaction in aromatics is thus substitution rather than addition, since substitution does not result in loss of the aromatic character.

24.21 (a) $CH_3CH_2CH_2CH_3$ or $CH_3CH_2CI_2CH_3$; (b) $CH_3CH_2CI{=}CH_2$

(c) $(CH_3)_2\overset{H}{\overset{|}{C}}-\overset{OH}{\overset{|}{C}}(C_2H_5)_2$ or $(CH_3)_2\overset{HO}{\overset{|}{C}}-\overset{H}{\overset{|}{C}}(C_2H_5)_2$ (d) $CH_2BrCHBrCH_3$

24.22 The small C-C-C angles in the cyclopropane ring, only 60°, cause strain that provides a driving force for reactions that result in ring-opening. There is no comparable strain in the five- or six-membered rings.

24.23 (a) $C_2H_6(g) + (7/2)O_2(g) \rightarrow 2CO_2(g) + 3H_2O(g)$

$\Delta H° = 2(-393.5) + 3(-241.8) - (-84.7) = -1427.7$ kJ

(b) $C_2H_4(g) + 3O_2(g) \rightarrow 2CO_2(g) + 2H_2O(g)$

$\Delta H° = 2(-393.5) + 2(-241.8) -52.3 = -1322.9$ kJ

(c) $C_2H_2(g) + (5/2)O_2(g) \rightarrow 2CO_2(g) + H_2O(g)$

$\Delta H° = 2(-393.5) + (-241.8) - 226.7 = -1255.5$ kJ

24.24

	ΔH
$C_{10}H_8(\ell) + 12O_2(g) \rightarrow 10CO_2(g) + 4H_2O(\ell)$	-5157 kJ
$-[C_{10}H_{18}(\ell) + (29/2)O_2(g) \rightarrow 10CO_2(g) + 9H_2O(\ell)]$	-(-6286) kJ
$C_{10}H_8(\ell) + 5H_2O(\ell) \rightarrow C_{10}H_{18}(\ell) + 5/2O_2(g)$	+1129 kJ
$(5/2)O_2(g) + 5H_2(g) \rightarrow 5H_2O(\ell)$	5(-285.8)
$C_{10}H_8(\ell) + 5H_2(g) \rightarrow C_{10}H_{18}(\ell)$	-300 kJ

Compare this with the heat of hydrogenation of ethylene: $C_2H_4(g) + H_2(g) \rightarrow C_2H_6(g)$; $\Delta H = -84.7 - (52.3) = -137$ kJ. This value applies to just one double bond. For five double bonds, we would expect about -685 kJ. The fact that hydrogenation of naphthalene yields only -300 kJ indicates that there must be some special stability associated with the aromatic system in this molecule.

24.25 We can first form an alkyl halide: $C_2H_4 + HBr \rightarrow CH_3CH_2Br$; then carry out a Friedel-Crafts reaction:

$$C_6H_6 + CH_3CH_2Br \xrightarrow{AlCl_3} C_6H_5CH_2CH_3 + HBr$$

24.26 (a) ketone; (b) carboxylic acid; (c) alcohol; (d) ester; (e) aldehyde; (f) ketone; (g) amide.

24.27 About each CH_3 carbon, 109°; about the carbonyl carbon, 120°, planar.

24.28 Alcohols are capable of hydrogen bonding with water as both electron pair donors, and as hydrogen sources (see Section 12.3). Ethers are able to act as electron pair donors toward an O-H bond, but this alone does not provide enough interaction to render them very soluble in water.

24.29 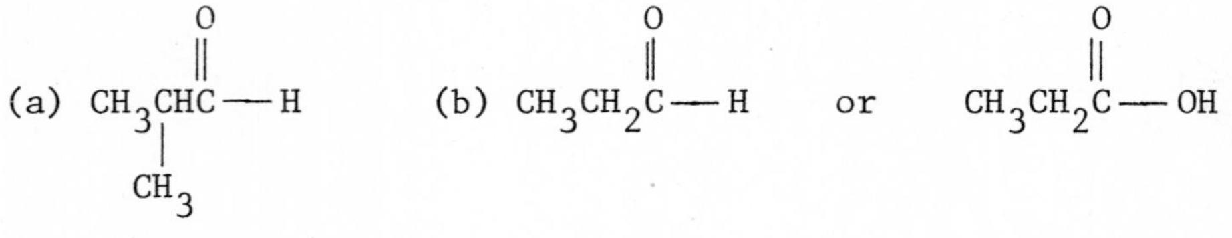

(c) No possibility for mild oxidation here. (d) Oxidation here would probably go all the way to CO_2 and H_2O, although possibly $CH_3\overset{\overset{O}{\|}}{O}-OH$ could be formed.

(e) $CH_3\overset{\overset{O}{\|}}{C}-OH$

<u>24.30</u> (a) $CH_3\overset{\overset{O}{\|}}{C}-O\overset{\overset{CH_3}{|}}{\underset{\underset{CH_3}{|}}{C}}H$ (b) $CH_3\overset{\overset{O}{\|}}{C}-OCH_2CH_2CH_3$ (c) $H\overset{\overset{O}{\|}}{C}-OCH_2CH_3$

<u>24.31</u> (a) 2-methyl-1-propanol; (b) 1-propanol; (c) 2-methyl-2-propanol (t-butyl alcohol); (d) acetone (dimethyl ketone); (e) acetaldehyde.

<u>24.32</u> (a) $CH_3CH_2\underset{\underset{OH}{|}}{C}HCH_3$ (b) $HOCH_2CH_2OH$ (c) $H\overset{\overset{O}{\|}}{C}-OCH_3$

(d) $CH_3CH_2\overset{\overset{O}{\|}}{C}CH_2CH_3$ (e) $CH_3CH_2OCH_2CH_3$

<u>24.33</u> (a) methanoic acid; (b) butanoic acid; (c) 3-methylpentanoic acid

<u>24.34</u> (a) $CH_3CH_2\overset{\overset{O}{\|}}{C}-H$ (b) $CH_3CH_2CH_2\overset{\overset{O}{\|}}{C}CH_3$

(c) $CH_3\underset{\underset{CH_3}{|}}{C}H\overset{\overset{O}{\|}}{C}CH_3$ (d) $CH_3CH_2\underset{\underset{CH_3}{|}}{C}H\overset{\overset{O}{\|}}{C}H$

<u>24.35</u> (a) $CH_2=CHCH_3 + H_2O \xrightarrow{H_2SO_4} CH_3\overset{\overset{OH}{|}}{C}HCH_3$

$CH_3\overset{\overset{OH}{|}}{C}HCH_3 + (0) \rightarrow CH_3\overset{\overset{O}{\|}}{C}CH_3 + H_2O$

(b) $CH_3CH_2CH_2OH + 2(0) \rightarrow CH_3CH_2\overset{\overset{O}{\|}}{C}-OH + H_2O$

(c) $CH_2=CH_2 + H_2O \xrightarrow{H_2SO_4} CH_3CH_2OH$

$CH_3CH_2OH + 2(0) \rightarrow CH_3\overset{\overset{O}{\|}}{C}-OH + H_2O$

$$CH_3\overset{\displaystyle O}{\overset{\|}{C}}\!\!-\!\!OH + NaOH(aq) \rightarrow CH_3CO_2^-(aq) + Na^+(aq) + H_2O(\ell)$$

(d) $CH_3\overset{\displaystyle O}{\overset{\|}{C}}\!\!-\!\!OH + HOCH_3 \rightarrow CH_3\overset{\displaystyle O}{\overset{\|}{C}}\!\!-\!\!OCH_3 + H_2O(\ell)$

(e) $CH_3CH_2\overset{\displaystyle OH}{\overset{|}{C}}HCH_3 + (O) \rightarrow CH_3CH_2\overset{\displaystyle O}{\overset{\|}{C}}CH_3 + H_2O$

24.36 (a) $CH_3COCH_3(aq) + OH^-(aq) \rightarrow CH_3\overset{\displaystyle O}{\overset{\|}{C}}\!\!-\!\!O^-(aq) + CH_3OH(aq)$

(b) $2CH_3OH \xrightarrow{H_2SO_4} CH_3OCH_3 + H_2O$

24.37 $HC \equiv C\!-\!CH_2CH_3$ $CH_3\!-\!C \equiv C\!-\!CH_3$ $\begin{array}{c} HC = CH \\ | \quad\quad | \\ CH_2\!-\!CH_2 \end{array}$

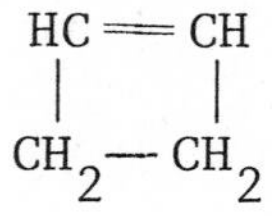

$CH_2 = CH\!-\!CH = CH_2$

24.38 (a) $C_{20}H_{42}$; (b) $C_{18}H_{36}$; (c) $C_{12}H_{22}$

24.39 $\begin{array}{c} H \\ \\ CH_3 \end{array}\!\!\!\!C = C\!\!\!\!\begin{array}{c} H \\ \\ CH_2\!-\!CH_3 \end{array}$ $\begin{array}{c} CH_3 \\ \\ H \end{array}\!\!\!\!C = C\!\!\!\!\begin{array}{c} H \\ \\ CH_2CH_3 \end{array}$

    cis   trans

To show complete Lewis structures all C-H bonds should be shown. We have
omitted that to save space. Cyclopentene does not show cis-trans isomerism
because the existence of the ring demands that the C-C bonds be cis to
one another.

24.40 (a) alkane; (b) alkyne (or diene); (c) alkene; (d) alkane

24.41 (a) 2-propanol; (b) dimethyl ether; (c) 3-butene-1-ol;
(d) 3-methyl-3-hexene; (e) 4-methyl-2-pentyne

24.42 (a) $CH_3C \equiv CH + HCl \rightarrow CH_3CCl = CH_2$

$CH_3CCl = CH_2 + HBr \rightarrow CH_3\!-\!\overset{\displaystyle Cl}{\overset{|}{\underset{|}{\underset{\displaystyle Br}{C}}}}\!\!-\!CH_3$

(b) $CH_2 = CH_2 + Cl_2 \rightarrow CH_2Cl\!-\!CH_2Cl$

(c) Water is eliminated between the alcohol and acid functions to form ester linkages along a chain:

$$\cdots CH_2O\overset{\overset{O}{\|}}{C}CH_2CH_2\overset{\overset{O}{\|}}{C}OCH_2CH_2O\overset{\overset{O}{\|}}{C}CH_2CH_2O\overset{\overset{O}{\|}}{C}\cdots$$

(d) $$C_6H_5\overset{\overset{O}{\|}}{C}-OCH_3 + Na^+(aq) + OH^-(aq) \rightarrow C_6H_5CO_2^-(aq) + Na^+(aq) + CH_3OH(aq)$$

(e) $$nH_2C{=}CHCH_3 \xrightarrow{\text{catalyst}} \cdots-CH_2-\overset{\overset{CH_3}{|}}{CH}-CH_2-\overset{\overset{CH_3}{|}}{CH}-CH_2-\overset{\overset{CH_3}{|}}{CH}-\cdots$$

In this reaction the double bonds are opened to react with one another. This addition polymerization reaction does not occur readily, and special catalysts are needed to prepare commercially useful materials.

(f) $CH_2{=}CH_2 + HBr \rightarrow CH_3CH_2Br$

$CH_3CH_2Br + C_{10}H_8 \xrightarrow{AlCl_3}$ 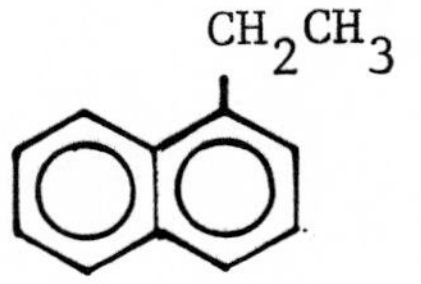 $+ HBr$

24.43 $CH_3CH_2\overset{\overset{OH}{|}}{CH_2}$ $\qquad CH_3\overset{\overset{OH}{|}}{C}HCH_3$ $\qquad CH_3-O-CH_2CH_3$

24.44 The C-Cl bonds in the trans compound are pointing in exactly opposite directions. Thus, the C-Cl bond dipoles cancel (Section 8.2). This is not the case in the cis compound, as can be seen by writing out the structure:

trans

net dipole moment
cis

24.45 Because of the strain in bond angles about the ring, cyclohexyne would not be stable. The alkyne carbons preferentially have a 180° bond angle. However, there are not enough carbons in the ring to make this possible without gross distortions of other bond lengths and angles:

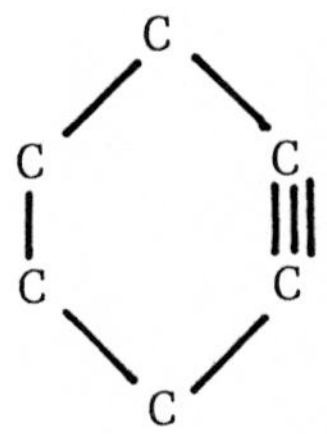

24.46 (a) ether, C-O-C; alkene, $-CH{=}CH_2$; (b) carboxylic acid, $-\overset{O}{\overset{\|}{C}}-OH$; ester, $CH_3\overset{O}{\overset{\|}{C}}-O-$; (c) ketone, $-\overset{O}{\overset{\|}{C}}-$; alkene, $-CH{=}CH-$; alcohol, $-\overset{|}{\underset{|}{C}}-OH$

24.47 (a) $CH_3OCH_2CH_3$ (b) $CH_3CH_2\overset{O}{\overset{\|}{C}}-H$ (c) $CH_3\overset{O}{\overset{\|}{C}}CH_3$

(d) $CH_3\overset{F}{\overset{|}{C}}HCH_3$ (e) $CH_3CH_2CH_2CH_2OH$ (f) $CH_3CH_2\overset{O}{\overset{\|}{C}}OH$

(g) $CH_3\overset{O}{\overset{\|}{C}}-OCH_3$

24.48 (a) $CH_3\overset{O}{\overset{\|}{C}}-OH$, C_6H_5OH; (b) $C_6H_5\overset{O}{\overset{\|}{C}}-OH$, CH_3OH

24.49 $$(0.512\text{ g }CO_2)\left(\frac{1\text{ mol }CO_2}{44.0\text{ g }CO_2}\right)\left(\frac{1\text{ mol C}}{1\text{ mol }CO_2}\right) = 1.16 \times 10^{-2}\text{ mol C}$$

$$(0.209\text{ g }H_2O)\left(\frac{1\text{ mol }H_2O}{18.0\text{ g }H_2O}\right)\left(\frac{2\text{ mol H}}{1\text{ mol }H_2O}\right) = 2.32 \times 10^{-2}\text{ mol H}$$

When we calculate the masses of these quantities and subtract from 0.256 g, we find that there are 0.093 g O in the sample, corresponding to 5.82×10^{-3} mol O. The empirical formula is thus C_2H_4O, with formula weight 44 g. To obtain the molecular weight from the gas density data, we use Equation [9.11]: MW = dRT/P.

$$= \frac{(0.155\text{ g}/0.100\text{ L})(0.082\text{ L-atm/mol-K})\ 400\text{K}}{1.16\text{ atm}} = 43.8\text{ g/mol}$$

The formula weight is thus the molecular weight, possibly CH_3CHO, acetaldehyde. This compound should readily oxidize to acetic acid: $CH_3CHO + (O) \rightarrow CH_3COOH$.

24.50 The linkages that have been formed are ester or amide type. These can be opened in a hydrolysis type reaction under strong base conditions, just as one hydrolyzes an ordinary ester linkage, Equation [24.40]. Obviously, however, commercially useful materials are designed to be resistant to this sort of process.

24.51 The compound is clearly an alcohol. Its slight solubility in water is consistent with the properties expected of a secondary alcohol with a five-carbon chain. The fact that it is oxidized to a ketone rather than to an aldehyde and on to a carboxylic acid tells us that it is a secondary alcohol:

$$CH_3CHCH_2CH_2CH_3 \qquad CH_3CH_2CHCH_2CH_3 \qquad CH_3CHCH(CH_3)_2$$
$$\quad\ \ | \qquad\qquad\qquad\qquad\ \ | \qquad\qquad\qquad\quad\ | $$
$$\quad\ \ OH \qquad\qquad\qquad\qquad OH \qquad\qquad\qquad\quad OH$$

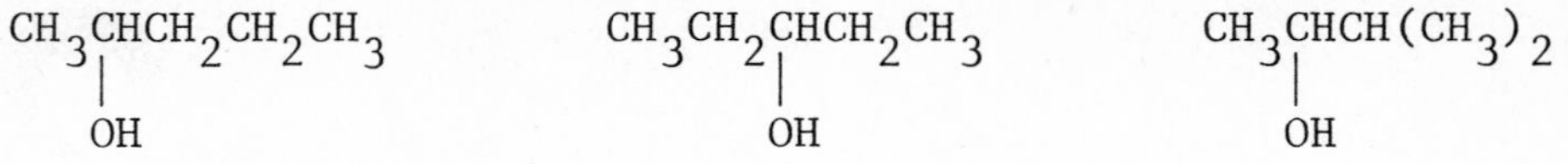

__24.52__ These compounds must be olefins. A is either $CH_3CH = CHCH_3$ $CH_2 = CHCH_2CH_3$, B is $(CH_3)_2C = CH_2$.

$$CH_3CH = CHCH_3$$
$$\text{or} \qquad\qquad + H_2O \xrightarrow{H_2SO_4} CH_3-\underset{\underset{OH}{|}}{C}HCH_2CH_3 \xrightarrow{(O)} CH_3\overset{\overset{O}{\|}}{C}CH_2CH_3$$
$$CH_2 = CHCH_2CH_3$$

$$(CH_3)_2C = CH_2 + H_2O \xrightarrow{H_2SO_4} CH_3 - \underset{\underset{CH_3}{|}}{\overset{\overset{CH_3}{|}}{C}} - OH \xrightarrow{(O)} \text{no reaction}$$

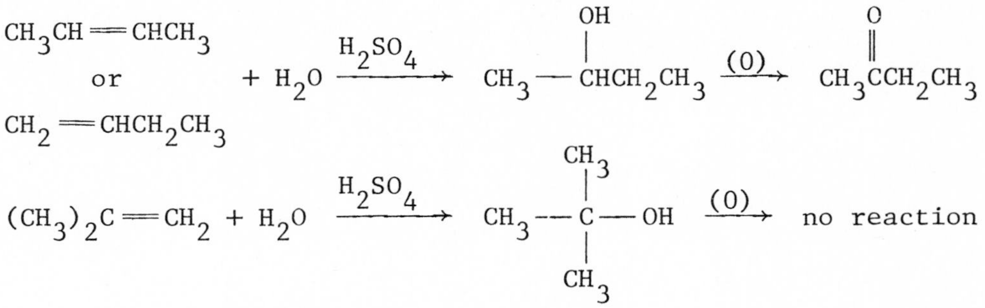

__24.53__ $(85.7 \text{ g C})\left(\dfrac{1 \text{ mol C}}{12.0 \text{ g C}}\right) = 7.14 \text{ mol C}$

$(14.3 \text{ g H})\left(\dfrac{1 \text{ mol H}}{1.01 \text{ g H}}\right) = 14.2 \text{ mol H}$

Empirical formula is CH_2. Using Equation [9.11]:

$$MW = \frac{(2.21 \text{ g/L})(0.082 \text{ L-atm/mol-K})(373 \text{ K})}{(735/760) \text{ atm}} = 69.9 \text{ g/mol}$$

The molecular formula is thus C_5H_{10}. Because the absence of reaction with aqueous Br_2 indicates that the compound is not an alkene, we can guess that it is the cycloalkane, cyclopentane. To verify this, we can compare the boiling point of the substance with the handbook value for cyclopentane.

__24.54__ $Br_2 \xrightarrow{h\nu} 2Br\cdot$

$Br\cdot + CH_3CH_2CH_2CH_3 \rightarrow HBr + CH_3CH_2CH_2CH_2$

$CH_3CH_2CH_2CH_2\cdot + Br_2 \rightarrow CH_3CH_2CH_2CH_2Br + Br\cdot$

Note that this is a radical chain process; the Br atom formed in the last reaction goes on to abstract hydrogen from another alkane molecule, and so forth. We have simplified matters here by assuming that the butyl radical formed in the hydrogen atom abstraction step is from the end carbon, whereas, in fact, abstraction from one of the two carbons inside the chain is found to be a bit favored. Thus, 2-bromobutane is also a product as are other, more highly brominated molecules.

25

The biosphere

25.1 Organisms require energy for the maintenance of metabolic processes within themselves, for mobility, for maintenance of temperature, and so forth. This energy is derived from some external source, which might be solar energy as in the case of green plants, or some food source, as in animals. The conversion of the external energy source to biochemical energy of some form is an overall spontaneous process. As such, when it occurs, the entropy of the universe increases. The entropy change within the organism may be negative, but the entropy change in the surroundings is sufficiently positive so that the process has an overall positive entropy change.

25.2 All growth processes, which require synthesis of new cell materials, require energy. For example, in plants transpiration of water requires energy. In animals, the action of the involuntary muscles such as the heart, maintenance of a uniform body temperature, synthesis of new cells, motions of limbs, maintenance of correct body fluid compositions, all require energy.

25.3 (a) Photosynthesis is just the reverse of the process shown. Therefore, ΔG°_{298} is +2878 kJ per mole of glucose formed. Because ΔG° is highly positive, the reaction is nonspontaneous.

(b) $\Delta G = \Delta G^\circ + 2.3\ RT \log Q$

$$= 2.878 \times 10^6 + 2.3(8.314)(298) \log \frac{(1.0 \times 10^{-3})(0.20)^6}{(3.1 \times 10^{-4})^6}$$

$$= 2878\ kJ + 79\ kJ = 2957\ kJ$$

25.4 Pigments are responsible for the initial absorption of solar radiation. One would expect that they would possess absorption bands in the visible region of the spectrum (not necessarily throughout the visible, however, since then they would be black). Further, since they must absorb a photon, then release this energy in some other form, they must be fairly robust molecules, not easily decomposed by an influx of energy. They must also be of a structural type that permits their incorporation into the largely organic matrix of the plant leaf.

25.5 $(50 \text{ g } CO_2) \left(\dfrac{1 \text{ mol } CO_2}{44.0 \text{ g } CO_2} \right) \left(\dfrac{1 \text{ mol } O_2}{1 \text{ mol } CO_2} \right) \left(\dfrac{32 \text{ g } O_2}{1 \text{ mol } O_2} \right) = 36 \text{ g } O_2$

This corresponds to 1.14 mol O_2. Since 1 mol O_2 at STP occupies 22.4 L, the STP volume of O_2 produced is 25.4 L.

25.6 $\left(\dfrac{1 \times 10^{17} \text{ g C}}{1 \text{ yr}} \right) \left(\dfrac{1 \text{ mol C}}{12 \text{ g C}} \right) \left(\dfrac{1 \text{ mol glucose}}{6 \text{ mol C}} \right) \left(\dfrac{2878 \text{ kJ}}{1 \text{ mol glucose}} \right)$

$= 4.0 \times 10^{18} \text{ kJ/yr}$

The percentage fraction of solar energy that is employed in photosynthesis is thus $100(4 \times 10^{18}/4 \times 10^{21}) = 0.1\%$.

25.7 Proteins serve as the structural elements in animal tissues (for example, muscles). They can act as catalysts for biochemical reactions (enzymes). They may be involved in transport functions, as in O_2 transport in the blood. They are the major constituent of antibodies that act against invading organisms.

25.8 (a) An α-amino acid contains an NH_2 function attached to the carbon that is bound to the carbon of the carboxylic acid function. (b) In forming a protein amino acids undergo a condensation reaction between the amino group and carboxylic acid:

$$ -\overset{\overset{\displaystyle O}{\|}}{C}-OH + H-\underset{\underset{\displaystyle H}{|}}{N}-CH_2\overset{\overset{\displaystyle O}{\|}}{C}-OH \rightarrow -\overset{\overset{\displaystyle O}{\|}}{C}-\underset{\underset{\displaystyle H}{|}}{N}-CH_2-\overset{\overset{\displaystyle O}{\|}}{C}-OH + H_2O $$

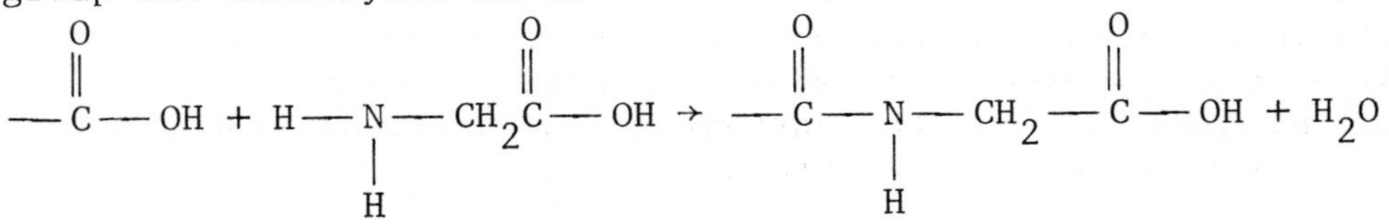

25.9 The side chains possess three characteristics that may be of importance. In the first place they may be bulky, and thus impose restraints on where and how the amino acid can undergo reaction. Secondly, the side chain may possess a polar group (e.g., the -OH group in serine), that will in part determine solubility. Finally, the side chain may contain an acidic or basic functional group that will partly determine solubility in acidic or basic medium, and that may become involved in interactions with other amino acids.

25.10 Two dipeptides are possible:

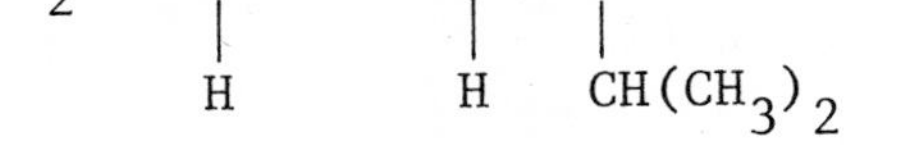

glycylvaline

and

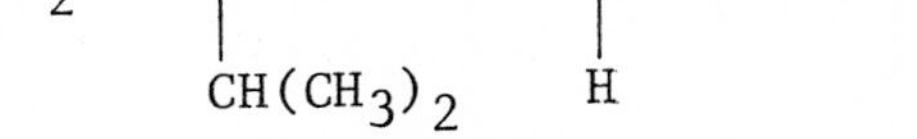

valinylglycine

25.11

$H_2NCH(CH_2CO_2H)CO_2H + H_2NCH(CH_2SH)CO_2H \rightarrow$ 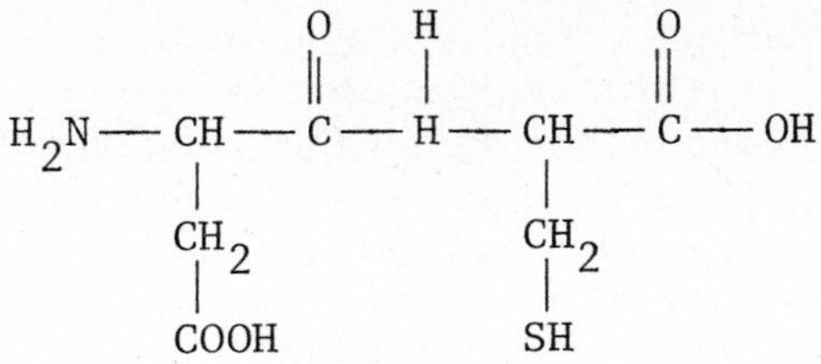

aspartylcysteine

25.12

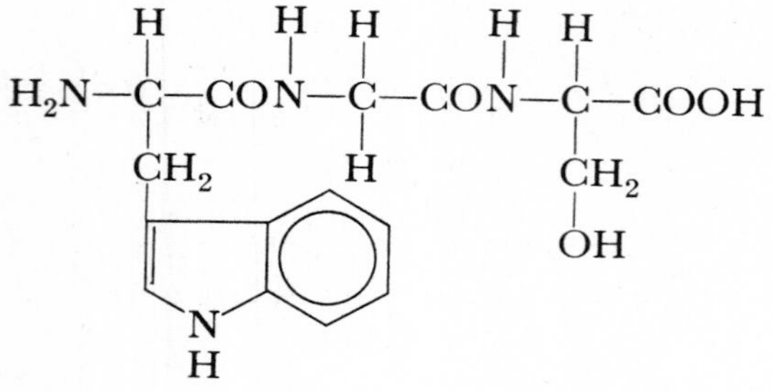

25.13 alanine, serine.

25.14 in basic solution as the anion: $H_2NCH_2CO_2^-$; in acidic solution as the cation: $H_3NCH_2CO_2H^+$.

25.15 gly-val-ala; gly-ala-val; val-gly-ala; val-ala-gly; ala-gly-val; ala-val-gly

25.16 Your drawing should look like Figure 25.4. The only difference is that the CH_3 group of the alanine is replaced by the CH_2COOH group of aspartic acid.

25.17 In glycine, the α carbon atom has two hydrogens. It therefore is not chiral in the sense illustrated in Figure 25.4. If you replace the CH_3 in that picture by H, the two mirror image structures become superimposable.

25.18 The primary structure of a protein refers to the sequence of amino acids in the chain. Along any particular section of the protein chain, the configuration may be that of a helical arrangement, or it may be an open chain, or arranged in some other way. This is called the secondary structure. The overall shape of the protein molecule is determined by the way the segments of protein chain come together, or pack. This overall shape aspect is referred to as the tertiary structure.

25.19 It is quite evident from Figure 25.5 that the hydrogen bonds between an NH group along the chain and the unshared electron pairs of a carbonyl group further along are responsbile for maintaining the helix. Indeed, the pitch, and general shape of the helix are determined by what makes for a good hydrogen bonding arrangement.

25.20 The biochemical properties are determined by all of the structural aspects of the protein, its primary, secondary and tertiary structure. The tertiary and secondary structures are maintained by relatively weak forces; for example, by hydrogen bonding interactions. A change in conditions that causes a change in the tertiary or secondary structure can drastically alter the chemical activity of the protein. This kind of change is referred to as denaturation. It may be brought about by a change in solvent, for example, from water to ethanol; or by a change in temperature. As the protein gains thermal energy, some of that energy goes into molecular motions that can cause rupture of the weak forces that maintain the secondary and tertiary structures. When this occurs, the protein loses activity.

25.21 All terms except the following are defined in the list of Key Terms: (b) An apoenzyme is an enzyme molecule without some essential component, usually a small molecule. (e) A holoenzyme is the entire enzyme system, consisting of a protein and one or more essential components referred to as coenzymes or cofactors. (g) A peptidase is an enzyme that promotes the hydrolysis, or breakup of a protein or polypeptide chain.

25.22 An enzyme inhibitor must possess sufficient similarity to the normal substrates of the enzyme so that it can fit into the active site and can compete with substrate molecules for the active sites. Once there, it must bind relatively strongly; if it did not, but merely came off and on with great rapidity, it would have no effect on the rate of normal reaction. It must, in effect, tie up the active site. There are different kinds of inhibitors. Some react with the protein and simply destroy the active site. Others bind so tightly that they do not, in effect, ever come off again. Others bind reversibly, and simply compete with the normal substrates for the active sites; these are reversible inhibitors.

25.23 1-(d); 2-(a); 3-(e); 4-(c); 5-(b).

25.24 The fact that an increase in rate is roughly proportional to the increase in substrate concentration tells us that the fraction of active sites involved at any one instant in the catalytic process is small. If a large fraction were involved, then increasing the concentration of substrate would have less than a proportional effect on rate.

25.25 These enzymes are peptidases; their function is to hydrolyze the amide functions along the polypeptide chains.

25.26 Holoenzyme: zinc plus protein. Apoenzyme: protein alone. Cofactor: the zinc (actually present as Zn^{2+}). Turnover number: 10^7 CO_2 molecules/sec.

25.27 Glucose exists in solution as a cyclic structure in which the aldehyde function on carbon 1 reacts with the OH group of carbon 5 to form what is called a hemiacetal. Carbon atom 1 carries an OH group in the hemiacetal form; in α-glucose this OH group is on the same side of the ring as an OH group on adjacent carbon atom 2 (illustrated in Figure 25.7). In the β (beta) form the OH groups are on opposite sides.

The final products look like this (only OH groups are labelled).

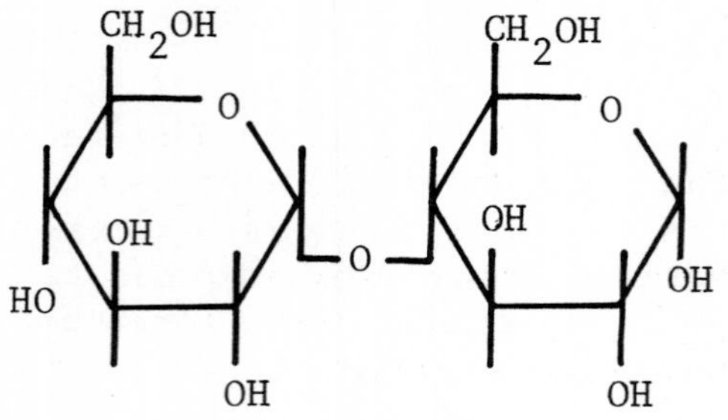

α-linkage

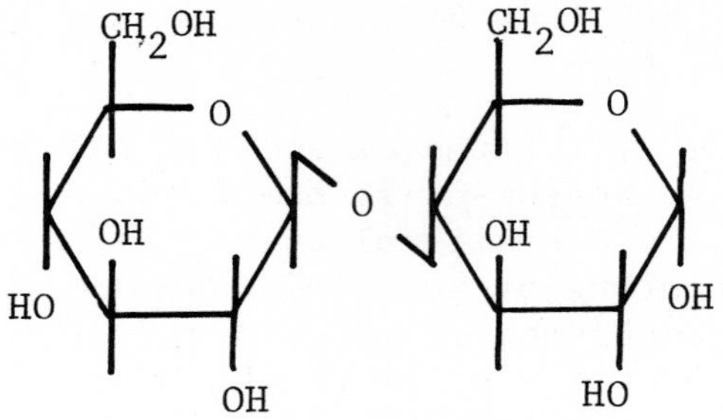

β-linkage

25.28 (a) α-form; (b) β-form; (c) α-form

25.29

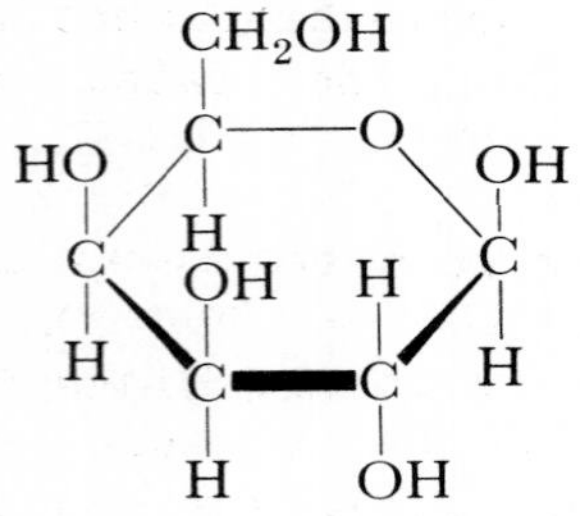

Galactose

The structure is best deduced by comparing galactose with glucose, and inverting the configurations at the appropriate carbon atoms.

25.30 Carbon atoms 2,3,4 and 5 (see numbering system in Figure 25.9) are chiral, because they carry four different groups on each.

25.31 The term hexose refers to the number of carbon atoms in the chain of the monosaccharide. Each carbon atom bears an alcohol or aldehyde function. The terms pyranose and furanose refer to the manner in which the ring forms the hemiacetal. If a six-membered ring is formed, the sugar is a pyranose. If a five-membered ring is formed, the sugar is a furanose. Glucose forms a pyranose because it is an aldehyde sugar. That is, the functional group that forms the hemiacetal is on the terminal carbon. In fructose, the functional group forming the ring is a ketone; it preferentially forms a five-membered ring.

25.32

$$\text{cis-CH}_3\text{(CH}_2\text{)}_7\text{CH}=\text{CH(CH}_2\text{)}_7-\overset{\displaystyle O}{\overset{\|}{C}}-O-CH_2$$

$$\text{cis-CH}_3\text{(CH}_2\text{)}_7\text{CH}=\text{CH(CH}_2\text{)}_7-\overset{\displaystyle O}{\overset{\|}{C}}-O-CH$$

$$\text{cis-CH}_3\text{(CH}_2\text{)}_7\text{CH}=\text{CH(CH}_2\text{)}_7-\overset{\displaystyle O}{\overset{\|}{C}}-O-CH_2$$

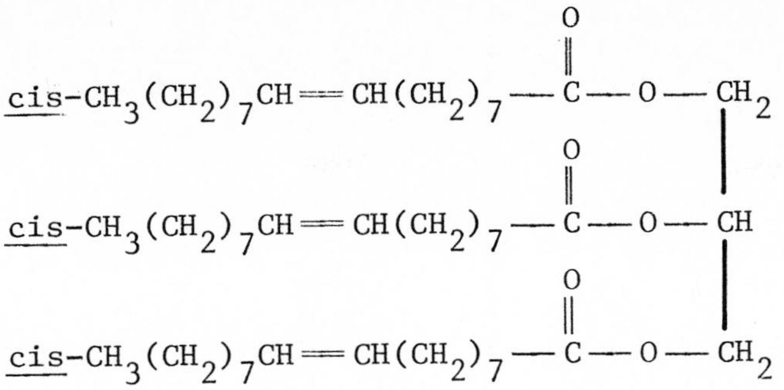

25.33 $C_3H_5(OH)_3 + 3C_{11}H_{23}\overset{\displaystyle O}{\overset{\|}{C}}-OH \rightarrow C_3H_5(OCC_{11}H_{23})_3 + 3H_2O$

25.34 $C_3H_5(O\overset{\displaystyle O}{\overset{\|}{C}}C_{13}H_{27})_3 + 3Na^+(aq) + 3OH^-(aq) \rightarrow 3C_{13}H_{27}\overset{\displaystyle O}{\overset{\|}{C}}-O^-(aq)$

$$+ 3Na^+(aq) + C_3H_5(OH)_3$$

25.35 $[CH_3(CH_2)_4CH=CHCH_2CH=CH-(CH_2)_7COO]_3C_3H_5 + 6H_2 \rightarrow [CH_3CH_2)_{16}COO]_3C_3H_5$

trilinolein   tristearin

Hydrogenation with only 4 mol of H_2 will result in fatty acid side chains with a total of two double bonds remaining. These double bonds might be in the same side chain, or one each in two different side chains. Furthermore, it turns out that the location of the single double bond that remains in a side chain might be in any of several places along the chain. Thus, many isomers are possible.

25.36 Fats contain as their only polar functional groups the ester linkage that ties the three fatty acid side chains to the glycerol. These groups are not sufficiently polar to interact strongly with a polar, hydrogen-bonding solvent such as water. By comparison, a disaccharide such as sucrose or a monosaccharide such as glucose bristles with polar hydroxyl groups, well situated to hydrogen bond to solvent water. These molecules are therefore quite soluble in water, whereas the fats are not. On the other hand, since fats consist mainly of hydrocarbon chains, they fit quite well into nonpolar organic solvents.

25.37 A nucleotide is a molecule consisting of a nitrogen-containing aromatic compound, a sugar in the furanose (5-membered) ring form, and a phosphoric acid molecule. The structure of deoxycytidine monosphosphate is shown at right.

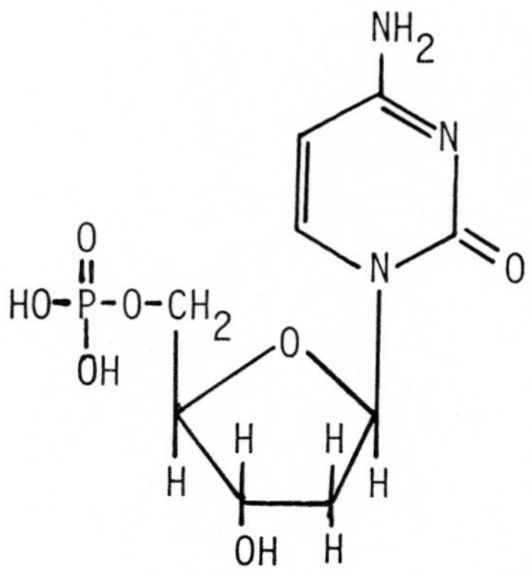

25.38

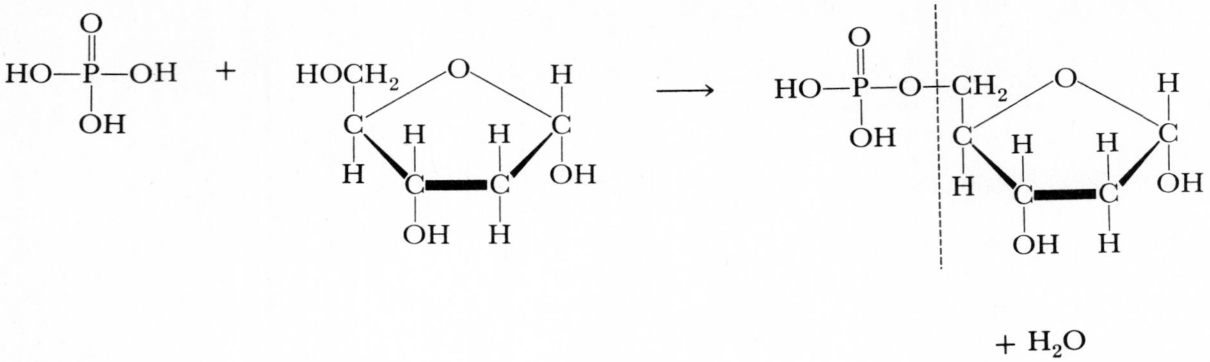

25.39 —A—C—T—C—G—A—

| | | | | |

—T—G—A—G—C—T— ← Complementary strand

25.40 In the helical structure for DNA, the strands of the polynucleotides are held together by hydrogen-bonding interactions between particular pairs of bases. It turns out that adenine and thymine form an especially effective base pair, and that guanine and cytosine are similarly related. Thus, each adenine has a thymine as its opposite number in the other strand, and each guanine has a cytosine as its opposite number. In the overall analysis of the double strand, total adenine must then equal total thymine, and total guanine equals total cytosine.

25.41 (a) $$\left(\frac{10{,}000\ \text{kJ}}{\text{day}}\right)\left(\frac{1\ \text{m}^2\text{-day}}{2.0\times 10^4\ \text{kJ}}\right)\left(\frac{1}{0.012}\right) = 42\ \text{m}^2$$

(b) This leaf area would result in conversion of about 10,000 kJ per day to plant matter. However, not all this energy is available for human diet. A large fraction will be in the form of cellulose, which is not digestible. Some of it goes into root formation, and some is consumed in the plant as energy is required for various functions within the living system.

25.42 (a) none; (b) The carbon bearing the secondary OH has four different groups attached, and is thus chiral. (c) The carbon bearing the $-NH_2$ group and the carbon bearing the CH_3 group are both chiral.

25.43 (a) $CH_3CH(NH_2)\overset{O}{\overset{\|}{C}}—OH + H_2NCH_2\overset{O}{\overset{\|}{C}}—OH$ (b) 3 mol $C_{17}H_{33}\overset{O}{\overset{\|}{C}}—OH + C_3H_5(OH)_3$

(c)

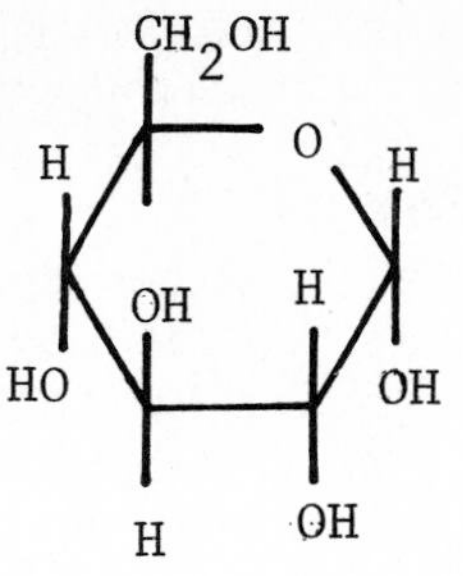

(d)

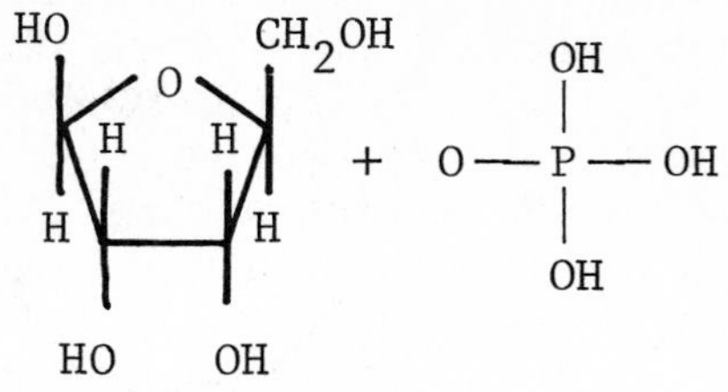

25.44 (a)

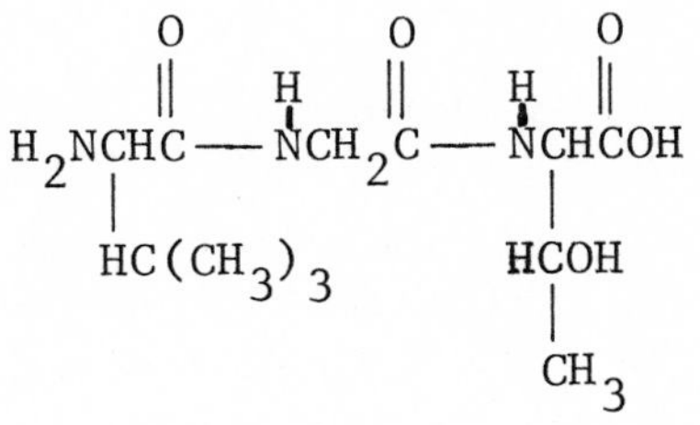

(b)

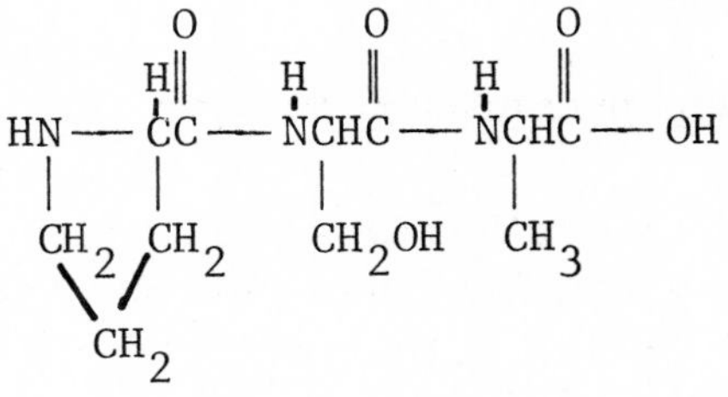

25.45 The VSEPR model (Section 8.1) tells us that the geometry about the carbonyl carbon should be planar with 120° bond angles. Notice that in the Lewis structure for the amide, or peptide, bond, there is an unshared electron pair on the amide nitrogen:

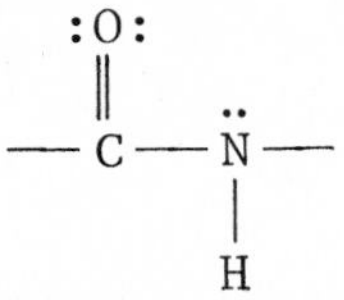

It is possible for the orbital containing this unshared pair to overlap with the $2p_{\pi}$ orbital on the carbonyl carbon, if the geometry about the nitrogen is also planar. This overlap makes possible resonance structures of the form:

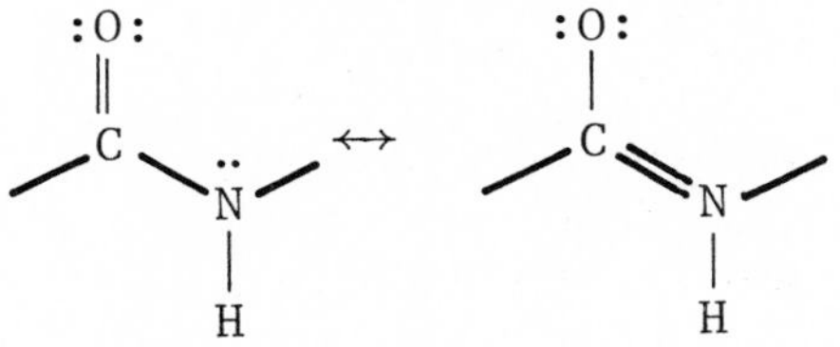

Because of the added stability that derives from such a resonance interaction, the entire peptide bond system is planar.

25.46 glu-cys-gly is the only possible structure.

25.47 The structures are shown in Figure 25.3. Note that the only difference between the amino acids is the OH group on the phenyl ring. The enzyme catalyzes the placement of this hydroxyl group, so is called a hydroxylase.

25.48 (a) An enzyme is thought in many cases to work by providing a kind of template on which reactants can come together in a certain arrangement to undergo reaction. The so-called "lock and key" model, illustrated in Figure 25.8, shows the idea in graphic form. (b) A substrate is any substance that undergoes the reaction catalyzed by a particular enzyme.

Enzyme inhibition occurs when the active site is blocked by some reagent which, while not a substrate, nevertheless interacts with the enzyme and prevents reaction from occurring. Most usually, the inhibitor occupies the active site and keeps out the substrate.

25.49 Water is nearly always involved in some way in the action of an enzyme. It may not be in the active site, but is important in determining the tertiary structure of the protein. Because it is capable of hydrogen bonding, alcohol can mix with water in cells (assuming it is not metabolized as yet) and thus play an active role in the interaction with enzyme proteins. The alcohol could bind in place of water in or near the active site, or could change the enzyme's tertiary structure, to prevent the enzyme from behaving properly.

25.50

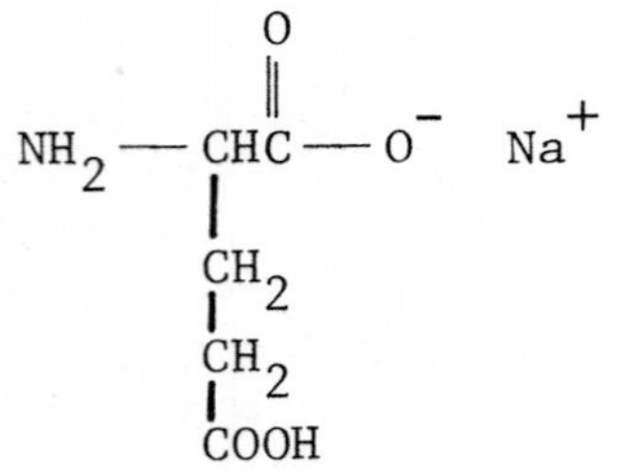

msg

The biological response that makes msg a flavor enhancer is doubtless related to functions that glutamic acid and its salt play in human metabolism. Since all naturally occurring glutamic acid is of the L form, the fact that only the L form works as a flavor enhancer is not surprising.

25.51 The reaction is:

$$2NH_2CH_2COOH(aq) \rightarrow NH_2CH_2CONHCH_2COOH(aq) + H_2O(\ell)$$

$$\Delta G = (-488) + (-285.8) - 2(-369) = -35.8 \text{ kJ}$$

25.52

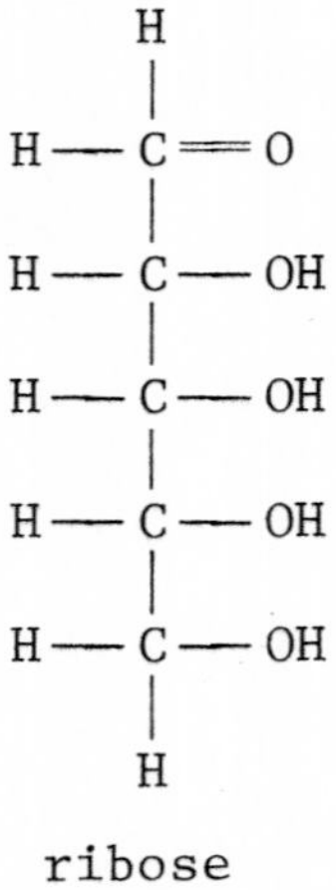

ribose

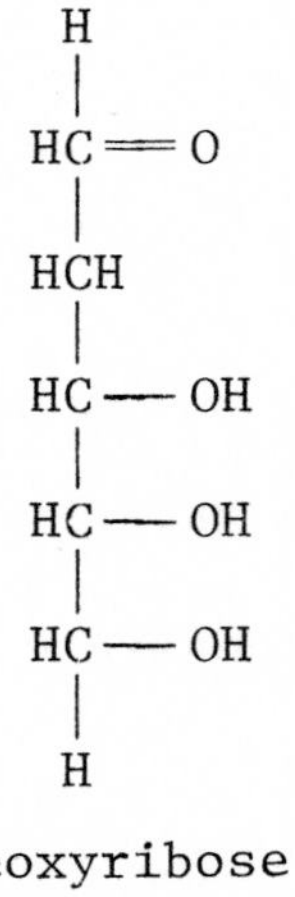

deoxyribose

25.53 The fact that the rate doubles with a doubling of the concentration of sugar tells us that the fraction of enzyme tied up in the form of an enzyme-substrate complex is small. A doubling of the substrate concentration leads to a doubling of the concentration of enzyme-substrate complex, because only a small fraction of enzyme molecules is tied up. The behavior of innositol suggests that it acts as a competitor with sucrose for binding at the active sites of the enzyme system. Such a competition results in a lower effective concentration of active sites for binding of sucrose, and thus results in a lower reaction rate.

25.54 Starch and cellulose differ in the geometry of the oxygen linkage joining monosaccharide units, as illustrated in Figures 25.11 and 25.12.

25.55 The reaction is approximately:

$$\underline{n}C_6H_{12}O_6(aq) \rightarrow -(C_6H_{11}O_5)_{\underline{n}}-(aq) + (n-1)H_2O(\ell)$$

$$\Delta G° = n(-662.3) + (n-1)(-285.8) - n(-917.2)$$

$$= n(-285.8 - 662.3 + 917.2) + 285.8$$

$$= n(-30.9) + 285.8 \text{ kJ}$$

25.56 A condensation reaction is one in which a small molecule, usually water, is eliminated between two molecular substances, forming a link that joins them. A hydrolysis reaction is one in which the reverse process occurs; water is added to split up two molecules. These reactions are central to biochemistry because all biopolymers are formed through condensation reactions, and many metabolic processes in living systems involve either condensation or hydrolysis.

25.57 (a) —T—A—T—G—C—A—
: : : : : :
—A—T—A—C—G—T— ← complementary strand

25.58 $AMPOH^-(aq) \rightleftarrows AMPO^{2-}(aq) + H^+(aq)$

$$K_a = \frac{[AMPO^{2-}][H^+]}{[AMPOH^-]} = 6.2 \times 10^{-8}$$

When pH = 7.40, $[H^+] = 4.0 \times 10^{-8}$.

Then $\frac{[AMPOH^-]}{[AMPO^{2-}]} = 4.0 \times 10^{-8}/6.2 \times 10^{-8} = 0.65$